Essentials

of

Musculoskeletal Care

American Academy of Orthopaedic Surgeons
American Academy of Pediatrics

Essentials

of *Musculoskeletal Care*

Robert K. Snider, MD Editor

Published 1997
by the American Academy of Orthopaedic Surgeons
6300 North River Road
Rosemont, Illinois 60018

First Edition
Copyright ©1997
by the American Academy of Orthopaedic Surgeons

The material presented in *Essentials of Musculoskeletal Care* has been made available by the American Academy of Orthopaedic Surgeons for educational purposes only. This material is not intended to present the only, or necessarily best, methods or procedures for the medical situations discussed, but rather is intended to represent an approach, view, statement, or opinion of the author(s) or producer(s), which may be helpful to others who face similar situations.

Some drugs and medical devices demonstrated in Academy courses or described in Academy print or electronic publications have Food and Drug Administration (FDA) clearance for use for specific purposes or for use only in restricted settings. The FDA has stated that it is the responsibility of the physician to determine the FDA status of each drug or device he or she wishes to use in clinical practice, and to use the products with appropriate patient consent and in compliance with applicable law.

Furthermore, any statements about commercial products are solely the opinion(s) of the author(s) and do not represent an Academy endorsement or evaluation of these products. These statements may not be used in advertising or for any commercial purpose.

ISBN 0-89203-162-x

Printed in the U.S.A.

Second printing, 1998

Contributors

William Anderson, MD
St. Vincent Hospital and Health Center
Billings, MT

David D. Aronsson, MD
University of Vermont
College of Medicine
Department of Orthopaedics
and Rehabilitation
McClure Musculoskeletal
Research Center
Burlington, VT

Daniel J. Berry, MD
Consultant, Orthopaedic Surgery
Assistant Professor, Orthopaedics
Mayo Clinic
Rochester, MN

James V. Bono, MD
Assistant Clinical Professor
of Orthopaedic Surgery
Tufts University School of Medicine
Staff Orthopaedic Surgeon
New England Baptist Hospital
Boston, MA

Richard F. Bruch, MD
Durham, NC

Robert M. Campbell, Jr, MD
Associate Professor, Orthopaedics
University of Texas Medical School
San Antonio, TX

Michael R. Clain, MD
Attending Orthopaedic Surgeon
Greenwich Hospital
Greenwich, CT
Clinical Instructor, Orthopaedics
Yale University
New Haven, CT

Frederick R. Dietz, MD
Department of Orthopaedics
University of Iowa
Iowa City, IA

James C. Drennan, MD
Professor, Orthopaedics and Pediatrics
University of New Mexico
School of Medicine
Medical Director/CEO
Carrie Tingley Hospital
University of New Mexico
Health Sciences Center
Albuquerque, NM

Joseph M. Erpelding, MD
Orthopedic Surgeons, P.S.C.
Billings, MT

Carol Frey, MD
Associate Clinical Professor of
Orthopaedic Surgery
University of Southern California
Director,
Orthopaedic Foot and Ankle Center
Manhattan Beach, CA

Mark J. Geppert, MD
Orthopaedic and Trauma Specialists
Somersworth, NH

W. L. Gorsuch, MD
Great Falls Orthopaedic Associates
Great Falls, MT

Charles D. Jennings, MD
Great Falls Orthopaedic Associates
Great Falls, MT

Jeffrey Kobs, MD
Raleigh Orthopaedic Clinic
Raleigh, NC

J. Bohannon Mason, MD
Charlotte Hip and Knee Center
Charlotte, NC

Charles L. Nance, MD
Wilmington Orthopaedic Group
Wilmington, NC

Brad Olney, MD
Associate Professor and Chairman
Department of Orthopaedic Surgery
University of Kansas Medical Center
Kansas City, KS

William L. Oppenheim, MD
Professor and Head,
Pediatric Orthopaedics
UCLA Medical Center
Los Angeles, CA

Mark E. Petrik, MD
Clinical Assistant Professor
University of Louisville
School of Medicine
Louisville, KY

Peter D. Pizzutillo, MD
St. Christopher's Hospital for Children
Philadelphia, PA

David A. Rendleman, III, MD
Bone and Joint Surgery Clinic
Raleigh, NC

Thomas L. Schmidt, MD
Professor, Orthopaedic Surgery
University of Missouri
Kansas City
Chief, Orthopaedic Surgery
The Children's Mercy Hospital
Kansas City, MO

James F. Schwarten, MD
Orthopedic Surgeons, P.S.C.
Billings, MT

George Sotiropoulos, MD
Clinical Assistant Professor
Department of Child Health
University of Missouri
Columbia, MO

Michael D. Sussman, MD
Chief of Staff
Shriners Hospital for Children
Portland, OR

George H. Thompson, MD
Rainbow Babies and Childrens Hospital
Case Western Reserve University
Cleveland, OH

Hugh Watts, MD
Clinical Professor
of Orthopaedic Surgery
University of California at Los Angeles
Los Angeles, CA

Reviewers

To my wife, Linda,

my children, Kristin, Erika, and Jon,

and my parents, Dale and Edna Snider,

who have allowed me

the privilege

of practicing medicine.

Preface

Essentials of Musculoskeletal Care is a guide for decision making in the care of approximately 300 common musculoskeletal conditions. As such, it does not and cannot cover all orthopaedic conditions. Specifically, in-depth coverage of fracture care and surgical treatment is beyond the scope of this text and constitutes text of its own. We hope that you will find this text a quick and easy reference for identifying common problems and their treatment options.

We realize that users of this text have varying levels of interest in and comfort dealing with musculoskeletal problems. Some treat common fractures while others never look at an orthopaedic radiograph. Some are comfortable managing conditions with potentially serious adverse outcomes, and others are not. The goal of this text is to present a practitioner with information that will allow him or her to deliver optimal care to a patient while using resources, such as diagnostic tests and specialty referral, in a wise and effective manner. To accomplish this, we asked many family practitioners, pediatricians, general surgeons, and internists to review the manuscript. They approved our approach to the coverage of this subject matter. Special thanks goes to the American Academy of Pediatrics which, as an organization, has reviewed and approved this text as a resource for their members. We hope you will find it useful as well.

The value of this book comes from the boundless energy and efforts of the section editors, who were responsible for the content of this text. They are Thomas R. Johnson, MD (Shoulder and Upper Extremity), Gregory S. McDowell, MD (Spine), Jay Lieberman, MD (Hip and Thigh), Scott Kelley, MD (Knee and Lower Leg), Glenn Pfeffer, MD (Foot and Ankle), and Walter Greene, MD (Pediatric Orthopaedics). Each of these editors has relied on the assistance of many dedicated contributors who have literally thousands of cumulative hours of practice experience to focus on this text. We are deeply indebted to the many reviewers who pored over this manuscript, offering suggestions and corrections that we have incorporated into the final version. We have relied heavily on their feedback to confirm the vision and scope of this text.

The text includes two general sections (General Orthopaedics and Pediatric Orthopaedics) and seven anatomic sections. The anatomic sections are as follows: Hand and Wrist, Elbow and Forearm, Shoulder, Spine, Hip and Thigh, Knee and Lower Leg, and Foot and Ankle. Each anatomic section begins with a diagram of surface pain patterns common to that region, a text overview of conditions that affect the region, an illustrated physical examination, and finally, an alphabetically arranged set of conditions that affect the region.

The "pain diagram" that opens each anatomic section shows common pain patterns that indicate underlying orthopaedic conditions. From a specific complaint of pain, you can move directly to the specific conditions that might be responsible for the complaint. The sections on general orthopaedics and pediatric orthopaedics differ from the anatomic region pain diagrams. Pain is often not a defining feature in many of the pediatric orthopaedic conditions, and in the general section, pain patterns refer to generalized or systemic conditions rather than specific joint problems.

The text overview uses common symptoms (such as pain, instability, or locking), and patient characteristics (such as gender or age) to pinpoint likely orthopaedic conditions. The purpose is to take the reader from a common complaint to a likely condition or conditions that would cause the complaint.

Following the overview is an illustrated physical examination of the anatomic region that details tests to help identify specific disease conditions. You can use this section to see how a specific test is done. The examination of pediatric patients is confined to common screening tests of orthopaedic disease. Since the section on pediatric orthopaedics is not subdivided anatomically, the specific tests for abnormalities are included in the description of that condition. In the same way, specific physical findings for conditions in the general orthopaedics section are listed with the specific condition.

Following the examination are the orthopaedic conditions common or significant to the section, arranged alphabetically. Each condition page begins with the name of the condition, synonyms for that condition, a definition of the condition, a description of the signs and symptoms, appropriate physical and diagnostic test findings, and differential diagnoses relevant to the condition. Next, we discuss the adverse outcomes of the disease (its untreated natural history), appropriate treatment options, possible adverse outcomes associated with the treatment, and any red flags that might indicate need for specialty referral. The condition pages contain many illustrations, including anatomic drawings, clinical photographs, and classic radiographic presentations illustrative of the condition. We have also included many summary tables as a quick reference for common types of bone tumors, hand tumors, causes of limping in children, as well as commonly used NSAIDs and their dosages, appropriate splinting materials, and CDC protocols for tetanus prophylaxis. We hope that you will find that these illustrations support the written material in a useful way.

Associated with some conditions are procedures that might be commonly employed in their treatment, such as injections, minor surgical procedures, and fabrication of splints or shoe inserts.

Essentials of Musculoskeletal Care has behind it the commitment of the Boards of Directors of the American Academy of Orthopaedic Surgeons and the American Academy of Pediatrics. In particular, James Strickland, MD, in his tenure as President, gave strong and vocal support to our efforts. Also, Walter Greene, MD, while chairman of the Publications Committee, championed this project in its early stages. This first edition would not have been possible without the unselfish contributions from many reviewers, contributors, and editors. We are also deeply indebted to the publications staff at the AAOS, particularly Lynne Shindoll, whose guidance, enthusiasm, and expertise have made this book such a delight to see and use. Thanks also to Anne Arnold for her help on the upper extremity text and to the staff of Orthopaedic Surgeons, PSC, for their help throughout the process.

We plan frequent revisions and need feedback from those who use this publication to make it a better tool for office management of musculoskeletal problems. We ask that you take the time to complete the enclosed comment card so that we can continue to improve this resource in future editions. Please also feel free to write to us at *Essentials of Musculoskeletal Care*, AAOS, 6300 River Road, Rosemont, IL, 60018. You may also send a fax to 847/823-8033, or send e-mail to bsnider@mcn.net or shindoll@aaos.org.

Robert K. Snider, MD
Chairman, Editorial Board

Foreword

Musculoskeletal injuries and symptoms are second only to respiratory illnesses as the most common reason patients seek medical care. Primary care physicians and non-physician health care providers currently evaluate and treat the largest percentage of these patients with musculoskeletal complaints. As we enter the 21st century, it appears that the musculoskeletal needs of populations of patients will be met by teams of providers. These teams will include primary care physicians, surgical and medical musculoskeletal specialists, physician assistants, physical therapists, nurses, and others. *Essentials of Musculoskeletal Care* has been written to assist those providing medical care to patients with musculoskeletal problems.

This publication was developed as a cooperative effort. The work of orthopaedists, pediatric orthopaedists, pediatricians and family practitioners is embodied within the pages of this text. Their common goal was to provide a sound basis for quality musculoskeletal care and health. An example of the authors' success is seen when The American Academy of Pediatrics recommended this text to their members as a definitive resource for treating pediatric musculoskeletal conditions.

The Board of Directors of the American Academy of Orthopaedic Surgeons congratulates Dr. Bob Snider and the Editorial Board of *Essentials of Musculoskeletal Care* for their insightful and outstanding effort in producing a resource that will assist practitioners in providing quality patient care for many years to come.

Douglas W. Jackson, MD
President
American Academy of Orthopaedic Surgeons

section one

g e n e r a l   o r t h o p a e d i c s

section two

s h o u l d e r

section three

e l b o w   a n d   f o r e a r m

section four

h a n d a n d w r i s t

section five

h i p a n d t h i g h

section six

k n e e a n d l o w e r l e g

section seven

section eight

section nine

p e d i a t r i c   o r t h o p a e d i c s

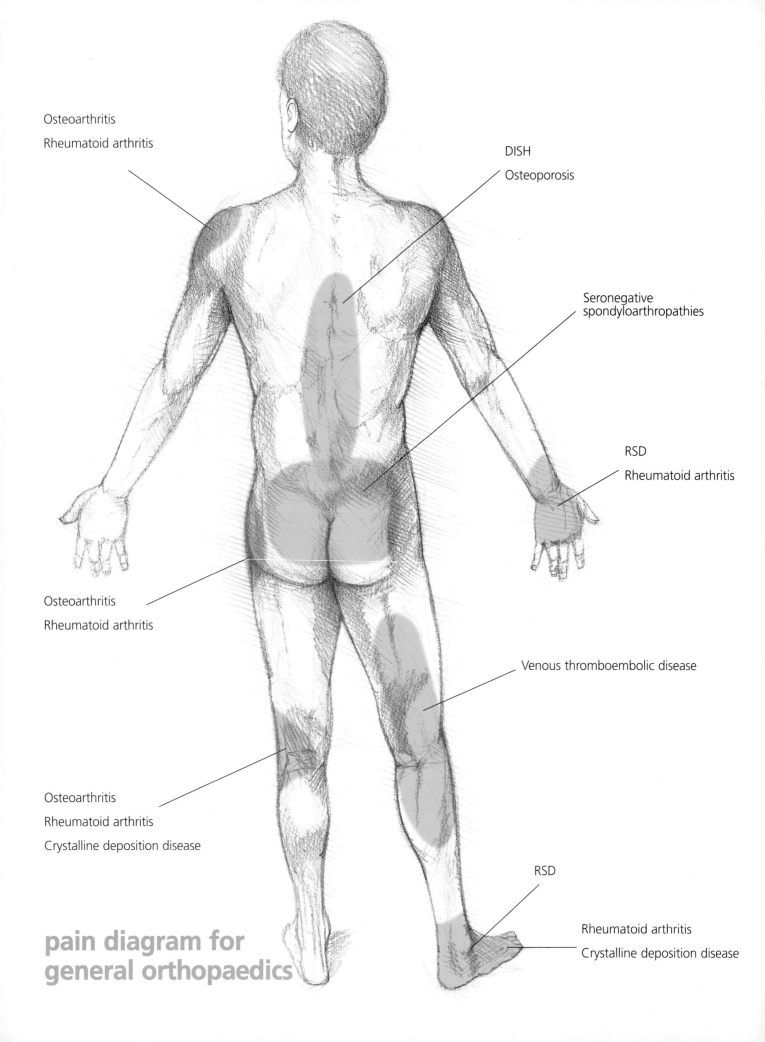

Osteoarthritis
Rheumatoid arthritis

DISH
Osteoporosis

Seronegative
spondyloarthropathies

RSD
Rheumatoid arthritis

Osteoarthritis
Rheumatoid arthritis

Venous thromboembolic disease

Osteoarthritis
Rheumatoid arthritis
Crystalline deposition disease

RSD

Rheumatoid arthritis
Crystalline deposition disease

**pain diagram for
general orthopaedics**

section one

general orthopaedics

Section Editor

Robert K. Snider, MD
Orthopedic Surgeons, P.S.C.
Billings, Montana

Richard F. Bruch, MD
Durham, North Carolina

James F. Schwarten, MD
Orthopedic Surgeons, P.S.C.
Billings, Montana

Thomas R. Johnson, MD
Orthopedic Surgeons, P.S.C.
Billings, Montana

Jay R. Lieberman, MD
Assistant Professor of Orthopaedic Surgery
UCLA Medical Center
Los Angeles, California

general orthopaedics

general orthopaedics— an overview

Some conditions involve multiple joints, multiple regions, and/or have systemic effects that cannot be classified by anatomic area. These conditions are the subject of this section of *Essentials of Musculoskeletal Care*. The purpose of this overview, as with all the others, is to highlight special conditions included in the section and to provide practical advice on how to diagnose and treat them. Also included at the end of this overview is a short glossary of common orthopaedic terms.

Pain disorders

Both arthritis and arthrosis are conditions that involve an inflammation of, or a diseased state of, a joint. The course of arthritis is as unpredictable as the weather. Even within arthritis types, the degree of impairment and suffering varies. The principle of rest, espoused by John Hunter in the 1700s, has been amended to allow protected function and movement, but remains the mainstay in the treatment of all types of arthritis.

Bursitis and tenosynovitis have a characteristic set of symptoms. The pain is worse on rising or during the night, and pain or a limp is severe until the bursa or tendon sheath is exercised. Patients may function well until they reach a limit of walking or arm use, and then the pain returns.

Arthritides
Of the arthritides, septic arthritis is the most urgent, as it demands immediate diagnosis and an efficacious treatment plan, typically involving surgical drainage or lavage. Some conditions, such as osteoarthritis and rheumatoid arthritis, affect multiple joints, with subsequent joint destruction and loss of function. Crystalline deposition diseases, such as gout, predominantly affect weightbearing joints. Spondyloarthropathies tend to affect a younger male population and often lead to significant loss of function by the time the patient reaches middle age.

Other conditions cause joint pain and sometimes swelling, such as Lyme disease, but joint destruction occurs in late stages, if at all. Hemophilic arthritis is a special condition, not covered in this text, that often leads to irreversible osteoarthritis-like changes at a young age, but can be ameliorated by aggressive management of the underlying disease, especially with newer treatments that do not bear the risk of transmitting HIV.

Medications and surgical procedures used in the treatment of arthritides vary in their effectiveness, potential to harm the patient, and cost. The current trend in arthritis treatment is to focus on obtaining a deeper understanding of the inflammatory process and using fewer invasive surgical techniques.

Use of corticosteroid injections in the treatment of arthritis and periarthritic conditions is well established. In general, repeated intra-articular injections are reserved for, and effective in, patients with rheumatoid arthritis. The course of osteoarthritis is not modified, nor well ameliorated, by intra-articular corticosteroid injections. Bursitis and tenosynovitis, however, often respond well to intermittent injections, and patients may obtain permanent or long-lasting relief.

Modalities such as manipulation, heat, massage, and special physical therapy techniques have not been scrutinized in a scientific fashion, but research addressing the effectiveness of these modalities is forthcoming. These modalities are generally reserved for short-term and intermittent use. Exercise is beneficial for most patients with arthritis, but only when the "crisis" or acute flare-up has been treated; rest is the key at that time, supplemented by short-term use of NSAIDs.

Other pain disorders

Three conditions, considered the "fuzzy areas" of arthralgias and periarticular pain, are discussed in this section: fibromyalgia, reflex sympathetic dystrophy (RSD), and cumulative trauma disorders (CTD). These processes are as difficult to treat as they are to understand, especially when issues of causation and compensation mix with issues of comfort. Like the above conditions, nonorganic behavior is often misinterpreted as a sign of nondisease, or even malingering; however, it is more likely a predictor of the patient's satisfaction with treatment outcomes than it is an indicator of "psychosomatic" behavior and often has a negative effect on the patient-physician relationship.

Trauma

Trauma is a principal cause of musculoskeletal disorders, including arthritis. Traumatic compartment syndrome is catastrophic if unrecognized and untreated. Appropriate splinting is necessary for all fractures, partly to reduce the likelihood of compartment syndromes, and partly to decrease soft-tissue injury and pain while the patient awaits definitive treatment. Early treatment of traumatic injuries must also focus on the risk of posttraumatic thromboembolic disease and its attendant risks of pulmonary embolism and death. There are accepted and standard principles of fracture care that form the basis of successful management of these injuries, both to reduce complications and to improve the likelihood of a good functional and structural outcome. These principles can be found in the chapter titled Fracture Principles.

Stiffness

Stiffness occurs in two varieties: that which improves with exercise (tendinitis, bursitis), and that which does not (joint effusion, end-stage joint degeneration).

Patients who have stiffness associated with tendinitis and bursitis usually feel better once the tendon or joint is moved. Think of it as mechanically squeezing out the edema and allowing free motion of the parts. However, with extended activity, inflammation develops and the patient notices increasing pain and the return of stiffness.

In end-stage degenerative joint disease, or with large joint effusions, patients often report stiffness that does not improve much with motion. One reason for this may be best explained by an engineering principle: one of the fundamental concepts of hydraulics is that water cannot be compressed. Thus, hydraulically, large effusions block joint motion; therefore, patients have stiffness rather than swelling. With chronic effusions, the joint capsule stretches, creating a larger cavity and less compression of the effusion. Stiffness improves, although the effusion is still large. When the capsule stretches, proprioception reduces as the joint changes position or as muscles contract. With this, patients often report buckling or collapse of the joint, especially at the knee.

Instability

Instability is typically caused by traumatic ligament tears. At the knee, the cruciate and collateral ligaments are commonly injured; at the shoulder, dislocations leave a torn capsule, which increases the likelihood that future dislocations or subluxations will occur.

Instability also occurs in association with rheumatoid arthritis. Patients with rheumatoid arthritis eventually have severely unstable joints in the extremities and spine as the ligaments and capsular structures stretch with chronic effusions and with erosion from the disease itself.

Patient age

Caring for musculoskeletal disorders in children and adolescents requires the practitioner to be familiar with a cornucopia of problems. Children may have congenital disorders, acquired conditions such as Legg-Calvé-Perthes disease, traumatic fractures and soft-tissue injuries involving primarily the upper extremities, and/or infectious and inflammatory diseases. Conditions that typically affect adolescents involve the accelerated growth phase, such as scoliosis, kyphosis, epiphyseal slips and fractures, and spondylolisthesis. Malignant primary bone tumors most commonly occur in this group. These topics are covered in depth in the section on Pediatric Orthopaedics.

With adults, the types of problems change. Trauma and conditions associated with overuse most commonly affect young adults. As adults reach their forties, degenerative conditions that affect tendons, intervertebral disks, and joints comprise the largest source of complaints. Common presenting problems in the elderly include fractures, metastatic tumors, and arthritis.

Abuse

Abuse involving children or the elderly is a complex social and medical problem. Failing to diagnose abuse (false negative) and overdiagnosing abuse (false positive) may lead to catastrophic consequences for the patient and the family.

Note, however, that it is essential that social service agencies be notified early if a patient presents with multiple bruises, lacerations, or fractures (especially torsional fractures, from twisting) in different stages of healing. Old, unattended lacerations and healing (or healed) fractures, as revealed on radiographs, strongly indicate neglect. In the section on Pediatric Orthopaedics, child abuse is discussed in a separate chapter.

Patients whose history is given wholly by a caregiver may feel unable to talk in their presence. In these circumstances, interview the patient and caregiver separately. A caregiver frustrated by an elderly patient's memory problems, behavior problems, alcoholism, or difficult personality may be abusing the patient regularly. Further, a financially stressed caregiver may be usurping the elderly patient's finances for his or her own benefit.

The complexity of these problems and the seriousness of the consequences demand familiarity with community resources and knowledge of the competence, compassion, and professionalism of those who will investigate the potential abuse.

Gender

While there are few conditions considered particular to men or women, some seem to affect one gender or the other with consistent frequency. Osteoporosis and trochanteric bursitis more commonly affect women, while diffuse idiopathic skeletal hyperostosis (DISH) and multiple myeloma more commonly affect men.

Table 1
Glossary of common orthopaedic terms

Arthrodesis	The surgical process of promoting bone growth across a joint to eliminate the joint and stop motion. The usual purpose is relief of pain or stabilization of an undependable joint.
Capsule	A complex ligamentous structure surrounding a joint like a sleeve. It is a composite of ligaments, tendon expansions, and tendon attachments. The capsule allows motion of joints while stabilizing them, especially in rotation.
Cartilage	A cellular tissue that, in the adult, is specific to joints, but in children forms a template for bone formation and growth. Hyaline cartilage is a cellular low-friction tissue that coats joint surfaces. Fibrocartilage is tough with high collagen content, such as found in the meniscus of the knee, or the annulus fibrosus portion of the intervertebral disk.
Closed fracture	A fracture in which the integrity of the surrounding skin is intact.
Closed reduction	A procedure in which normal relationships are restored to a fractured bone or dislocated joint; no incision is needed.
Condyle	The knobby portion of the end of a long bone that articulates with another bone through a joint and serves as a point of attachment for tendons and ligaments.
Cox-, Coxa	Hip. Coxa vara is a varus, or adduction, deformity of the hip.
Cubitus	Elbow. Cubitus varus is a bow or adduction deformity of the elbow.
Delayed union	A delay in normal fracture healing; not necessarily a pathologic process.
Diaphysis	The shaft of a bone; the portion between the metaphyses at either end.
Dislocation	A disruption in the relationship of two bones forming a joint in which the bones have (at least momentarily) completely moved out of their normal positions. Usually, bones lie alongside one another, locked in that position until replaced by a closed or open reduction.
Epiphysis	The end of a bone; the portion that articulates with an adjacent bone to form a joint.
Extensor	Relating to the extensor side of a limb, usually the posterior surface, but is anterior on the leg from embryonic limb rotation.
External fixation	Surgical insertion of pins through the skin that are then attached to an external frame to stabilize a fracture or joint.

Flexor	Relating to the flexor side of a limb, usually the anterior surface, but is posterior on the leg from embryonic limb rotation.
Fracture	A disruption in the integrity of a bone.
Fracture-dislocation	A fracture of bone associated with a dislocation of its adjacent joint.
Fusion	Arthrodesis; a biological "welding" process in which adjacent joint surfaces are excised to facilitate bony growth from one side of the joint to the other and stop painful joint motion. In the spine, bone graft material "welds" two vertebral bodies to create a single bone and stop motion of the adjacent vertebrae.
Genu	Knee. Genu valgum is knock-knee deformity; genu varum is a bowleg deformity.
Internal fixation	Surgical insertion of a device that stops motion across a fracture or joint to encourage bony healing or fusion.
Kyphosis	A deformity of the spine in which the spine bends forward.
Ligament	A collagenous tissue that either binds together two bones to form a joint, or acts to suspend a structure in a certain position.
Lordosis	A deformity of the spine in which the spine sways, or bends backwards.
Malunion	Healing of a fracture in an unacceptable position. Malunions are worst when they angulate the bone away from the plane of motion of the adjacent joints.
Meniscus	A fibrocartilage structure in the knee, interposed between the femur and tibia on the medial and lateral sides.
Metaphysis	The broad portion of the bone adjacent to a joint.
Myelopathy	An abnormal condition of the spinal cord, whether through disease or compression. The usual consequences are spasticity, impairment of sensation, and impairment of bowel and bladder function.
Neuropathy	An abnormal condition involving a peripheral nerve.
Nonunion	Failure of healing of a fracture. With continued motion through a nonunion, a pseudarthrosis will form.
Open fracture	A fracture that communicates with air via a disruption of the skin.
Open reduction	An open surgical procedure in which normal or near-normal relationships are restored to a fractured bone or dislocated joint.
Osteomyelitis	Infection of bone, either bacterial or mycotic. These infections may be life long and are difficult to treat.
Osteonecrosis	Literally, death of bone. This term is also known as aseptic necrosis, or death of bone in the absence of infection. In children, it occurs in Legg-Calvé-Perthes disease, and in adults may occur in the femoral head (in association with alcoholism) or in the medial

tibial plateau. Removing dead bone and laying down new bone is an incredibly slow process; the surrounding bony architecture usually collapses before the process occurs, leaving irreversible bony deformity and arthritis.

Osteosynthesis	The process of bony union, as in fracture healing. A biological welding process, sometimes facilitated with grafts of bone from the iliac crest and insertion of fixation devices.
Palmar	The anterior surface of the forearm, wrist, and hand.
Percutaneous pinning	Insertion of pins into bone through small puncture wounds in the skin for stabilization of a fracture or a dislocated joint that was realigned by closed reduction.
Physis	The growth plate. It is interposed between the metaphysis and epiphysis and is cartilaginous in structure.
Plantar	The sole, or flexor surface of the foot.
Pseudarthrosis	A false joint produced when a fracture fails to heal. It develops a pseudocapsule and synovial-like fluid.
Scoliosis	A deformity of the spine in which the spine bends to the right or left. With double curves, one may be to the right and the other to the left. Curves to the left in the thoracic region may be from an intraspinal tumor.
Septic arthritis	Infection of a joint, either bacterial or mycotic.
Spondylolisthesis	A slippage or subluxation of one vertebral body on the one below; the slippage can be anterior, posterior, or to either the right or left side. The common causes are degenerative changes in the disk and facet structures in adults, and a specific defect in the lamina called spondylolysis, which appears at or before early adolescence, creating a deformity that persists throughout life.
Subluxation	An incomplete disruption in the relationship of two bones forming a joint. The joint surfaces are still related to one another but are not perfectly aligned.
Synovium	The lining of a joint, which produces synovial fluid for joint lubrication.
Tendon	A highly collagenous tissue attached to muscle at one end and to a bone at the other. It transmits forces of muscular contraction to cause motion across a joint.
Tenosynovium	The sheath within which a tendon glides as it transmits muscle forces across joints.
Trochlea	A groove in a bone that articulates with another bone, or serves as a channel for a tendon to track in.
Valgus	Abduction of a distal bone in relation to its proximal partner. Valgus of the knee is a knock-knee deformity, with abduction of the tibia in relation to the femur.
Varus	Adduction of a distal bone in relation to its proximal partner. Varus of the knee is a bowleg deformity, with adduction of the tibia in relation to the femur.
Volar	The anterior surface of the forearm, wrist, and hand.

Acute Compartment Syndrome

Synonyms

Volkmann's ischemic contracture or necrosis

Definition

Compartment syndrome occurs when perfusion of muscle and nerve tissues decreases to a level inadequate to sustain the viability of these tissues. In compartment syndrome, the intracompartmental tissue pressure becomes elevated and produces a secondary elevation in venous pressure that obstructs venous outflow. An escalating cycle of continued increase in intracompartmental tissue pressure occurs. Necrosis of muscle and nerve tissues can develop in as few as 4 to 8 hours. Thus, treatment of compartment syndrome constitutes a surgical emergency.

Compartment syndrome most commonly follows trauma, especially in association with fractures of the long bones of the leg or forearm. However, it may also develop in long-distance runners, new military recruits, or others involved in a major change in activity level. Patients with peripheral vascular disease are also at risk, as their poor tissue perfusion leads to mild ischemia and transudation of fluid, which increases the local compartment pressure.

The volar aspect of the forearm and the anterior compartment of the leg (shin area) are the most commonly affected. If untreated, tissue necrosis can develop with secondary muscle paralysis, muscle contracture, and sensory impairment. This process is called Volkmann's ischemic necrosis and results in claw hand or foot caused by contracture of the ischemic muscle.

Clinical symptoms

Intractable pain and, sometimes, sensory hypesthesia distal to the involved compartment (top of foot, median distribution in the hand) are the most important early symptoms.

The "**P**" characteristics are **p**ain, **p**aresthesias, and **p**aralysis.

Tests

Exam

Examination of a patient who is wearing a cast or dressing begins by removing the cast and any padding in order to carefully evaluate the muscle compartments. Palpation of increased compartment pressure is subjective, but normally there should be some softness to the compartment. Manometric measurement of suspicious increases in compartment syndrome is mandatory.

The most important physical sign is extreme pain on stretching the long muscles passing through a compartment. Full extension of the fingers or plantar flexion of the toes will stretch the muscles of the forearm or shin. The inability to actively contract these muscles voluntarily, as when making a fist or dorsiflexing the toes, is an indication of paralysis. While passive or active stretching of muscles around fractured bones is always painful, it is usually tolerable; if it is not, the patient may have a compartment syndrome.

Pulselessness indicates arterial trauma—not compartment syndrome—although a compartment syndrome may exist in combination with vascular ischemia. Pulses are typically completely normal in compartment syndrome since the intracompartmental pressure rarely exceeds systolic or mean arterial pressure levels.

Diagnostic

Manometric techniques are available to measure compartment pressure directly. General guidelines indicate the presence of a compartment syndrome when the diastolic pressure minus the intracompartmental pressure is less than or equal to 20 mm Hg. Hypovolemic shock is a predisposing factor for reduced perfusion in a traumatized compartment that has elevated intracompartmental pressures and decreased arterial perfusion.

Differential diagnosis

Arterial injury (pulse deficit)

"Shin splints" (exercise-induced weakness and pain)

Adverse outcomes of the disease

Without immediate treatment, compartment syndrome may result in permanent loss of function. The muscles die, scar, and shorten; fingers and toes are often clawed and have little motion. The wrist is held in flexion, and sensation is impaired. Late reconstructive surgery has little chance of restoring original, normal function.

Treatment

Because muscle necrosis may develop within as few as 4 to 8 hours, there is little time to delay treatment. Even a suspicion of compartment syndrome probably requires treatment, especially if intracompartmental pressure measurements support the diagnosis. Surgical fasciotomy of the compartment is essential. The wound is left open, with delayed closure or skin grafting performed after the edema subsides.

For outpatient fractures, providing strict instructions to the patient or guardian to call at any hour, day or night, is necessary if pain is unbearable or if the patient is unable to actively extend the long extensors of the fingers or toes. Patients must be able to actively extend their fingers or toes before leaving the emergency department or office so that either they or the parents/guardians know what is acceptable motion and usage.

The use of splints is preferred over that of circumferential casts in the treatment of acute trauma. Elevating the lower extremity on a single pillow placed under the calf will help reduce edema. Elevating the upper extremity with a pillow is also appropriate. Excessive elevation should be avoided since this reduces hydrostatic pressure and may lower arterial pressure enough to decrease compartment perfusion.

Adverse outcomes of treatment

There is very little negative risk in treating an acute compartment syndrome with fasciotomy, except that the scar may be unsightly. However, failure to perform a fasciotomy could be disastrous for the patient.

Referral decisions/Red flags

Even a tentative diagnosis of compartment syndrome requires urgent evaluation so that surgical decompression by emergency fasciotomy can be considered.

Adult Seronegative Spondyloarthropathies

Synonyms

Enthesopathy

Behçet's disease

Reiter's syndrome

Ankylosing spondylitis

Marie-Strumpell arthritis

Rheumatoid spondylitis

Inflammatory bowel disease

Psoriatic arthropathy

Reactive arthritides (related to *Yersinia enterocolitica, Salmonella,* or *Shigella* infections)

Definition

Seronegative spondyloarthropathies are arthritides that involve the spine and peripheral joints but are not associated with presence of rheumatoid factor or antinuclear antibodies. They are often associated with HLA-B27 and other genetic markers, but the B27 marker is neither necessary nor sufficient alone to define the disease. These conditions often affect young men and are associated with back pain and stiffness.

Clinical symptoms

Spine, sacroiliac, and major weightbearing joint pain is common. Enthesopathy, or pain and tenderness at the insertion of ligaments and tendons into bone, is common. Morning stiffness is a common complaint, with patients needing more than 30 minutes to "warm up" when arising.

Tests

Exam

Reiter's syndrome involves an asymmetric inflammatory arthritis of the lower extremities associated with urethritis, conjunctivitis, skin rash, or mucous membrane ulceration. Enthesopathy and unilateral sacroiliitis are common.

In ankylosing spondylitis, tenderness over the sacroiliac joint, enthesopathy, and stiffness are common, especially in young men.

Diagnostic

HLA-B27 antigen is present in 90% of patients with ankylosing spondylitis and 85% of patients with Reiter's syndrome, in addition to a negative rheumatoid factor (RF) and antinuclear antibody (ANA). The HLA test is used to confirm the clinical and radiologic diagnoses. In ankylosing spondylitis, the erythrocyte sedimentation rate is usually normal or mildly elevated unless the disease is highly symptomatic; it may be as high as 100 mm/hr in association with Reiter's syndrome.

Radiographs show sacroiliitis and subsequent fusion of the sacroiliac joint (Figures 1 and 2). There is typically ankylosis or ossification of the anterior longitudinal ligament in the spine, and spontaneous fusion of the posterior facet joints leading to the classic "poker spine," which is fused from top to bottom.

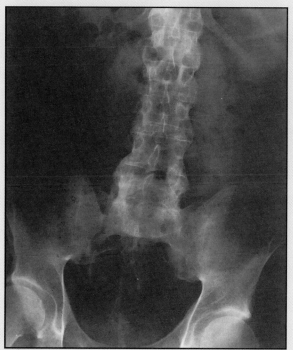

Figure 1
AP radiograph of the pelvis demonstrating advanced sacroiliitis

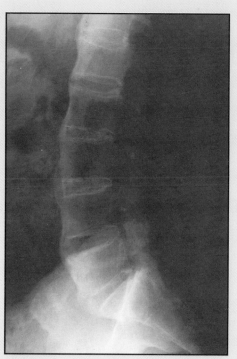

Figure 2
Lateral radiograph of the lumbar spine demonstrating bridging syndesmophytes

Differential diagnosis

Rheumatoid arthritis (positive RF, peripheral joint involvement)

Adverse outcomes of the disease

Severe spinal deformity in which patients are unable to see the horizon is common. Loss of function and motion in the major weightbearing joints of the lower extremities may also occur. Uveitis affects 10% to 20% of patients with ankylosing spondylitis. In addition, restrictive pulmonary disease and pneumonia develop in some patients due to poor chest expansion.

Treatment

NSAIDs are effective in controlling symptoms for many patients. Sulfasalazine may be of value if further studies confirm its efficacy, and tetracycline is appropriate for suspected Chlamydia infections. Surgery is reserved for patients who require joint replacement or spinal osteotomy for severe kyphosis.

Adverse outcomes of treatment

Postoperative infection and/or loosening of implants are possible. Heterotopic ossification may complicate total joint replacement. NSAIDs can cause gastric, renal, or hepatic complications.

Referral decisions/Red flags

Patients with kyphosis, pain at rest, or pain at night in a weightbearing joint need further evaluation.

Bone Tumor

Synonyms

Malignancy

Neoplasm

Bone lesion

Bone cancer

Definition

Bone tumors are classified as either benign or malignant. Malignant tumors can metastasize, whereas most benign tumors rarely do, with the following exceptions: giant cell tumors, osteoblastomas, and chondroblastomas. Note that metastatic foci associated with these benign tumors are usually not life threatening.

Bone tumors are further classified by their tissue of origin. There are both benign and malignant bone-forming and cartilage-forming tumors. In addition, nonosseous tumors may be classified as fibrous, fibrous histiocytic, vascular, and bone marrow in origin.

Age is an important factor in predicting which bone tumor is most likely present. Any malignant lesion of bone in a patient older than age 40 years must be considered as a possible skeletal metastasis, although age is not always a critical factor (Table 1).

Table 1
Bone tumors and tumor-like conditions by age

1 to 5 years	6 to 18 years	19 to 40 years	40+ years
Osteomyelitis	Simple bone cyst	Ewing's sarcoma	Metastases
Metastatic neuroblastoma	Aneurysmal bone cyst	Giant cell tumor	Multiple myeloma
Leukemia	Nonossifying fibroma	Osteosarcoma	Chondrosarcoma
Eosinophilic granuloma	Ewing's sarcoma		Fibrosarcoma
Simple bone cyst	Osteomyelitis		Malignant fibrous histiocytoma
	Osteosarcoma		Chordoma
	Enchondroma		
	Chondroblastoma		
	Chondromyxoidfibroma		
	Osteoblastoma		
	Fibrous dysplasia		
	Osteofibrous dysplasia		

general orthopaedics

Clinical symptoms

Mild pain and persistent swelling are the usual presenting complaints that may be helpful in identifying the character and location of a lesion. Constant pain or pain that prevents the patient from sleeping is characteristic of both benign and malignant tumors. This is especially true of tumors such as osteoid osteomas, osteoblastomas, aneurysmal bone cysts, or giant cell tumors.

Fever, malaise, weakness, and other constitutional symptoms often occur in association with Ewing's sarcoma. Weakness and anemia often accompany multiple myeloma.

Tests

Exam
Physical examination should focus on identifying masses, bony tenderness, joint involvement with reduced range of motion, a limp, or deformity. In patients older than age 40 years, careful examination of the prostate in men and breasts in women is essential to rule out a primary carcinoma.

Diagnostic
AP, lateral, and oblique radiographs help identify the location of the lesion within the bone, the presence of calcification within the lesion, and periosteal reaction adjacent to the lesion. Radio-graphs will also identify whether the lesion is eccentric or central, lytic or blastic, and whether its border is well circumscribed or not well delineated (Figures 1 and 2). Primary bone tumors do not cross a joint to involve bones on both sides, in contrast to infections.

Radioisotope studies are helpful in determining the presence or absence of other lesions and the relative activity of a lesion. Lesions associated with multiple myeloma generally will not demonstrate increased activity on bone scan. A skeletal survey is required to define the extent and number of metastases.

A CT scan is excellent for demonstrating bony changes and degree of calcification within the lesion. An MRI scan is particularly helpful in delineating soft-tissue lesions and marrow involvement. In addition, these studies are helpful in accurately staging the tumor. The goal in staging a bone neoplasm is to determine the extent of the disease before performing a biopsy and initiating definitive treatment. If a malignant tumor is suspected, a chest radiograph and CT scan of the chest should also be obtained. Biopsy evaluation of the tumor with imaging and other diagnostic studies will decrease the risk of tissue contamination or tumor spread.

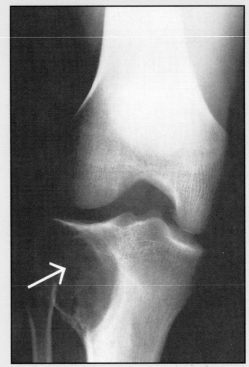

Figure 1
Lytic tumor destroying bone of proximal lateral tibia

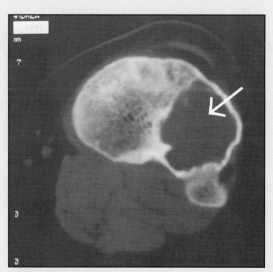

Figure 2
CT scan of tumor shown in Figure 1

Routine laboratory tests may help narrow the differential diagnosis and provide baseline values for patients who need chemotherapy. The following laboratory tests are usually obtained: CBC with differential, serum electrolytes, blood urea nitrogen, creatinine, calcium, phosphorus, and alkaline phosphatase. Tests for patients age 40 years and older include the following: urinalysis, urine and serum protein electrophoresis, and prostate specific antigen (in men).

A biopsy is performed to determine if the lesion is benign or malignant, to determine the cell type of lesion, and to determine the grade of the lesion. The biopsy should be performed by the surgeon who may be required to perform the definitive surgical procedure.

Differential diagnosis

Infection (fever, constitutional symptoms may involve joint)

Tumors metastatic to bone (lung, breast, prostate, thyroid, and renal)

Benign bony lesions

Aneurysmal bone cyst (metaphyseal, in children and teens)

Chondroblastoma (epiphyseal, in children and teens)

Chondromyxoid fibroma (metaphyseal, in young adults)

Enchondroma (central metaphyseal, in children and teens)

Fibrous dysplasia (diaphyseal, deformity, limb shortening, in children and teens)

Giant cell tumor (epiphyseal, metaphyseal, in adults younger than 40 years)

Nonossifying fibroma (metaphyseal, in children and teens)

Osteochondroma (bony cap on a stalk that points away from adjacent joint, in children, teens, and adults)

Osteoid osteoma (night pain, pain relieved by aspirin, small size, in children and teens)

Simple bone cyst (metaphyseal, in children and teens)

Malignant bony lesions

Chondrosarcoma (central metaphyseal, pelvis, and flat bones, in adults older than 40 years)

Chordoma (typically sacral with mass on rectal exam, coccygeal pain, in adults older than 40 years)

Ewing's sarcoma (round cell tumor that may simulate osteomyelitis, in children and teens)

Fibrosarcoma/malignant fibrous histiocytoma (metaphyseal, non-bone-producing, in adults)

Myeloma (spine, marrow bones like pelvis, bone destruction, in adults older than 40 years)

Osteosarcoma (metaphyseal, long bones, in children and teens)

Paget's sarcoma (substantial increase in pain suggests malignant transformation of Paget's disease, in older adults)

Adverse outcomes of the disease

Pathologic fracture (minimal trauma, poor healing), disability, and even death are possible.

Treatment

Simple observation is appropriate for many benign bone tumors in which the diagnosis is clear on radiographs. However, others may require complete excision, as they can behave in a malignant manner by causing repeated pathologic fractures and disability. Patients with minimal bone destruction and minimal symptoms are observed for changes that might weaken the cortical bone.

In malignant primary bone lesions, surgery in conjunction with radiation therapy, chemotherapy, and immunotherapy offers varying responses dependent on tumor type, age, and location. Limb salvage surgery (internal amputations with replacement using allografts or prostheses) often preserves function lost with amputation techniques, such as above-knee amputation or hemipelvectomy.

Adverse outcomes of treatment

Disability, disfigurement, pathologic fracture, infection, failure of implant, and serious nerve and vascular deficit may accompany these often heroic procedures.

Referral decisions/Red flags

Suspicious bone or soft-tissue masses, unusual pain or night pain, constitutional symptoms in association with bone pain or lytic or blastic changes of bone, soft-tissue calcification, or periosteal reaction on radiography all require further evaluation.

Acknowledgements

Figures 1 and 2 and Table 1 are reproduced with permission from Kasser JR (ed): *Orthopaedic Knowledge Update 5.* Rosemont, IL, American Academy of Orthopaedic Surgeons, 1996, pp 133–148.

Crystalline Deposition Diseases

Synonyms

Gout

Podagra

Calcific periarthritis

Calcium pyrophosphate deposition disease (CPPD)

Pseudogout

Definition

Crystalline deposition diseases, such as gout, induce a synovitis and secondary inflammatory process within a synovial joint. The inflammation is caused by substances released during leukocyte phagocytosis of the crystals. While the affected crystals and joints differ among the various types of disease, the clinical pathway of joint inflammation is similar. The onset is rapid, usually of finite duration, and is associated with pain and erythema. The prevalence of this condition increases with age.

Clinical symptoms

Patients with gout often report sudden pain at the base of the great toe. Other weight-bearing joints, especially the knee, are affected in all forms of crystalline deposition disease, including pseudogout. Onset is often linked to trauma or surgery, but most episodes are not associated with remembered trauma.

Patients may have a fever—up to 103°F (39.4°C)—especially with the polyarticular form of the disease. Up to 40% of patients with gout also have renal stones.

Masses may appear, especially around the feet, as gouty tophi form as a chronic manifestation of the disease. Note that not all cases of gout are associated with symptoms; some patients have asymptomatic tophi.

Tests

Exam
Redness, limited motion, and swelling are the principal findings. With long-standing gout, tophaceous deposits of urate crystals create masses around the joints and in the cartilage of the outer ear. In gout, the foot is most often involved, possibly because of repetitive trauma. Other common sites include the olecranon and patellar bursae. In pseudogout, the knee is most often involved.

Diagnostic
Normal values for serum uric acid levels range from 3 to 7 mg/dL in adult men and 2 to 6 mg/dL in women and in children. Acute episodes occur more frequently if patients have chronic hyperuricemia (levels greater than 10 mg/dL), but an elevated uric acid level is not required for an acute episode. Lowered renal excretion of urates is the most common cause of elevated serum uric acid levels. Measurement of 24-hour urinary uric acid excretion helps differentiate patients who "overproduce" uric acid from those who "under excrete."

Differential diagnosis

Charcot foot (diabetic neuropathy, bony deformity, or enlargement)

Leukemia (peripheral blood smear)

Lyme arthritis (chronic fatigue, memory loss, history of rash)

Reiter's syndrome (urethritis)

Rheumatoid arthritis (positive RF, hand and shoulder involvement)

Septic arthritis and infection (elevated erythrocyte sedimentation rate)

Trauma (history, evidence of fracture on radiograph)

Adverse outcomes of the disease

Tophaceous deposits and joint destruction are associated with chronic gout.

Treatment

For an acute episode of gout, drug therapy of 150 mg of oral indomethacin for the first dose, followed by 50 mg every 8 hours is the usual regimen; however, the initial dose may be lowered to 50 mg, followed by 25 mg every 8 hours, especially if gastrointestinal intolerance is a concern with the higher initial dose. An alternate treatment uses colchicine orally, 0.5-mg tablets, 1 to 2 initially followed by 1 tablet every hour until symptoms abate or gastrointestinal side effects (stomach pain, vomiting, or diarrhea) occur.

Each patient should learn the dosage needed to abort their acute attack and keep an appropriate supply of tablets on hand.

Corticosteroids may be necessary for polyarticular or refractory gout. Behavior modification may help reduce the frequency of attacks.

Allopurinol, a xanthine oxidase inhibitor, is given in dosages of 100 to 300 mg/day. It is usually reserved for patients whose serum uric acid levels remain chronically greater than 10 mg/dL despite dietary restriction of protein and calories. Side effects include rashes, drug fevers, hepatitis, and bone marrow depression. Allopurinol should be avoided in patients with renal insufficiency. Prophylaxis is reserved for individuals who have chronic tophaceous or intercutical forms, or frequent episodes of gout. Hyperuricemia alone is not a reason for treatment.

Probenecid is usually given to patients with chronic gout and hyperuricemia, as it increases urate excretion; the typical dose is 0.25 g once daily for a week, followed by 0.5 g twice daily. Sulfinpyrazone is a stronger uricosuric agent, but its use can result in symptoms of gastric distress.

Joint aspiration, especially in patients with pseudogout, may relieve acute attacks. Splint the affected joint for a day or two, if needed for pain control.

Adverse outcomes of treatment

Drug fevers, renal failure, hepatitis, and/or marrow depression are all possible.

Referral decisions/Red flags

Patients with signs of joint destruction need further evaluation.

Cumulative Trauma Disorders of the Upper Extremity

Synonyms

Overuse syndrome

Repetitive strain injury

Repetitive trauma disorder

Occupational cramp

Occupational arm pain

Occupational stress syndrome

Work-related pain disorder

Pain dysfunction syndrome

Regional musculoskeletal disorder

Definition

Cumulative trauma disorders (CTDs) are either caused by or aggravated by repetitive motion of or sustained exertion of a particular body part. CTD is an umbrella term that encompasses specific conditions such as carpal tunnel syndrome, epicondylitis, flexor tendinitis, as well as generalized myofascial pain. While the exact cause of CTDs remains controversial, several factors likely contribute to or aggravate these conditions. Repetitive tasks, forceful exertions, vibration, cold temperatures, awkward postures in the workplace, the ergonomic environment, the patient's job satisfaction, the state workers' compensation laws, the patient's psychological makeup, and the social environment all contribute to the development of cumulative trauma disorders. CTDs reflect an interplay of physical, psychosocial, and sociopolitical factors.

Clinical symptoms

Unfortunately, there is no consensus on the definition of cumulative trauma disorders nor any agreement concerning the diagnostic criteria. The many synonyms described above illustrate the confusion that surrounds these conditions. CTDs develop over time—from a few weeks to years. Frequently, the onset is insidious, and patients may not report their problems early on in anticipation that the condition will improve.

Typical complaints include pain, fatigue, numbness, or any combination of the above. While patients often have difficulty pinpointing the location of the pain, they may have specific complaints related to carpal tunnel syndrome, lateral elbow epicondylitis, or flexor tenosynovitis in the palm, associated with vague and nonanatomic discomfort elsewhere in the arm. In addition to specific numbness related to a carpal tunnel syndrome or ulnar nerve entrapment at the elbow, they may report numbness in a nonanatomic or nondermatomal distribution. Patients may also report a sensation of swelling in the extremity, although it is not generally apparent on examination.

Certain individuals are "at risk" for CTDs, principally those with exposure to physical stresses (repetition, force, awkward postures, temperature extremes, and vibration) and those with psychosocial stresses (fast work pace, inflexibility in the workplace, monotonous demanding tasks, and patient depression). Women with serious psychosocial problems, such as physical and sexual abuse, are at increased risk for experiencing unexplained musculoskeletal pain. Individuals who are poorly educated or in mundane, low-paying jobs also are at risk. Note, too, that highly skilled individuals who believe that they are overworked, underpaid, overstressed, and unappreciated are also at higher risk.

general orthopaedics

Tests

Exam
Patients should be asked specific job-related questions, including questions about job satisfaction, working conditions, the relationship with the supervisor and co-workers, exposure to repetitive and forceful exertions, vibration, cold temperature, and job harassment. Identifying the type of industry and specific job is also important, as the incidence of claims for CTDs is higher for certain industries and jobs, such as meat packers, assembly line workers, grocery checkers, and clerical workers.

Examination may reveal evidence of localized tenderness, weakness, swelling, and skin discoloration. The patient may have findings of carpal tunnel syndrome, deQuervain's disease, flexor tenosynovitis in the hand, "tennis elbow," or ulnar nerve entrapment at the elbow.

In addition, the examination also may reveal multiple tender point areas, exaggerated response to gentle palpation, loss of sensation in nondermatomal distributions, increased sweating, and changes in skin temperature. Frequently, patients appear anxious or depressed.

Many of the conditions listed in the differential diagnosis may be present concomitantly.

Diagnostic
Radiographs are indicated if there is a clear history of trauma, but they are often normal. Likewise, nerve conduction studies can be ordered to rule out carpal tunnel syndrome or ulnar nerve entrapment at the elbow, but results of these studies are usually normal.

Differential diagnosis

Carpal tunnel syndrome (positive Phalen's test, weakness in opponens muscle)

deQuervain's tenosynovitis (positive Finkelstein's test, local tenderness, and swelling)

Elbow epicondylitis (local tenderness, decreased grip strength)

Flexor tenosynovitis (local palmar tenderness, trigger finger)

Ulnar nerve entrapment (numbness at the ulnar side of the fingers, Tinel's sign at the medial elbow)

Adverse outcomes of the disease

With these disorders, patients often lose time from work and experience psychological changes. They may even change occupations or simply never return to work.

general orthopaedics

Treatment

Cumulative trauma disorders are best treated by a team of health care professionals skilled in dealing with work-related injuries. A satisfactory outcome for the patient is dependent on the cooperative efforts of the employer, the insurance carrier, and a physician-directed health care team that might include physical and occupational therapists, occupational health nurses, and vocational rehabilitation counselors. A case manager, such as an occupational health nurse, is invaluable in coordinating the efforts of these groups to return the patient to work. Although surgery may be indicated, it is seldom urgent. While the outcome of surgical intervention in these patients is quite often disappointing, carpal tunnel release appears to be effective if carpal tunnel syndrome is indeed the principal problem.

Initial treatment should include ice and rest of the affected part, along with a progressive exercise program of subsymptomatic stresses to strengthen the limb. At the workplace, modify tasks and work schedules, and consider job change if the symptoms persist.

Lack of job satisfaction and depression are also important predictors of recovery. Poor prognosis is associated with both long-standing disability (longer than 6 months) and litigation.

Adverse outcomes of treatment

Drug dependence from narcotics or antidepressants is possible. If the patient has had multiple, ill-advised surgeries, persistent tender scars may be left.

Referral decisions/Red flags

Once a diagnosis of CTDs is suspected, a team of health care professionals may be the best way to manage these difficult problems.

Diffuse Idiopathic Skeletal Hyperostosis

Synonyms

Ankylosing hyperostosis

Spondylitis ossificans ligamentosa

Vertebral osteophytosis

Definition

Diffuse idiopathic skeletal hyperostosis (DISH) is an idiopathic disease of spinal ligamentous structures that often leads to stiffness and pain. It does not produce the severe flexion deformity of ankylosing spondylitis. The hallmark is confluent ossification spanning three or more intervertebral disks, most commonly in the thoracic and thoracolumbar spine. The disease primarily affects Caucasian men (male to female ratio is 2:1) who are age 60 years or older. DISH may be a multisystem disorder since it is often associated with diabetes mellitus, obesity, hypertension, and coronary artery disease. However, these conditions also affect patients in the same age group, and there may be no causal relationship.

Clinical symptoms

Nonradicular back pain, especially in the lumbar and thoracolumbar junction area, is the most common symptom (Figure 1). Patients may report stiffness, especially in the morning and evening, and those with cervical spine involvement may notice dysphagia related to a large anterior cervical osteophyte located behind the esophagus. Other weightbearing joints may be painful, but spinal pain is the principal symptom. Note that patients may report that they have had symptoms for 10 years or longer.

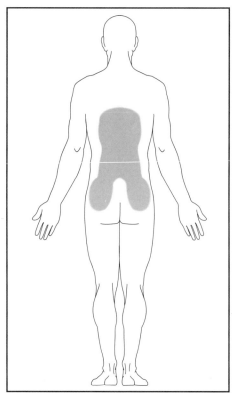

Figure 1
Distribution of pain in DISH

Tests

Exam

Examination reveals stiffness in the spine on forward flexion and on extension. In addition, there may be reduced hip motion or associated knee arthritis.

Diagnostic

Radiographs of the thoracic and lumbar spine, especially the lateral view, show confluent ossification spanning the intervertebral disks of at least four contiguous vertebral bodies (three disks) (Figure 2). The intervertebral disk height is preserved in those segments fused. The posterior apophyseal joints and sacroiliac joints are normal as opposed to the findings in ankylosing spondylitis.

In the cervical spine, ossification of the posterior longitudinal ligament occurs and is the second most common cause of cervical myelopathy (after cervical spondylosis).

The pelvis often shows "whiskering" or shaggy, hyperostotic bone at the pelvic rim. There is often hyperostotic change in the ribs, and DISH is often detected first on a chest radiograph.

There is no clear human leukocyte antigen (HLA) association.

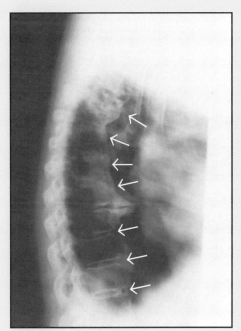

Figure 2
Confluent ossification anteriorly spanning multiple disk levels

Differential diagnosis

Acromegaly (facial and phalangeal changes)

Ankylosing spondylitis (sacroiliac and apophyseal joint involvement, positive HLA-B27)

Degenerative disk disease (reduced disk height)

Early rheumatoid arthritis

Multiple myeloma (30% of patients also have DISH)

Paget's disease (32% of patients also have DISH)

Polymyalgia rheumatica

Adverse outcomes of the disease

Spinal stiffness is common. With widespread involvement, a single mobile segment may remain but it may become unstable and painful.

Treatment

Walking and exercise programs are the most common initial treatment. Intermittent NSAIDs may help, but pain is usually mild and tolerable.

Adverse outcomes of treatment

Heterotopic ossification occurs five times more often following hip replacement surgery in patients with DISH. NSAIDs may cause gastric, renal, or hepatic complications.

Referral decisions/Red flags

Symptoms of neurogenic claudication, myelopathy, or dysphagia indicate the need for further evaluation.

Fibromyalgia

Synonyms

Fibrositis

Fibromyositis

Definition

Fibromyalgia is a common noninflammatory, nonarticular musculoskeletal condition characterized by generalized muscular pain and fatigue; the joints, however, are spared. Stiffness and sleep disorders are common, and women between the ages of 20 and 60 years are at greatest risk. The cause is unknown.

Clinical symptoms

In 1990, the American College of Rheumatology proposed the following criteria for the diagnosis of fibromyalgia: 1) widespread pain that has been present for 3 months; and 2) tenderness at 11 or more of 18 tender point sites (Figures 1 and 2). Widespread pain is defined as pain on both sides of the body, pain above and below the waist, and axial skeletal pain. Sleep disturbances and stiffness are often present. More than 90% of patients with fibromyalgia report fatigue that is worse in the morning and late in the day. Mood changes include depression and anxiety and often develop concurrently with multiple somatic complaints, such as migraine and tension headaches, irritable bowel syndrome, urinary frequency, and paresthesias in the hands and feet.

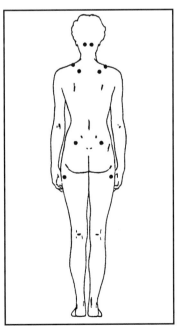

Figure 1
Posterior trigger points

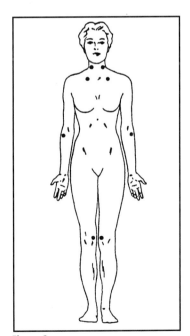

Figure 2
Anterior trigger points

Tests

Exam
Examination reveals tenderness to palpation over several of the 18 tender point sites. Any indication of pain to gentle pressure is significant. Examination of the joints is normal.

Diagnostic
No radiographs or laboratory tests are diagnostic of fibromyalgia.

Differential diagnosis

Bursitis or tendinitis (usually single joint or limb)

Carpal tunnel syndrome (paresthesias, positive Phalen's test, positive nerve conduction studies)

Hypothyroidism

Multiple sclerosis

Tenosynovitis (single focus, associated with tendon motion)

Adverse outcomes of the disease

The chronic pain associated with fibromyalgia can result in depression, anxiety, and inactivity. The battery of multiple tests often ordered to make the correct diagnosis can be expensive.

Treatment

Patients should be advised that fibromyalgia is not a life-threatening or progressive disease. No permanent changes have been observed in the musculoskeletal system from this condition.

Administration of amitriptyline in doses of 10 to 50 mg or cyclobenzaprine in doses of 10 to 40 mg, taken at bedtime, can be useful in treating the symptoms. Fluoxetine is useful as well. Short-term or intermittent NSAIDs may help diminish pain, but corticosteroids and narcotic analgesics are not indicated.

Patients should be instructed in a stretching program to increase flexibility. In addition, use of an aerobic exercise program to increase cardiac fitness is recommended. Many patients will benefit from participation in fibromyalgia support groups, which are a useful source of information and encouragement.

Adverse outcomes of treatment

Patients can become dependent on narcotics and tranquilizers. Medications also have side effects, including drowsiness, dry mouth, change in appetite, and constipation.

Referral decisions/Red flags

The presence of severe symptoms that interfere with patients' ability to work or the presence of serious psychiatric problems may indicate the need for evaluation at a center with a multi-disciplinary team approach.

Acknowledgements

Figures 1 and 2 are reproduced with permission from the Arthritis Foundation, Atlanta, GA.

general orthopaedics

Fracture Principles

Definition

A fracture is a disruption in the continuity of a bone. Some fractures occur at a microscopic level (stress fractures) and may not appear on radiographs. These fractures are usually stable; therefore, treatment is often only activity restriction.

Young bone is less brittle than mature bone and can actually bend, producing a "torus" fracture with buckling of one cortex but no visible fracture on the opposite side. These fractures require only observation and short-term splinting for comfort. Further bending of the bone will produce a greenstick fracture, with a break in one cortex and bending of the other; these fractures are also stable, but may be quite deformed and require reduction.

Fracture of a mature bone completely interrupts the bone, but the fracture may or may not be displaced. Treatment of undisplaced fractures usually consists of immobilization in a cast or splint until symptoms improve and radiographic healing is obvious. However, most displaced fractures are unstable and often require surgery. Some displaced fractures can be reduced and may be maintained in proper position with a splint or cast until union occurs. Others may require surgical reduction and internal fixation.

Clinical symptoms

Swelling, tenderness, pain, and deformity are the classic signs of acute fractures. With an undisplaced fracture, the bone may not be angulated with obvious deformity, but there will be local swelling. With stress fractures, there may be mild swelling and tenderness or pain with weightbearing.

Tests

Exam
Examination reveals local tenderness, deformity, and swelling. Muscle and nerve function should be tested both proximal and distal to the fracture.

Diagnostic
Radiographs usually indicate the presence of an acute fracture. However, some fractures cannot be diagnosed for 1 to 2 weeks until bone absorbs at the fracture site, making the injury radiographically apparent. This is especially true with stress fractures and fractures of the carpal navicular. Splinting is usually necessary to treat symptoms until radiographs can confirm the presence (or absence) of a fracture.

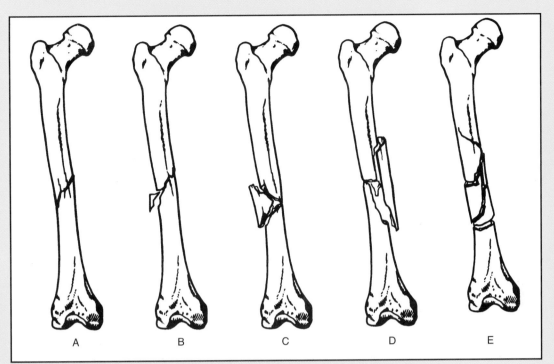

Figure 1
Common fracture patterns

Fractures are classified in several ways. The most common involves the configuration of the fracture itself, as follows:

- intra-articular (fracture line crosses articular cartilage and involves the joint);
- displaced (expressed in mm or cm in the direction of displacement of the distal fragment);
- nondisplaced (Figure 1A);
- angulated (expressed in degrees in the direction of the apex);
- comminuted (more than one piece at a single fracture location) (Figures 1B, C, D);
- compression (impacted or depressed);
- linear (straight line along the axis of the bone);
- segmental (two or more fractures in a single bone) (Figure 1E);
- oblique (Figures 1A, B);
- spiral;
- transverse.

A fracture is considered open if the fracture communicates with air through a skin wound; otherwise, it is closed. Open fractures are classified as follows:

- type I, in which the wound is clean and less than 1 cm in size;

- type II, in which the wound is greater than 1 cm in size, but without extensive soft-tissue damage;

- type IIIA, in which there is extensive soft-tissue damage, but enough tissue remains to potentially cover the bone;

- type IIIB, in which there is extensive soft-tissue loss, periosteal stripping, and grossly exposed bone;

- type IIIC, in which there is an arterial injury that requires repair.

Differential diagnosis

Infection (fever, elevated erythrocyte sedimentation rate)

Tumor (bony destruction)

Adverse outcomes of the disease

With all fractures, delayed union (slow healing, not a pathologic process), nonunion (failure to heal, a pathologic process), and malunion (healing with deformity at the fracture site) are always a possibility. With open fractures, osteomyelitis may develop. Excessive shortening of the limb can also occur. Associated muscle and nerve damage may be irreversible; ischemic compartment syndrome may develop.

Treatment

The principal goal of fracture treatment is to achieve union, especially in weightbearing bones. Also important are preserving function of the joints adjacent to the fracture, especially in nonweightbearing bones, restoring alignment in the plane of joint motion in weightbearing bones, and preserving bone length.

Reducing a fracture
The first step in reducing a fracture is to align the distal fragment to the proximal fragment. The distal fragment is the one that is to be moved; the one you have control over. To reduce a fracture, exaggerate the deformity (recreate the injury), hook the distal fragment on the fractured end of the proximal fragment, and lift the distal fragment onto the end of the proximal fragment.

Factors increasing fracture stability or favoring union

- Young patient age (a thick periosteum that remains partially intact and helps stabilize the fragments)

- Single bone broken in the forearm (radius or ulna), lower leg (fibula or tibia), or pelvis

- Presence of an adjacent bone or body part for support (single phalanx fractured, adjacent finger to splint to; ribs adjacent to thoracic vertebrae)

- Transverse fractures in phalanges

- Spiral fractures

Factors decreasing fracture stability

- Skeletal maturity (periosteum is thin and tears easily)
- Both bones fractured in forearm or lower leg
- Comminuted pelvis fracture
- Oblique fractures
- Single long bones (humerus, femur) or cervical/lumbar spine
- Comminuted fractures (single fracture site, multiple pieces)
- Segmental fractures (multiple fracture levels in the same bone)
- Fractures involving a joint

Adverse outcomes of treatment

Infection, failure of surgical fixation, and vascular or nerve damage are all possible.

Referral decisions/Red flags

Patients with open fractures, unstable fractures, irreducible fractures, or those with nerve, vascular, or muscle damage need further evaluation. Most patients who have fractures require further evaluation, as even fractures that appear innocuous are associated with poor outcomes.

Acknowledgements

Figure 1 is reproduced with permission from Poss R (ed): *Orthopaedic Knowledge Update 3*. Park Ridge, IL, 1990, American Academy of Orthopaedic Surgeons, p 514.

Imaging Techniques

Surgical specialists often have preferences regarding special imaging, such as MRI and CT scans. Some perform their own myelograms and some rely on special procedures performed in conjunction with advanced imaging. Since special imaging is costly, be sure you know the preferences of the surgical specialist who is part of your health care team.

Radiography

Plain radiographs are the mainstay of bone and joint imaging.

Principles
Radiographic examination of a long bone for fracture should meet the following criteria:

- Any film of a long bone should include the joint below and the joint above to avoid missing a dislocation associated with a fracture, often at the opposite end.

- Radiographs of a joint may be confusing if the limb is in a nonstandard position. Comparison views of the opposite extremity may help you to understand the patient's anatomy, especially with fractures (Table 1).

Indications
Deformity of a bone or joint

Inability to use a limb or joint

Unexplained pain in a bone or joint

MRI scan

Principles
Magnetic resonance imaging offers the advantage of seeing soft-tissue detail and often identifies the extent of sarcomas. MRI scans are excellent in spine, joint, and soft-tissue imaging. Combining MRI scans with contrast materials demonstrates a distinction between scar and avascular tissues.

Indications
In the musculoskeletal system, it is often valuable in *preoperative planning* to confirm a diagnosis or *anticipate anatomic lesions*; however, the underlying diagnosis can usually be made by less expensive means.

Table 1
Standard radiographic joint views

Region	Views	Special considerations
Hand	PA and lateral	
Wrist	PA and lateral	
Elbow/forearm	AP and lateral	Comparison views are helpful (opposite elbow)
Shoulder	AP of shoulder, AP of glenohumeral joint, axillary	Transscapular lateral view if unable to do axillary view
Cervical spine	AP and lateral	Include from C1 through C7 in the lateral view; swimmer's view may help visualize C7
Thoracic	AP and lateral	Swimmer's view may help visualize C7–T5
Lumbar spine	AP and lateral	Spot lateral of L5 if not seen well on standard lateral; try to see from T12-sacrum
Pelvis	AP	
Hip	AP and groin lateral or true lateral	
Knee	Weightbearing AP in full extension and 30° flexion in patients older than 40 years; otherwise standard AP. Lateral and bilateral Merchant view	
Ankle	AP and lateral	
Foot	Weightbearing AP and lateral	

Diagnostic dilemmas may require MRI scans if the potential diagnoses involve significant adverse health consequences.

Issues of *liability* (such as personal injury or workers' compensation claims) may require an MRI scan to determine the anatomic extent of injury or disease for purposes of claim settlement.

CT scan

Principles
Computed axial tomography offers axial visualization of bone, muscle, and fat tissues with the same limitations of conventional radiographs. Bone visualization is usually excellent, and soft-tissue structures less so.

Indications
Preoperative planning for bony procedures to assist in localizing lesions and appreciating the scope of bony changes.

Visualizing *complex fracture patterns,* especially those with joint involvement. Often used after myelography in complex degenerative spine disease.

Arthrography

Principles

Arthrography is a technique of injecting contrast material into the joint to evaluate joint capsule and articular surface integrity.

Indications

Rotator cuff tears in the shoulder, interosseous ligament tears at the wrist, and knee meniscal tears are conditions well evaluated by this less expensive but invasive technique.

Bone scan

Principles

A bone scan is a radioisotope technique indicative of blood flow and thereby of bone formation or destruction.

Indications

The bone scan is useful for identifying infection, tumor, and fractures.

Tomography

Principles

Tomograms are conventional radiographs imaged with a moving tube or plate to isolate a specific plane in skeletal structure.

Indications

Failed spine fusions, nonunions of fractures, and certain bone tumors may be localized in this manner. In lumbar spondylolisthesis, tomography may identify the laminar defect responsible for the slippage. Another name for these studies is a "blur-o-gram," indicating its primary fault: lack of a distinct image. Still, this technique provides excellent information in regions with low scatter from other tissues.

Injections and Corticosteroids

General guidelines for use of corticosteroids

Corticosteroids have an accepted role in the treatment of chronic inflammatory diseases, but indications for their use in acute and chronic musculoskeletal and soft-tissue disorders are not clear. In general, injections work well to relieve pain or inflammation and to improve function; however, they may be coupled with significant adverse effects.

There are many long-acting corticosteroid ester preparations that enhance anti-inflammatory actions and reduce undesirable hormonal side effects. The most widely used compounds are as follows:

- hydrocortisone and methylprednisolone tertiary butyl acetate (TBA esters);
- methylprednisolone acetate;
- triamcinolone acetonide;
- triamcinolone hexacetonide;
- betamethasone phosphate and acetate;
- dexamethasone acetate.

The triamcinolone compounds appear to have the greatest potency and longest duration of action. For local injections, the less-soluble TBA esters seem to have greater efficacy and duration of response.

How corticosteroids work

Corticosteroids suppress the initial events in inflammation. They decrease collagenase and prostaglandin formation and formation of granulation tissue. They are catabolic promoters that block glucose uptake in the tissues, enhance protein breakdown, and decrease new protein synthesis in muscle, skin, bone, connective tissue, and lymphoid tissue (predominantly T cells).

The intra-articular dose of corticosteroids is not proportional to the effective oral dose, because only the excess not retained by synovial tissue is absorbed systemically. The optimal dose for local injection is the maximum amount that can be held locally. Usual doses of a methylprednisolone preparation are 5 mg in the proximal interphalangeal joints, 20 mg in the wrist, 20 to 30 mg at the elbow, and 40 to 80 mg for the shoulder or knee. Some of the locally injected steroid is absorbed systemically and may produce transient systemic effects.

general orthopaedics

Use of intra-articular injections

Rheumatoid arthritis

In patients with active rheumatoid arthritis, use of an intra-articular injection produces symptomatic improvement in the injected joint about 50% of the time, lasting from several days to several weeks.

Repeated injections to suppress rheumatoid synovitis are generally effective in the knees, elbows, and interphalangeal joints. Long-term studies are scarce, but do not show accelerated destructive changes in the injected joints when compared with control joints.

Osteoarthritis

Intra-articular corticosteroid injection is less effective and its effects are of shorter duration in patients with osteoarthritis than in those with rheumatoid arthritis. The joints most often helped are the knee and the interphalangeal or metacarpophalangeal joints of the hand.

Crystal-induced arthritis

Use of an intra-articular corticosteroid injection in patients with gout or pseudogout may be especially helpful for those whose comorbid conditions or allergies prohibit use of systemic medications.

Tenosynovitis and bursitis

Both flexor tenosynovitis in the hand (trigger finger), and deQuervain's tenosynovitis at the wrist respond well to injections of corticosteroids into the tenosynovial sheath. Bursitis associated with shoulder impingement often responds well to a single injection. Trochanteric bursitis usually responds well to injection, and this bursa may be injected repeatedly with few reported adverse effects for most patients.

Injections into ligamentous structures carry the risk of spontaneous rupture of the ligament and are usually quite painful. The Achilles and patellar tendons should not be injected in the substance of the tendon since pain in these structures usually indicates interstitial tears, which already have reduced their tensile strength. Tennis elbow and plantar fasciitis may also respond well to corticosteroid injection.

Entrapment neuropathies

Carpal tunnel syndrome is often treated with injections into the carpal canal, but there is a substantial relapse rate and a chance of intraneural injection.

Ganglia

Injection of corticosteroids into ganglia is not necessary. Use a large-bore needle and make multiple punctures to decompress the ganglion. Recurrence of the ganglion is common.

Injection principles

Prior to injecting a corticosteroid, review the following guidelines:

1. Follow CDC guidelines for sterile technique and universal precautions.
2. Cleanse the top of the solution vial with an antiseptic solution.
3. Use an 18- or 20-gauge needle for easy withdrawal of solutions. Discard this needle.
4. Consider anesthetizing the injection site with ethyl chloride or a small amount of local anesthetic given with a 25- or 27-gauge needle.

5. Do not use local anesthetics that contain epinephrine when injecting the hand or foot; these may cause arterial constriction and infarction of a digit.

6. Corticosteroid and local anesthetic solutions can be mixed in the same syringe, usually in a 1:2 ratio. Large joints (knee and shoulder) may need an additional 4 mL of local anesthetic.

7. Short- and long-acting local anesthetics may also be mixed in the same syringe.

8. Inject anesthetic or corticosteroid preparation with a sterile 22- to 25-gauge needle.

9. Do not inject anesthetic or corticosteroid into a nerve, a tendon, or into subcutaneous fat.

10. Use multiple injections only if clear improvement has occurred. Limit the number to three injections.

11. Following the injection, have the patient rest the limb for 24 hours and avoid the precipitating cause of the problem.

Improper uses of injections

Acute trauma

Injection into a tendon or nerve

Injection into an infected joint, tendon, or bursa

Multiple injections (except as above) in conditions other than rheumatoid arthritis

Adverse outcomes of the drug

Adverse systemic effects include transient serum cortisol suppression and transient hyperglycemia (a special problem in patients with diabetes). Significant effects are uncommon with doses of 25 to 50 mg of methylprednisolone. Postinjection infectious arthritis is uncommon (perhaps 1/13,000 injections or fewer), but potentially catastrophic.

Local side effects after injection include lipodystrophy, loss of skin pigmentation, tendon rupture, and possible accelerated joint degeneration (although recent evidence suggests this is unlikely). Up to 10% of treated patients may have a transient flare or increased pain for 24 to 48 hours following an injection.

Nerve injection injuries can be catastrophic. Extrafascicular injection usually results in no permanent damage, but intrafascicular injection can be disastrous depending on the specific corticosteroid used. Injection of dexamethasone results in minimal damage; triamcinolone acetonide and methylprednisolone result in moderate damage, and hydrocortisone hexacetonide causes severe damage. The damage has been blamed on the carrier agent, but these agents have not been separated from the therapeutic agent.

It is not possible to establish a safe dose of corticosteroid because of individual variability regarding the sensitivity to the drug.

Olecranon and prepatellar intrabursal injections carry a significant risk for infection. Inject these structures only when the patient's problem has not resolved with time, and there is clearly no evidence of underlying infection.

Patients with diabetes mellitus are at risk for serious infection and for systemic effects of absorbed corticosteroids.

Lyme Disease

Synonyms

Lyme arthritis

Tick-borne Borreliosis

Definition

Lyme disease is a multi-system illness with acute and chronic manifestations caused by the spirochete *Borrelia burgdorferi* and is borne by the deer tick, *Ixodes dammini.* Lyme disease is named after a town in Connecticut where, in 1975, several children developed a mysterious arthritis of unknown cause, subsequently found to be due to this spirochete. We now know this multi-system disease is the most prevalent vector-borne illness in the United States; nearly 50,000 cases have been reported since 1982. The highest numbers of reported cases occur in the Northeast (Maryland to northern Massachussetts), the upper Midwest (Wisconsin and Minnesota), and the Far West (northern California and Oregon). Lyme disease has been reported in 48 states, as well as in Asia and Europe, where its existence has been recognized since the early 20th century.

Clinical symptoms

The acute symptoms, including fatigue, fever, headache, nausea, myalgias, and arthralgias, usually accompany a distinctive skin lesion named erythema migrans (EM) or erythema chronicum migrans. Its chief characteristic is a reddened expanding lesion at the site of the tick bite.

The acute phase can evolve if not treated. Arthralgias and arthritis develop in up to 80% of patients. Typically, only one or two joints are involved, most commonly the knee. Rarely are multiple joints affected. Cardiovascular involvement consists of conduction abnormalities with varying degrees of AV block, which occur in 4% to 8% of patients. Neurologic involvement occurs in 15% of patients. The most common neurologic syndromes are meningitis and neuropathies of the peripheral and cranial nerves. Bell's palsy or facial nerve paralysis develops in 50% of patients with neurologic involvement. Likewise, a peripheral nerve in the extremity may be affected and mimic common conditions such as carpal tunnel syndrome. In rare cases, myositis or hepatitis may develop, or the eyes or lungs may be affected.

The chronic stage of Lyme disease may follow by several months or years. Chronic arthritis and recurrent arthralgias are particularly troublesome. Other symptoms include chronic fatigue, polyradiculopathy, and encephalopathy with loss of memory and concentration.

Tests

Exam

Patients with a history of a tick bite, especially in a region known to have a high incidence of Lyme disease, should be examined for skin lesions (EM). A red macule or papule develops anywhere from 3 to 21 days after a tick bite. Subsequently, the rash may expand to a large (5 to 15 cm) circular red lesion with a clearing in the center, resembling a bull's eye. Several lesions may be present, and patients commonly report burning pain at the site of the skin lesions. The rash usually fades within a month.

Diagnostic

Lyme disease is a clinical diagnosis. Serologic testing can be helpful to confirm the diagnosis. The ELISA assay test detects antibodies against *Borrelia burgdorferi* and can be confirmed by a Western blot test. Serologic test results are negative early in the illness. An ECG should be done to rule out a conduction defect. A lumbar puncture is indicated if the patient has signs and symptoms of meningitis.

Differential diagnosis

Acute rheumatic fever

Complete heart block (no rash or systemic symptoms)

Idiopathic Bell's palsy (no rash or systemic symptoms)

Juvenile rheumatoid arthritis (no history of tick bite)

Meningitis (no rash or history of tick bite)

Multiple sclerosis (no rash or history of tick bite)

Peripheral neuritis

Reiter's syndrome (no rash or history of tick bite)

Rheumatoid arthritis (serologic testing, no rash or history of tick bite)

Adverse outcomes of the disease

Lyme disease can be complicated by arthritis in major weightbearing joints, facial paralysis, chronic fatigue, concentration deficits, cardiac conduction block, and peripheral neuritis.

Treatment

Lyme disease is effectively treated with antibiotics when the diagnosis is made early. Oral antibiotics given early in the course of the disease do seem to prevent some of the long-term sequelae. Oral doxycycline, 100 mg two times a day for 10 to 30 days, or amoxicillin, 500 mg three times a day for 10 to 30 days are accepted treatment regimens. For children younger than age 8 years, amoxicillin in a dose of 20 mg/kg/day in divided doses is indicated. Ceftriaxone in a dose of 2 g once a day for 2 to 3 weeks is recommended for late-stage Lyme disease.

Clinical trials are presently under way to test vaccines that will prevent the development of Lyme disease. However, until these vaccines are available, prevention may be the best treatment. People living in high-risk areas should avoid or minimize walking in the woods. Those living in heavily wooded areas should wear long-sleeved shirts tucked into trousers, with the trouser legs tucked into socks. Most importantly, check for ticks on the skin and clothing. If the tick is removed within 24 to 36 hours, the chance of acquiring Lyme disease is minimal.

Adverse outcomes of treatment

Allergic reactions or adverse drug interactions related to antibiotics can occur.

Referral decisions/Red flags

Patients who have the classic skin rash and multiple organ system involvement, but do not respond to a 2- to 3-week course of antibiotics may need further evaluation.

Nonorganic Physical Findings

Synonyms

Psychosomatic illness

Functional overlay

Definition

Patient responses or complaints that do not fit known patterns of illness or injury are considered nonorganic physical findings or symptoms. These findings do not indicate malingering; rather, they are often the way patients communicate, "See how bad I am?" or "I need more help than I'm getting." True malingering is rare and is often a manifestation of bizarre social behavior.

Nonorganic findings may be present in patients with significant underlying pathology and should not be construed as indicating lack of concomitant disease. Nonorganic findings occur three to four times more often in situations where compensation is an issue than in situations where it is not.

Clinical symptoms

Nonsegmental numbness, global pain, and pain or symptoms that travel from one side or area of the body to another in a nonanatomic fashion are characteristic.

Tests

Exam

Exaggerated responses: Light touch causes a jerk or withdrawal. Other findings include grimacing, groaning, and grabbing the affected leg or arm during examination when there is no obvious trauma or medical problem.

Axial loading (low back pain): With the patient standing, place both hands on the head and push down, asking if it causes pain. Low back pain elicited in this position is a nonorganic finding; however, neck pain may be a legitimate finding.

Axial rotation (spine): With the patient's hands on the iliac crests, grasp and rotate the pelvis, asking the patient if this causes back pain. This maneuver should not elicit back pain, since the motion occurs at the hips, not in the back. Since this test may be positive in 20% of patients, it is not as sensitive in indicating nonorganic behavior as are exaggerated responses and axial loading.

Flip sign: With the patient seated leaning slightly forward with his or her hands on the edge of the exam table, lift the foot and extend the knee as you ask if the patient has knee problems. Patients with sciatica will involuntarily "flip" back against the wall, complaining of back and leg pain, as this maneuver increases tension on the sciatic nerve. In the low back examination, a negative flip sign along with a positive supine straight leg raising test that produces leg and back pain at less than 45° of leg elevation is a significant nonorganic finding, as these maneuvers are the same test.

Distraction: A provocative test (such as palpation) may be negative while you distract the patient with conversation, but positive when you draw attention to the test or body part.

Giving way: During muscle testing, the patient may let go with the affected side; this is usually "ratchety" and uneven. With coaxing, the patient may intermittently contract the muscle, then let go.

"Stocking" or nonanatomic numbness: Some patients report hypesthesia to pinprick or light touch that covers non-adjacent dermatomes, or sensory patterns that do not fit the distribution of peripheral sensory nerves. However, patients with diabetes mellitus or multiple sclerosis might exhibit peripheral neuropathy in a "stocking" distribution.

Pain diagram: On a diagram of the body, ask the patient to draw representations of symptoms, using dashes, slashes, Xs, etc. Bizarre drawings do not indicate malingering or mental disease; many people have perceptual disorders that blunt the scientific validity of these drawings, but they often yield insight into how pain is perceived by patients, and what areas of the body they feel are related in their current problem (Figure 1).

A thorough neurologic exam is required to put the above tests in perspective and to rule out concomitant disease.

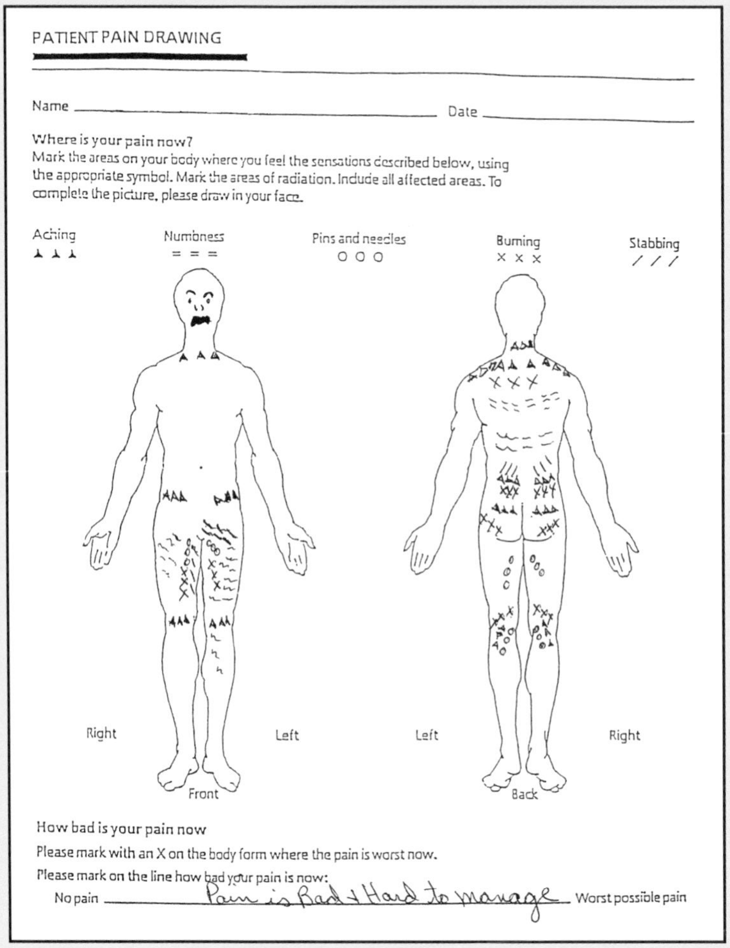

Diagnostic

None

Figure 1

Pain diagram for patient with nonorganic physical findings

Differential diagnosis

Acute injury (withdrawal from a painful examination maneuver may be an appropriate response)

Diabetes mellitus ("stocking" type peripheral neuropathy)

Multiple sclerosis (may present with bizarre sensory patterns that are not segmental)

Stroke or other central lesions with altered sensory appreciation

Treatment

Rule out serious disorders that may be masked by these symptoms and discuss them with the patient. Since the patient is often indirectly indicating that he or she is getting less support or help than is needed, discuss psychological and social support interventions and inquire about contributing factors such as anxiety, stress, marital difficulties, etc.

Adverse outcomes of treatment

Failure to identify an associated serious condition is possible.

Referral decisions/Red flags

True diminished pinprick, light touch with or without areflexia, clonus, or spasticity usually indicates neurologic involvement.

Nonsteroidal Anti-Inflammatory Drugs

Nonsteroidal anti-inflammatory drugs (NSAIDs) are a diverse group of drugs used to treat inflammatory conditions such as arthritis, bursitis, and tendinitis. They vary in chemical structure and mechanism of action, but have in common the ability to suppress inflammation, decrease pain, and reduce fever (Table 1). Although aspirin is a nonsteroidal anti-inflammatory drug, the term NSAID is usually reserved for the newer aspirin-like agents that were developed to decrease the severity of aspirin's effects on the gastric mucosa.

Mechanism of action

Inflammation is an essential, normal protective mechanism associated with injury to the musculoskeletal system. The pain associated with inflammation is the "gift that nobody wants." Inflammation may be triggered by trauma, infection, allergy, or an autoimmune response, as in rheumatoid arthritis. The inflammatory response is mediated by chemicals, such as prostaglandins, which are released from mast cells, granulocytes, and basophils. Prostaglandin synthesis is blocked by NSAIDs and aspirin, as these interfere with the action of the enzyme cyclooxygenase. Cortico-steroids also block the formation of prostaglandins but do so at a different step in the production chain.

Other properties of NSAIDs

In addition to the anti-inflammatory properties, all NSAIDs exhibit analgesic and antipyretic activity, decrease platelet adhesiveness, and inhibit the production of prothrombin, which may precipitate bleeding problems. Nephrotoxicity may also develop early in treatment and is being observed with increasing frequency; however, it is not dose related or duration dependent. Because of these potential complications, checking renal and liver functions every 6 months seems reasonable, although arbitrary. Fluid retention can increase blood pressure in susceptible patients. The inhibition of prostaglandin synthesis may also be responsible for hives and asthma attacks in some patients. In addition, NSAIDs should be used with caution in the elderly because there is a much higher risk that gastrointestinal hemorrhage and perforated ulcers will develop in this population.

Table 1
Currently available NSAIDs and their dosages *

Drug	Strength (mg)	Trade name	Typical dosage
Aspirin	300, 325, 600, 650	Several	325 mg qid
Choline magnesium trisalicylate	500, 750, 1000	Trilisate	1500 mg bid
Diclofenac K (immediate release)	50	Cataflam	50 mg bid or tid
Diclofenac sodium			
(delayed release)	25, 50, 75	Voltaren	50 mg tid, 75 mg bid
(extended release)	100	Voltaren-XR	100 mg q day
Diflunisal	250, 500	Dolobid	250 mg bid or tid
			500 mg bid
Etodolac	200, 300, 400, 500	Lodine	200–400 mg tid
			500 mg bid
Etodolac (extended release)	400, 600	Lodine XL	400–1000 mg q day
Fenoprofen	300, 600	Nalfon	300–600 tid or qid
Flurbiprofen	50, 100	Ansaid	100 mg bid
Ibuprofen (OTC)	200	Motrin IB, Advil, Rufen, Nuprin	200–400 mg bid or tid
Ibuprofen (prescription)	400, 600, 800	Motrin	400–800 mg bid or tid
Indomethacin	25, 50	Indocin	25–50 mg tid
Indomethacin (sustained release)	75	Indocin SR	75 mg q day
Ketoprofen	25, 50, 75	Orudis	50 mg qid, 75 mg tid
Ketoprofen (extended release)	100, 150, 200	Oruvail	200 mg q day
Ketorolac tromethamine	10	Toradol Oral **	10 mg qid, maximum 5 days (oral and IM combined)
Nabumetone	500, 750	Relafen	1000–2000 mg q day
Naproxen	250, 375, 500	Naprosyn	250, 375, or 500 mg bid
Naproxen sodium	220	Aleve	220 mg bid
(controlled release)	375, 500	Naprelan 375, 500	750–1000 mg q day
	275, DS 550	Anaprox	275 mg bid or tid
			550 mg bid
Oxaprozin	600	Daypro	600–1200 mg q day
Piroxicam	10, 20	Feldene	20 mg q day
Salsalate	500, 750	Disalcid	1500 mg bid
			1000 mg tid
Sulindac	150, 200	Clinoril	150–200 mg bid
Tolmetin sodium	200, 400, 600	Tolectin 200 Tolectin DS Tolectin 600	400 mg tid (starting dose) for 1 week; 200-600 mg tid (control dose)

* The recommended doses of NSAIDs may change as new data becomes available. Therefore, the clinician should refer to the specific drug's information package insert or to a current *Physicians' Desk Reference* (PDR) for the most current dosage recommendations.

**Toradol Oral should be given only as a continuation to IM Toradol and is usually used to treat moderate to severe acute pain, most often following surgery.

Side effects

All NSAIDs cause similar side effects, but in varying degrees. All cause gastric and duodenal irritation both by local and systemic effects, most commonly during the first few weeks of therapy. Even rectal NSAIDs can cause upper gastrointestinal tract ulceration. However, injury to the gastrointestinal tract can be decreased by applying a "mucosal protection strategy," which involves giving the patient a prostaglandin E analogue along with the NSAID. Misoprostol (Cytotec) given in doses of 100 to 200 μg four times a day (qid) offers protection for both the stomach and the duodenum. Ranitidine (Zantac) in a dose of 150 mg two times a day (bid) offers protection against duodenal ulcer formation, but may be less effective in preventing gastric ulcers.

Choice of therapeutic agents

Aspirin remains the gold standard in the treatment of inflammatory joint disorders, and enteric coating helps with gastric side effects. If there is no contraindication for its use, aspirin is the initial drug of choice. Note that the use of acetaminophen for primary management with intermittent use of NSAIDs for break-through pain may help reduce gastric symptoms. In general, all NSAIDs should be taken with food to decrease gastric irritation.

For the 15% to 20% of patients who cannot tolerate aspirin, one of the newer NSAIDs can be used, based on cost, side effects, and the dosage schedule involved. However, newer NSAIDs may cost 10 to 20 times more than a generic form of aspirin. Once or twice daily dosage schedules are optimal for patients with poor compliance.

Using NSAIDs with patients who are already taking oral hypoglycemic medication or warfarin (Coumadin) may potentiate the effects of these agents. Under these circumstances, ibuprofen is a suitable choice. Aspirin or other NSAIDs should not be taken at the same time as another NSAID.

For patients who need an NSAID, but are at risk for gastric complications, one of three alternatives may help. A nonacetylated NSAID such as salsalate (Disalcid) appears to cause less mucosal injury. NSAIDs that are nonacidic (such as nabumetone) and prodrugs (drugs that must undergo biotransformation to an active metabolite) are reported to decrease gastric injury. NSAIDs of the pyranocarboxylic acid class, such as etodolac (Lodine), are appropriate in high-risk patients.

Osteoarthritis

Synonyms

Wear and tear arthritis

Degenerative joint disease

Definition

Osteoarthritis is a progressive, currently irreversible condition with loss of articular cartilage that leads to pain and sometimes deformity, principally in the weightbearing joints of the lower extremities and the spine. It is the most common type of arthritis and is most often associated with age, obesity, and trauma. Less common associations include chondrocalcinosis, hemophilia, hemochromatosis, epiphyseal dysplasia, and hydroxyapatite degenerative disease.

Clinical symptoms

Stiffness, pain, and deformity are the common complaints. Stiffness, rather than swelling, is the most common complaint associated with joint effusions. The pain associated with osteoarthritis is usually relieved by rest until the arthritis is advanced. With advanced arthritis, the osteophytes (spurs) themselves may block joint motion as well.

Tests

Exam

At the knee, ankle, elbow, hand, and foot, osteophytes are often palpable at the joint margins. There is sometimes an effusion, especially if the patient has noted stiffness. At the knee particularly, there is often a genu varum (bowleg) or genu valgum (knock-knee). At the hip joint, the patient often "toes out" and walks with the limb externally rotated, like Charlie Chaplin.

Diagnostic

Radiographs of the affected joint usually demonstrate loss of joint space, sclerosis, and osteophytosis or spurs at the joint margin.

Differential diagnosis

Charcot joint (primarily foot and ankle, diabetic neuropathy)

Chondrocalcinosis (crystals in joint aspirate)

Degenerative changes secondary to inflammatory arthritis (positive RF)

Epiphyseal dysplasia (short stature)

Hemochromatosis (abnormal liver studies)

Hemophilia (bleeding tendency)

Adverse outcomes of the disease

Pain, deformity, loss of joint motion, loss of limb function, and joint instability are possible.

Treatment

Reassurance, patient education, and protection of the joint from overuse are critical. Simply calling patients from time to time can often help them cope with this irreversible disorder. Gentle, regular joint exercises help maintain function and manage pain. Water exercise, cycling, and nonweightbearing range of motion exercises all help reduce symptoms and preserve muscle support for the affected joints. Isometric exercises help if patients are unable to tolerate exercises involving joint motion.

Weight loss is important, especially in the weightbearing joints of the lower extremity, not only to reduce symptoms, but to improve the survival of joint implants placed when symptoms are intolerable.

Pain management with acetaminophen, propoxyphene, or similar medications is often helpful. Many elderly patients are better able to cope with symptoms if they can manage their pain and get adequate rest at night. Simple medications may allow them to postpone or avoid the risks associated with surgery.

Salicylates and NSAIDs often help, but may cause renal, hepatic, or gastric problems. In addition, NSAIDs may inhibit joint repair by interfering with prostaglandin synthesis. This is currently a theoretic consideration of unknown clinical relevance.

Intra-articular corticosteroids often relieve symptoms, but the duration of relief is often short (1 to 2 weeks) in weightbearing joints of the lower extremities, shoulder, or elbow. In the hand, intra-articular steroids may relieve symptoms for months. No more than one or two injections should be given in any weightbearing joint.

Surgery, typically joint replacement or arthrodesis, is appropriate when patients have pain at rest, pain at night, or unacceptable loss of joint function. Hip and knee replacement are especially effective in reducing pain and increasing function.

Adverse outcomes of treatment

Removal of prosthetic implants due to infection or loosening is possible.

Referral decisions/Red flags

Patients who have pain at rest, pain at night, or unacceptable loss of joint function need further evaluation.

Osteoporosis

Synonyms

Brittle bone disease

Osteopenia

Definition

Bone is a metabolically active organ system, constantly tearing down old bone and replacing it with new. Osteoporosis is an imbalance in this process, whether by inadequate production, excessive removal, or both; the result is too little bone, though what is there is qualitatively normal. In older women, there is often an associated nutritional osteomalacia.

Clinical symptoms

Fractures, whether "microscopic" or obvious, produce pain in patients with osteoporosis. Back pain, primarily in the thoracolumbar area, sometimes begins after a "pop" and usually radiates around the ribs or along the iliac crest. Hip pain may indicate an impending intertrochanteric, femoral neck, or pubic ramus fracture. Wrist pain and wrist fractures are common after a fall. Other sites vulnerable for fracture include the ankle and shoulder.

Dizziness, stroke, syncope, medication changes, alcoholism, use of psychotropic medication, age-related loss of proprioception, dementia, and external factors such as loose rugs and electrical cords compromise balance and predispose patients to falls.

Tests

Exam
Wrist deformity (Colles fracture) following a fall is often the first sign of osteoporosis, especially in women younger than age 65 years. Kyphosis (dowager's hump) is often severe as the disease advances; this collapse of the spine with preservation of limb length produces an appearance of a shortened trunk (Figure 1). The lower rib cage may come to rest on the iliac crest. Fractures about the hip precipitate most hospital admissions for osteoporosis.

Diagnostic
There are several techniques for estimating bone mass and, with repeated measurements, the rate of bone loss. However, none of these techniques will establish an accurate rate of bone loss without multiple (usually more than three) annual measurements. Single measurements are valuable only to estimate the patient's bone mass relative to an age-matched population.

The following techniques are available for estimating bone mass:

- Singh index: AP radiograph of the hip, counting trabeculae; high variability

- DEXA: computerized-assisted densitometry

- CT scan: quantification of radiograph

- Single and dual photon absorption: quantitative absorption by bone

Results of laboratory studies are usually normal; abnormalities, such as elevated alkaline phosphatase, are commonly associated with fractures. Urinary calcium excretion may be high in rapid bone resorption states. Untreated hyperparathyroidism may cause osteopenia in younger patients.

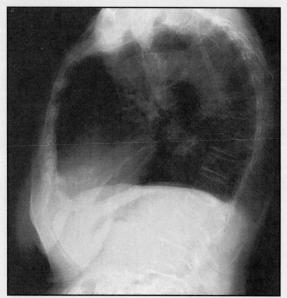

Figure 1
Severe kyphosis

Differential diagnosis

Hyperparathyroidism (younger patients)

Metastatic cancer with spinal fracture (breast in women, prostate or multiple myeloma in men)

Renal osteodystrophies (usually osteomalacia resulting in bony deformity [bowing] rather than fracture)

Adverse outcomes of the disease

Severe kyphosis, multiple fractures with attendant pain, disability, and deformity are the most common findings in association with osteoporosis. Spinal cord lesions are rare. Death may occur within 6 months after a hip fracture.

Treatment

Prevention is the optimal treatment. Emphasize adequate calcium intake for teenage women and regular exercise to build bone mass. Oral calcium taken in doses of 1,000 mg/day in premenopausal years and 1,500 mg/day in postmenopausal years is appropriate. Note that skim milk has the highest calcium content of the milk group. Exercise is crucial to preserving bone mass and should be started as soon as possible following diagnosis. Following these two simple steps puts "money in the bank" by creating a large peak bone mass for those later years when bone loss exceeds bone formation.

Advise the patient to remove loose rugs, trailing electrical cords, and other obstacles that contribute to falls. Bath seats and rails and shower seats are useful. Use of bifocal lenses at night may be less safe than a single prescription lens. Patients who have problems with balance should use a walking aid, such as a walker or cane.

Estrogen-progesterone regimens are the gold standard in treating postmenopausal bone loss. Vitamin D supplements are appropriate, especially if there is a malacic component to the osteopenia. Use of salmon calcitonin may be effective for short-term pain management; however, the nasal spray is the only current route for administration. Cyclical bisphosphonates appear useful in early clinical trials.

Treatment of the first fracture should be aggressive; however, by this time, the patient has already lost substantial bone mass, perhaps up to 50%, and it takes years to gain back enough bone to withstand falls.

Corsets may help during the 2 to 3 months of significant pain following a spinal fracture; however, many women object to the chafing that occurs under their breasts, and severe kyphosis makes fitting these devices difficult. Braces are often tolerated better in the acute phase following a vertebral fracture. A Jewett-type brace is lightweight, reasonably well tolerated, and provides good immobilization.

Adverse outcomes of treatment

There is controversy about the relationship between endometrial and breast cancers and postmenopausal estrogen-progesterone supplementation. However, benefits in reduction of heart disease and fracture morbidity outweigh these potential risks in the population at large. Breakthrough bleeding and spotting associated with estrogen-progesterone treatment usually subsides within a few months.

Referral decisions/Red flags

A limp associated with hip (groin or anterior thigh) or knee pain on weightbearing, even with normal radiograph, may indicate an impending fracture.

Reflex Sympathetic Dystrophy

general orthopaedics

Synonyms

RSD

Causalgia

Sympathetically
maintained pain (SMP)

Shoulder-hand syndrome

Major causalgia

Sudeck's atrophy

Minor dystrophy

Definition

Reflex sympathetic dystrophy (RSD) is a condition of unknown etiology, characterized by severe pain (usually localized to an extremity), swelling, discoloration, and excessive perspiration. The hallmark of RSD is severe pain that is out of proportion to the original injury. Patients with nerve injuries are particularly susceptible, but RSD may also occur as a result of fractures and soft-tissue injuries. RSD more commonly affects the upper extremity, but can also occur in the lower extremity.

Clinical symptoms

A bone or soft-tissue injury most commonly precipitates RSD. The injury may be innocuous, and 30% of patients have no injury at all. RSD affects 5% to 8% of patients with incomplete nerve injuries, either from a cut or a stretch. Patients often indicate they cannot use the extremity at all.

Patients with RSD often have significant anxiety and depression, even to the point of suicide.

Tests

Exam
RSD typically occurs in three stages. Stage 1, which can last up to 3 months, is characterized by severe pain that is out of proportion to the injury. Autonomic nervous system dysfunction in this stage is manifested by swelling in the affected extremity, increased sweating, a change in skin color from red to cyanotic, temperature changes, an increased amount of hair growth, and excessive nail growth. Stage 2 occurs after 3 to 4 months and is marked by loss of skin lines (causing the skin to look pale and waxy), joint stiffness, brittle nails, muscle spasms, and persistent pain. Stage 3 occurs after 8 to 9 months, at which time there are irreversible changes. The extremity becomes atrophic with loss of muscle and skin, permanent joint contractures, loss of motion, and persistent pain becomes severe (Figure 1).

Diagnostic
In late stage 1 through stage 3, plain radiographs should be ordered as they may show spotty areas of osteoporosis or demineralization in the bones of the affected extremity. Bone scans may show increased uptake in the limb, usually distal to the point of injury. The bone scan findings usually correlate with the patchy osteopenia seen on the radiograph.

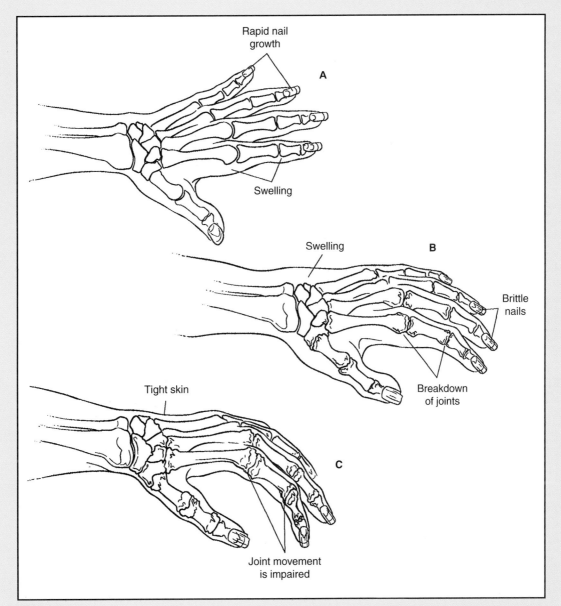

Figure 1
Early RSD stage 1 (A); Middle RSD stage 2 (B); Late RSD stage 3 (C)

Upper extremity pain relieved by a cervical sympathetic anesthetic block supports the diagnosis of RSD. Similarly, a lumbar sympathetic block may relieve pain, at least temporarily, in the lower extremity.

Differential diagnosis

Nerve injury (Morton's neuroma, incomplete or complete nerve laceration)

Peripheral nerve entrapment (carpal tunnel syndrome, tarsal tunnel syndrome)

Adverse outcomes of the disease

Chronic, possibly debilitating pain, joint contractures and stiffness, and skin and muscle atrophy can develop. More serious consequences include loss of function in the affected extremity. Psychiatric problems, including depression, anxiety, and even the potential to commit suicide, have also been noted.

Treatment

Early recognition and prompt treatment appear to decrease the severity of the symptoms, although the cure may be incomplete. Use of sympathetic anesthetic blocks can be therapeutic as well as diagnostic. If an anesthetic block provides pain relief, then a series of three to six blocks over a 2- to 3-week period is indicated. Note that a single block does not distinguish the placebo effect (in which up to one third of patients report temporary improvement following an injection of normal saline) from true pain relief.

Immediately after each block, range of motion and stress loading programs are begun, usually under the direction of a physical or occupational therapist. When RSD affects the hand, a "home" stress loading program is begun. This type of program includes performing routine chores, such as scrubbing a 3-foot-square area of kitchen floor (if tile or wood) for 20 minutes three times a day, bearing as much weight through the wrist and hand as tolerated. A home program also includes lifting and carrying household objects, such as a 25-lb suitcase three times a day, and generally using the affected hand as often as practical.

Oral steroids and NSAIDs can decrease pain, but they do not alter the course of the disease. In stage 1, narcotic analgesics may be required for pain control; however, use of narcotics requires special caution as these can result in drug dependency in addition to the RSD. Tricyclic antidepressants (amitriptyline) in doses of 25 to 50 mg taken at night allow patients to sleep and may offer partial pain relief during the day.

A transcutaneous electrical nerve stimulation (TENS) unit applied to the extremity, under the supervision of a therapist, is a useful adjunct for the control of pain and may decrease the need for stronger narcotic pain medications. Biofeedback can also be tried. With biofeedback, patients are taught to control the autonomic functions of the body that regulate sweating, skin temperature, and blood flow.

Adverse outcomes of treatment

Side effects from cervical sympathetic blocks include hoarseness, weakness or numbness in the arm, pneumothorax, and possible seizure from vertebral artery injection. Use of narcotics can lead to addiction, and excessive use of NSAIDs can result in gastric, hepatic, and renal symptoms. Fractures can also occur from too vigorous manipulations.

Referral decisions/Red flags

Severe pain and swelling associated with any type of injury to an extremity indicate the need for further evaluation at a center where a multi-disciplinary team is available to treat this difficult problem. Also, patients with severe, unexplainable pain following even a trivial injury may need further evaluation.

Rheumatoid Arthritis

Synonyms

None

Definition

Rheumatoid arthritis is a systemic polyarthritis with a multi-factorial etiology and a wide spectrum of clinical presentations. At its worst, it is a debilitating and disabling disease with disastrous multi-system involvement. There is no specific biologic marker to characterize the disease, so the diagnosis is often made by observing the patient over a period of time. The primary joint disorder involves a chronic synovitis with erosive pannus formation at the articular margins.

Clinical symptoms

Pain, stiffness, swelling, and systemic symptoms are common. The pain may move from joint to joint. Many patients have joint deformities and contractures, which are the presenting complaints. Foot pain, bunion, and hammer toes are common with long-standing disease, as are severe ulnar drifting and deformities of the metacarpophalangeal joints in the hand. The small joints are most often affected and involvement is typically symmetrical.

Tests

Exam

Joint contractures (inability to fully extend or flex because of tight capsular structure), joint effusions, deformity (genu valgum or knock-knee), and painful motion are common. Phalen's test is often positive, and an associated carpal tunnel syndrome may begin suddenly, accompanied by severe pain.

Most joints show increased warmth and synovial bogginess. Joint aspirations often produce less fluid than expected because much of the enlargement comes from synovial hypertrophy.

Rheumatoid nodules often appear, especially along the extensor aspect of the arm. Other findings include episcleritis and pulmonary interstitial disease.

Diagnostic

Results of rheumatoid factor (RF) test are positive in 75% to 90% of patients. IgM is the rheumatoid factor most commonly present, though IgG and IgA are sometimes present. However, testing is often deferred during the first 4 to 6 weeks of symptoms to allow resolution of confounding self-limiting conditions such as postviral arthropathy.

general orthopaedics

Radiographs often show periarticular osteopenia and bony erosion at the joint margin (correlated with pannus formation). Lateral flexion and extension views of the neck may demonstrate C1–C2 instability secondary to erosion of the ligaments that hold the odontoid in place. This information is important if the patient is anticipating any surgery that involves intubation or manipulation of the neck, since it could lead to quadriparesis or death.

Differential diagnosis

Hepatitis

Lyme disease

Seronegative arthropathies

Systemic lupus erythematosus

Adverse outcomes of the disease

Joint contractures, pain, loss of function, loss of ambulation, and multi-system disorders are possible. Osteoporosis is common, in part related to the disease, inactivity, and steroid use.

Treatment

Medical treatment for this condition is beyond the scope of this text. Commonly, salicylates, NSAIDs, splinting, and oral and intra-articular corticosteroids are used to treat the polyarthralgias and pain. Disease-modifying agents, such as hydroxychloroquine, methotrexate, and gold are being used at earlier stages of the disease. Corticosteroid injections into the carpal tunnel may relieve acute carpal tunnel syndrome associated with a flare of synovitis. Splints often help manage acute episodes of pain associated with synovitis. Custom shoes are helpful with the severe foot deformities seen in chronic disease.

Adverse outcomes of treatment

Infection secondary to surgery or injection; gastric, hepatic, or renal complications of NSAIDs; and osteonecrosis of bone associated with steroid use are possible. Skin rashes and other side effects related to the various medications used can also develop.

Referral decisions/Red flags

Patients whose symptoms persist for more than 3 months, or who have uncontrollable joint pain at rest, or who have severe joint deformity need further evaluation. Patients with foot deformities not responsive to custom shoes also need additional evaluation, as do those who develop extra-articular manifestations.

Splinting Principles

Splinting most fractures, dislocations, or tendon ruptures of the upper or lower extremities is often required as part of initial emergency management. A well-applied splint reduces pain, bleeding, and swelling by immobilizing the injured part. Splinting also helps prevent the following:

- further damage of muscles, nerves (including the spinal cord), and blood vessels by the sharp ends of fractured bones;
- laceration of the skin by sharp fracture ends (One of the primary indications for splinting is to prevent conversion of a closed fracture to an open fracture);
- constriction of vascular structures by malaligned bone ends.

General principles of splinting

1. In most situations, remove clothing from the area of any suspected fracture or dislocation to inspect the limb for open wounds, deformity, swelling, and ecchymosis.
2. Note and record the pulse and capillary refill and neurologic status distal to the site of injury.
3. Cover all wounds with a dry, sterile dressing before applying a splint. Notify the receiving hospital or orthopaedist of all open wounds, if further evaluation is necessary.
4. Ensure that the splint immobilizes the joint above and the joint below the injury and suspected fracture.
5. With injuries in and around the joint, ensure that the splint immobilizes the bone above and the bone below the injured joint.
6. Pad all rigid splints to prevent local pressure.
7. During application of the splint, use your hands to minimize movement of the limb and to support the injury site until the limb is completely splinted.
8. Align a limb severely deformed from a fracture of the shaft of a long bone with constant gentle manual traction so that it can be incorporated into a splint.
9. If you encounter resistance to limb alignment when you apply traction, splint the limb in the position of deformity.
10. When in doubt, splint.

general orthopaedics

Materials

Although prefabricated metal splints are available, they are generally unsatisfactory except for very brief periods of emergency treatment. If a splint is expected to be effective and to remain in place for more than a few hours, custom application of a well-padded plaster or fiberglass splint is preferred. Because plaster is cheaper, more readily available, and can be more easily molded to the extremity, it is usually preferred. See Table 1 for the materials needed for splinting.

Table 1
Splinting materials

Thumb/finger	Arm	Short leg splint
1 to 2 rolls 4″ cast padding (adults) or 3″ cast padding (children)	2 or 3 rolls 4″ cast padding (adults) or 3″ cast padding (children)	2 rolls of 4″ to 5″ wide cast padding
4″ × 15″ splints, six thicknesses (adults) or 3″ roll folded into splint of appropriate length (children)	5″ × 30″ splints six thicknesses (adults) or 4″ roll folded to necessary length (children)	12 to 14 thicknesses of 5″ × 30″ or 5″ × 45″ plaster strips
3″ or 4″ elastic bandage	3″ or 4″ elastic bandage	One 3″ to 4″ wide roll of plaster
Tepid water	Tepid water	One roll of 4″ wide elastic bandage
Nonsterile gloves	Nonsterile gloves	One bucket of tepid water
		Nonsterile gloves
Wrist and forearm	**Long leg splint**	
2 rolls 4″ cast padding (adults) or 3″ cast padding (children)	3 to 4 rolls of 6″ wide cast padding	
5″ × 30″ splints, six thicknesses (adults) for "sugar tong" or 4″ × 15″ six thicknesses splints (children) for simple dorsal or volar splint	3 to 4 rolls of 5″ or 6″ wide plaster or 5″ × 45″ plaster splints	
3″ or 4″ elastic bandage	One roll each of 4″ and 6″ wide elastic bandages	
Tepid water	One bucket of tepid water	
Nonsterile gloves	Nonsterile gloves	

Splinting the upper extremity

Fractures or injuries distal to the distal forearm

Position the patient supine or sitting and have an assistant hold the patient's thumb and/or index fingers. Loosely wrap cast padding from the palm to the elbow, making sure that there are three layers of padding at any bony prominence. Place a 4″ × 15″ preassembled splint in the palm and carry it up the volar aspect of the forearm to just below the elbow (Figure 1). If the injury involves the thumb, wrap it separately with 2″ or 3″ of cast padding. Place the splint on the volar or radial aspect and fold the plaster around the thumb, extending across the wrist to the proximal forearm. Leave the dorsal or ulnar side open for swelling (Figure 2). Wrap the cast padding loosely over the plaster, then wrap an elastic bandage loosely over the cast padding as you mold the splint. Trim the palmar portion of the splint back to the distal palmar flexion crease, proximal to the metacarpophalangeal (MP) joint.

Fractures or injuries to the elbow and wrist

With the patient sitting or supine, have an assistant support the patient's hand with the elbow flexed to 90°. If sitting, the patient should lean slightly to the affected side so that the elbow falls away from the body. Loosely wrap cast padding from the palm to above the elbow, taking care to avoid creating a constriction in the antecubital fossa. Make sure that there are three layers of padding at any bony prominence, such as the wrist and elbow. Begin the splint in the palm, carry it up the forearm to the elbow, around the posterior elbow, then distally on the extensor aspect of the forearm to the dorsum of the hand (Figure 3). Use multiple 4″ × 15″ preassembled splints, or a 5″ × 30″ preassembled splint if that size is appropriate. Wrap cast padding loosely over the plaster, then wrap an elastic bandage loosely over the cast padding as you mold the splint. Trim the palmar portion of the splint back to the distal palmar flexion crease, proximal to the MP joint.

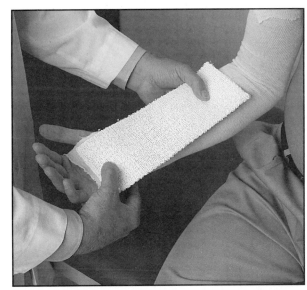

Figure 1
Begin in the palm and extend up the volar surface of the forearm to below the elbow

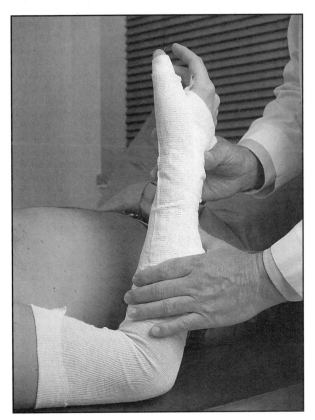

Figure 2
Apply the splint along the volar aspect of the thumb, extending across the wrist to the proximal forearm

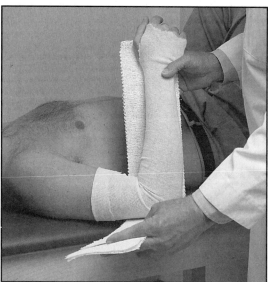

Figure 3A
Begin in the palm and extend proximally around the posterior elbow

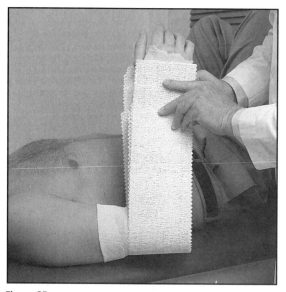

Figure 3B
Complete the splint distally on the extensor aspect of the forearm to the dorsum of the hand

Fractures or injuries at or above the elbow

With the patient sitting, have an assistant support the patient's hand with the elbow flexed to 90°. The patient should lean slightly to the affected side so that the elbow falls away from the body. With an elbow injury, loosely wrap cast padding from the palm to the upper arm. Begin the splint below the axilla, carry it under the elbow, then up the lateral aspect of the arm (Figure 4). For unstable humeral fractures, continue the splint over the top of the shoulder, cover the plaster with a layer of cast padding, and then loosely wrap the entire arm with an elastic bandage (elephant ear splint). For lower humeral fractures or elbow injuries, end the splint below the lateral shoulder, cover the plaster with a layer of cast padding, and then loosely wrap the entire arm with an elastic bandage (coaptation splint) (Figure 5).

Provide the patient with a strap sling that loops around the wrist, then around the neck, and back to the wrist (Figure 6). The sling should be long enough to allow the elbow to be maintained at 90°. Ensure that the sling has padding at the neck and wrist; these straps do not slide at night and can be adjusted for different arm lengths.

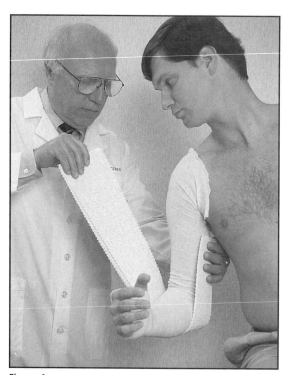

Figure 4
Begin the splint below the axilla, extending it under the elbow then up the lateral aspect of the arm

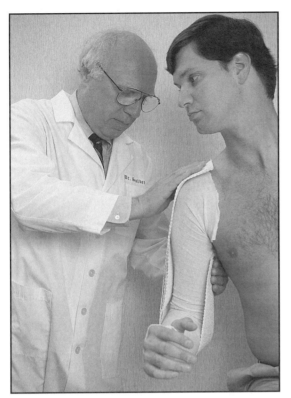

Figure 5
For lower humeral or elbow fractures, end the splint below the lateral shoulder

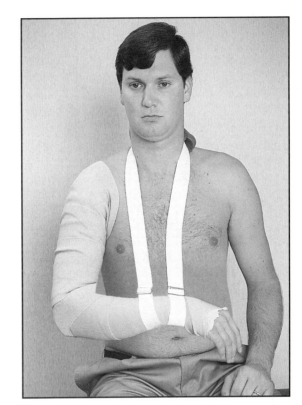

Figure 6
Proper positioning of a sling to maintain the elbow at 90°

Patient instructions

Advise the patient to take the following steps when caring for the injured upper extremity at home, after the splint has been applied.

Protect your splint for 24 hours, until the plaster cures and hardens (fiberglass splints harden faster than plaster splints).

Do not place your splinted arm on any plastic covered surfaces (including pillows) until the plaster has cooled.

Splinting the lower extremity

Long leg splint

For unstable fractures of the leg or ankle, use a long leg splint with the knee flexed 30° and the ankle 90°. With the patient supine, move his or her buttocks to the edge of the table, allowing the entire leg to hang suspended with an assistant holding the patient's forefoot. Ask the patient to allow the heel to "sink" so that the foot will be maintained at 90° during splinting.

Use either a "stirrup" type splint or a long posterior splint. For a stirrup splint, have an assistant hold the patient's forefoot as you wrap the leg with three layers of 6" cast padding. Place extra padding over the kneecap and lateral knee (fibular head). Begin the 5" × 45" plaster splint (10 to 12 thicknesses) on the lateral aspect of the thigh, extend it down the lateral aspect of the leg, under the heel, and then back up the medial side (Figure 7). Start a second splint medially, and extend it beneath the foot and up the lateral side. Apply a layer of cast padding over the plaster, and wrap a 5" or 6" elastic bandage over the padding as you mold the splint. Ensure that the knee is positioned in 25° to 30° of bend and the foot positioned at 90° (Figure 8). Avoid folds in the plaster over the area of the peroneal nerve below the lateral knee (fibular head) or around the ankle. Preassembled foam padded splints are convenient, but use them with caution as they may develop folds or ridges in critical areas.

Use tepid or cool—never hot—water when applying the splint. The heat generated by the reaction of the plaster, if coupled with the use of hot water, can seriously burn the skin. For the same reason, place the leg on a cloth (not plastic) pillow and leave it uncovered for about 10 minutes following application to allow better convection of the heat.

Hold the splint in place with a loosely applied elastic bandage or bias-cut stockinet, rolled on with almost no tension. As the splint hardens, maintain the ankle at 90°. Mold or support the splint with the flat of the hand—only while it hardens—to avoid causing dents. Dents not only make the splint uncomfortable, but can cause cast sores or peroneal nerve palsy and foot drop.

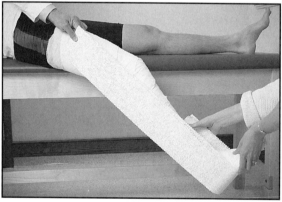

Figure 7

Begin the splint on the lateral aspect of the thigh, extending it down the lateral aspect of the leg, under the heel, and then back up the medial side

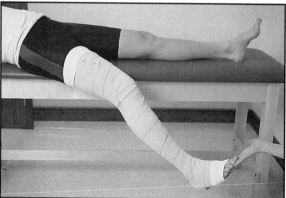

Figure 8

Maintain the knee in 25° to 30° of bend, maintaining the ankle at 90°

Short leg splint

With the patient sitting, have an assistant hold the forefoot to maintain the ankle at 90°, wrap the foot, ankle, and leg loosely with three thicknesses of cast padding. Use either 5″ × 45″ cast padding or fashion a splint with 4″ or 5″ rolls folded to length. Begin the splint laterally, three fingers breadths below the knee flexion crease, and extend it down and wrap it under the heel and then up the medial side of the leg (Figure 9). Apply the splint like a stirrup, extending material under the foot, covering the heel and arch. Place a single layer of cast padding over the splint, and loosely wrap a 4″ elastic bandage to secure the splint as you mold it to the extremity. Maintain the ankle at 90° as the splint hardens (Figure 10). An additional splint may be placed posteriorly if needed. Leave the plaster open in front and/or back for swelling, so the patient can unwrap the elastic bandage and spread the splint if needed.

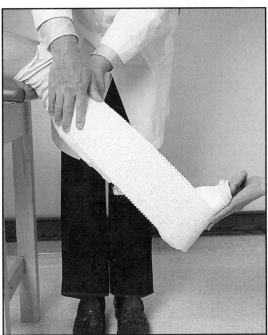

Figure 9
Begin the splint laterally, three fingers breadths below the knee flexion crease, and extend it down and wrap it under the heel and then up the medial side of the leg

Figure 10
Maintain the ankle at 90° as the splint hardens

Patient instructions

Advise the patient to take the following steps when caring for the injured lower extremity at home, after the splint has been applied.

Keep your injured leg elevated to the level of your heart as much as possible. A reclining chair with a pillow beneath your leg is useful for this.

Keep ice bags on your injured leg as much as possible for the next 2 to 3 days. This will help reduce pain and minimize swelling.

If the pain becomes a lot worse, your foot begins to feel numb or like it is "going to sleep," or you cannot move your toes up and down, loosen your splint by unwrapping the elastic bandage and tearing the padding down the front of your leg. If your leg does not feel better in 20 to 30 minutes, contact your doctor. You may be developing problems with the circulation to your leg, which can have serious consequences.

Keep your splint dry. To bathe, place a plastic bag or commercially available cast cover over your leg, prop your leg on the side of the tub, and fill the tub around you, keeping the splinted leg out of the water. Do not shower.

Contact your doctor if you notice any places where the splint feels as though it is chafing or digging into your skin.

Adverse outcomes of treatment

Compartment syndrome, burns, and pressure sores can occur in injuries of both the upper and lower extremities. Plantar flexion contractures of the ankle can develop if the ankle is splinted for prolonged periods with the ankle plantar flexed beyond the neutral position.

Venous Thromboembolic Disease

Synonyms

Blood clot

Thrombophlebitis

Phlebothrombosis

Pulmonary embolism

Definition

Venous thromboembolic disease is characterized by an altered state of coagulation, obstruction of venous outflow, and/or endothelial trauma that precipitates venous clot formation, which makes this condition potentially fatal. Major musculoskeletal procedures (total hip arthroplasty, total knee arthroplasty, hip fracture surgery), multiple trauma, and spinal cord injuries increase the likelihood of clot formation and subsequent embolism. Other risk factors include previous thrombosis, stroke, congestive heart failure, inflammatory bowel disease, malignancy, presence of antiphospholipid antibody, and familial thrombophilia.

Clinical symptoms

Most patients with venous thrombosis are asymptomatic, but some report pain or swelling. Patients with pulmonary embolism may complain of dyspnea, pleuritic or substernal chest pain, and hemoptysis.

Tests

Exam
Clinical signs of venous thrombosis are often nonspecific. Swelling, whether painful or not, is common, but embolism may occur before swelling manifests. Patients with superficial saphenous vein thrombophlebitis have a tender, warm, ropy vein, which may be serious if it extends into the common femoral vein.

The clinical signs of pulmonary embolism include dyspnea, hemoptysis, tachycardia, pleural rub, tachypnea, and sometimes circulatory collapse.

Diagnostic
If patients report swelling and pain, duplex ultrasound is highly accurate in diagnosing proximal clot formation. However, the role of a duplex scan in screening for asymptomatic thrombi remains controversial. Venography remains the gold standard, but is an invasive procedure that may cause pain, allergic reactions to contrast medium, or even a deep vein thrombosis.

Differential diagnosis

Cellulitis

Contusion

Hematoma

Lymphedema

Adverse outcomes of the disease

Both distal and proximal clots can lead to pulmonary embolism and/or death. Symptomatic proximal leg vein thromboses may have silent pulmonary emboli in as many as 50% of patients. Milk leg, or severe edema, which is often painful, may also occur. Other possible problems include postthrombotic syndrome with venous stasis ulceration, chronic edema, venous claudication, and recurrent thromboses.

Treatment

Most patients who die of a pulmonary embolism do so within 30 minutes of the acute event, which is too soon for therapeutic anticoagulation to be effective. Consequently, prophylaxis is needed to reduce the incidence of thromboembolism. There is no universally accepted regimen (agent or duration) for prophylaxis in orthopaedic surgery conditions. The effect of early hospital discharge on the duration of prophylaxis and the impact on the prevalence of symptomatic thromboembolism following discharge is unclear. Effective prophylaxis may be obtained with both pharmacologic and mechanical methods.

The ultimate goal of prophylaxis is to prevent development of a symptomatic pulmonary embolism. With warfarin, 10 mg is generally given the night of surgery (although some prefer to give 5 mg the night before surgery), but nothing is ordinarily given on the first postoperative day, and then the dose of warfarin is adjusted to keep the international normalized ratio (INR) between 1.8 and 2.5. Warfarin should continue for at least 2 weeks; if a shorter course of prophylaxis is used, then screening with a duplex scan or venogram should be considered. However, the overall duration of prophylaxis and the use of screening studies for asymptomatic thrombi remains controversial.

The prophylaxis regimen used with low-molecular-weight heparin depends on the particular drug being prescribed. These drugs have a very short half-life and should not be given until at least 18 to 24 hours after surgery. Monitoring of INR or partial thromboplastin time (PTT) is not necessary. The recommended duration of prophylaxis is approximately 2 weeks. Compared to warfarin, low-molecular-weight heparins are more effective in reducing the overall deep vein thrombosis rates associated with hip and knee surgery, but there has been no difference noted in the prevalence of pulmonary embolism.

When using standard unfractionated (adjusted-dose) heparin, give the first subcutaneous injection of 5,000 USP units 1 to 2 hours preoperatively or within 12 hours after the procedure. Subsequent doses are given every 8 to 12 hours and are adjusted to achieve an activated PTT of 1.5 to 2.0 times more than the upper limit of normal. The blood sample for the measurement of the activated PTT is obtained 4 to 6 hours after the morning dose of heparin. Although this regimen provides effective prophylaxis, it is associated with higher bleeding rates than other modalities.

Pneumatic compression boots or plantar arch compression devices reduce the overall risk of deep vein thrombosis and should be applied during and after surgery. The advantage to using these devices is that they do not require laboratory monitoring, and there is no risk of bleeding. However, patient compliance can be a problem, and the efficacy of these devices in limiting proximal clot formation in patients undergoing total hip arthroplasty is questionable. These devices can serve as an adjunct to pharmacologic prophylaxis.

Effective prophylaxis after total hip arthroplasty includes low-dose warfarin, low-molecular-weight heparin, and adjusted-dose heparin. Effective prophylaxis after total knee arthroplasty includes low-molecular-weight heparin, low-dose warfarin, pneumatic compression boots, and plantar compression devices. Low-molecular-weight heparin also provides effective prophylaxis for some multiple-trauma patients.

The purpose of treatment of a pulmonary embolism or proximal or popliteal thrombus is to reduce the likelihood of recurrence or propagation of the thrombus. When a patient is diagnosed with a proximal venous thrombosis or a pulmonary embolism, start IV heparin with a bolus dose of 5,000 USP units. Follow this with a continuous infusion beginning at 30,000 USP units per day (which is adequate for most patients) to maintain a PTT between 1.5 and 2.0 times normal while starting warfarin therapy. Stop the heparin when the INR has been maintained between 2.0 and 3.0 for 2 consecutive days. Continue warfarin therapy for at least 3 months.

Contraindications of long-term warfarin therapy include pregnancy, liver insufficiency, severe liver disease, noncompliance, severe alcoholism, uncontrolled hypertension, active major hemorrhage, and inability to return for monitoring.

Intravenous heparin is currently the mainstay for early therapy of acute pulmonary embolism or proximal thrombus. However, low-molecular-weight heparins have been demonstrated to be effective in treating acute thromboembolic disease in clinical trials. These agents may eventually be the treatment of choice for acute thromboembolic disease as hospital stays and costs can be reduced.

Adverse outcomes of treatment

Bleeding has been associated with all anticoagulants. Thrombocytopenia occurs with low-molecular-weight heparin or adjusted-dose heparin. However, bleeding risks are low and it is important that high-risk patients receive prophylaxis despite these potential problems. Skin necrosis may occur with warfarin, but is uncommon.

Referral decisions/Red flags

None

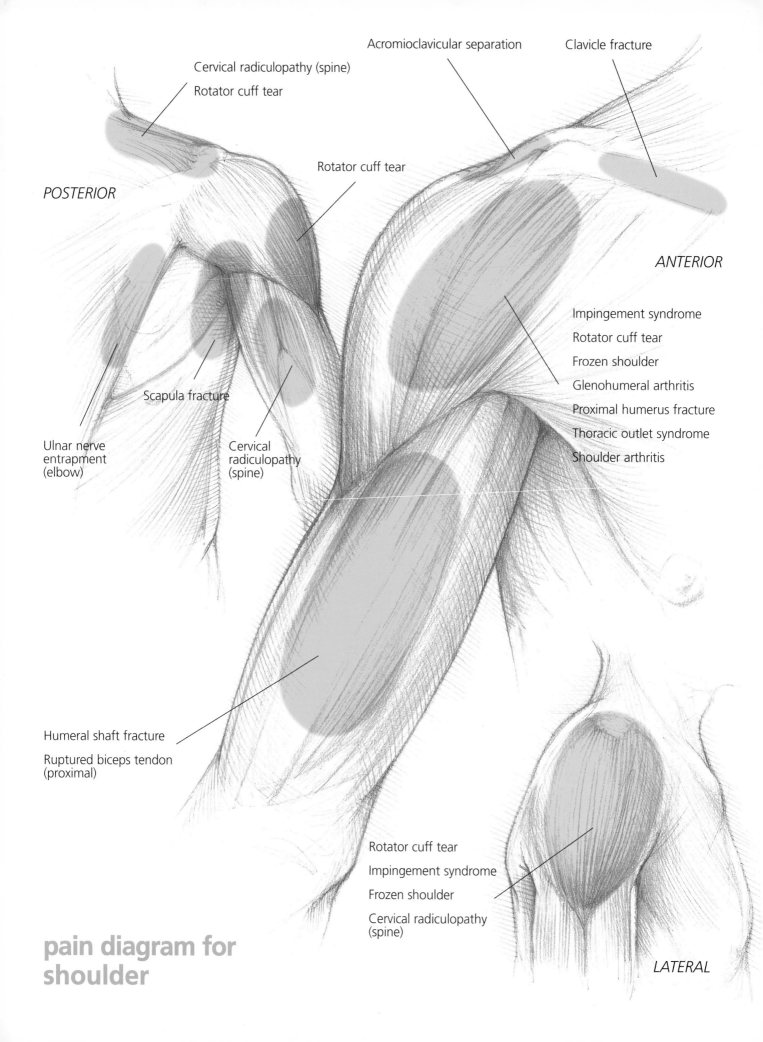

Cervical radiculopathy (spine)

Rotator cuff tear

Acromioclavicular separation

Clavicle fracture

POSTERIOR

Rotator cuff tear

ANTERIOR

Scapula fracture

Impingement syndrome

Rotator cuff tear

Frozen shoulder

Glenohumeral arthritis

Proximal humerus fracture

Thoracic outlet syndrome

Shoulder arthritis

Ulnar nerve
entrapment
(elbow)

Cervical
radiculopathy
(spine)

Humeral shaft fracture

Ruptured biceps tendon
(proximal)

Rotator cuff tear

Impingement syndrome

Frozen shoulder

Cervical radiculopathy
(spine)

**pain diagram for
shoulder**

LATERAL

section two

shoulder

Section Editor

Thomas R. Johnson, MD
Orthopedic Surgeons, P.S.C.
Billings, Montana

shoulder sidebar label

shoulder— an overview

This section of *Essentials of Musculoskeletal Care* emphasizes common shoulder problems in adults, primarily fractures and conditions that affect motion. There are three valuable clues in diagnosing shoulder disorders: chronicity, components of the chief complaint, and patient age. Most disorders of the shoulder can be diagnosed from a careful patient history, physical examination, and routine plain radiographs. A standard trauma plain radiograph series reveals most fractures and dislocations. Some conditions, such as locked posterior dislocations, are easily missed on AP radiographs alone; therefore, an axillary view should be added routinely to aid in diagnosis. MRI scans are seldom needed to make a correct diagnosis.

Type of pain

Patients with acute symptoms (of less than 2 weeks duration) usually have a history of a recent injury, typically a dislocation, fracture, or rotator cuff tear. Common signs and symptoms of these conditions include discoloration, deformity, and swelling. Knowing the mechanism and magnitude of the injury helps in making the diagnosis. For example, a football player who has severe pain and deformity about the tip of the shoulder after falling onto that shoulder most likely has an acromioclavicular (AC) joint separation.

In the primary care setting, most patients with shoulder problems have had chronic symptoms of more than 2 weeks duration. Two elements in the patient's history—age and chief complaint—serve as valuable keys in making the correct diagnosis of chronic shoulder problems.

Patients with pain due to a rotator cuff problem report pain at night in which sleeping on the affected shoulder is impossible. Pain localized at the top of the shoulder suggests arthritis or an AC joint separation. Pain from an inflamed bursa or torn rotator cuff originates in the deltoid region and radiates to the lateral upper arm. Occasionally, patients with rotator cuff problems will have a tingling sensation into the forearm and hand that may be mistaken for carpal tunnel syndrome; however, this tingling likely represents traction on portions of the brachial plexus from drooping of the shoulder. Pain from rotator cuff problems is aggravated by lifting the arm overhead, especially as the arm passes through an arc of 60° to 120°, the so-called "painful arc." A throwing motion also causes shoulder pain if cuff disease is the problem. In contrast, shoulder pain arising from cervical nerve root irritation is sometimes relieved by placing the forearm and hand on the top of the head.

Instability

"My shoulder slips out of joint" is a common complaint among patients younger than age 30 years. This slipping sensation may be transient (ie, a subluxation), and the shoulder reduces itself when the patient allows the arm drop to the side. A patient who has had multiple previous dislocations may experience a complete dislocation following relatively minor trauma. The position of instability is the "throwing position" (ie, with the arm abducted and externally rotated). This is also the position of apprehension. If the examiner moves the symptomatic arm into the throwing position, the patient will experience pain and a sensation that the arm will "slip out of joint." Over 95% of instability problems involve anterior subluxations or dislocations. Posterior instability is rare. Posterior dislocations most commonly occur after seizures.

Stiffness

Stiffness or limited motion can be secondary to soft-tissue contracture, such as a tight shoulder capsule or alterations in bony architecture from trauma or osteoarthritis. A middle-aged woman with the insidious onset of shoulder pain and progressive loss of motion most likely has a frozen shoulder. A 70-year-old male rancher who reports shoulder pain and loss of active and passive range of motion most likely has advanced glenohumeral osteoarthritis.

Patient Age

Patient age is a key variable in evaluating shoulder problems. Diagnoses typically vary, depending on patient age; patients are categorized as follows: young (younger than 30 years), middle-aged (30 to 50 years), and older (over 50 years).

Young patients

Patients younger than age 30 years typically have symptoms of instability, in which they describe the shoulder as "slipping out," "giving out," or "going numb." These episodes represent subluxations or dislocations of the glenohumeral joint. The second most common problem in this age group is subluxation or dislocation of the AC joint from a direct fall on the point of the shoulder. These injuries are commonly called shoulder separations. Complete rotator cuff tears rarely occur in a person younger than age 30 years.

Middle-aged patients

Patients between ages 30 and 50 years can have any of the common shoulder problems described above. In addition, impingement syndrome, associated with rotator cuff tears, is common in this age group. Frozen shoulder commonly develops in slender females, as well as in patients with diabetes. Remember that patients with a frozen shoulder rarely have an associated tear of the rotator cuff. The converse is also true; patients with a rotator cuff tear seldom lose passive shoulder motion.

Older patients

Patients older than 50 years are likely to have a complete rotator cuff tear (most likely), degenerative arthritis, or a frozen shoulder. Acute pain following trauma may indicate a fracture of the surgical neck of the humerus in an older patient.

shoulder—physical exam

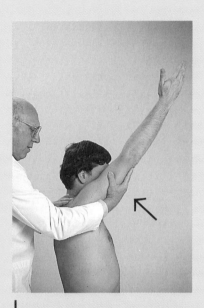

1

Figure 1
Neer impingement sign

With the patient seated, forcibly flex the arm to an overhead position to produce pain from impingement of the humerus against the coracoacromial arch.

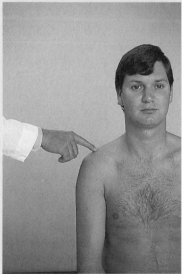

2

Figure 2
Acromioclavicular (AC) joint

With the patient seated, palpate the end of the clavicle and look for tenderness or spurs at the clavicular end. Tenderness is usually most pronounced at the posterior margin of the AC joint and is exaggerated when the patient places that hand on the opposite, unexamined shoulder.

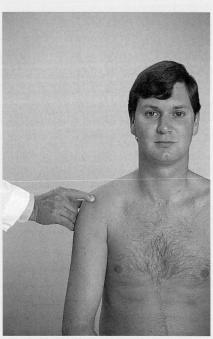

3

Figure 3
Subacromial bursa

With the patient seated, palpate the anterolateral portion of the acromion, moving down toward the deltoid until you feel the acromiohumeral sulcus. Tenderness here is usually related to subacromial bursitis or a supraspinatus tendon tear.

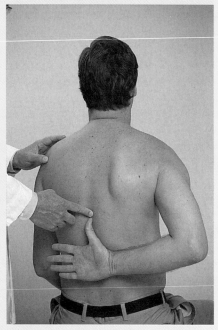

4

Figure 4
Internal shoulder rotation

Ask the patient to reach the thumb as far up the back as possible, while seated. Note which vertebra is reached (the most prominent lowest cervical spinous process is C7). Repeat the test with the other hand and compare the results.

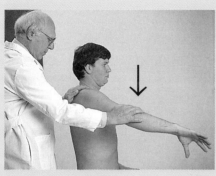

5

Figure 5
Supraspinatus strength test

With the patient seated, place both arms in a position of 90° abduction and 30° forward flexion with the thumbs pointing down. Push down on both arms as the patient resists this pressure. Lack of resistance or presence of pain suggests a rotator cuff tear.

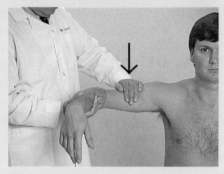

6

Figure 6
Inferior instability

With the patient seated with arm abducted to 90°, push directly downward on the midhumerus to demonstrate inferior subluxation of the glenohumeral joint. A slippage as the humeral head slips over the inferior rim of the glenoid indicates a positive test result, and the patient may try to drop the arm to the side.

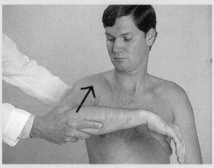

7

Figure 7
Posterior instability

With the patient seated with shoulder flexed to 90°, push posteriorly on the shoulder to demonstrate posterior instability. A slippage as the humeral head rides over the posterior rim of the glenoid indicates a positive test result.

shoulder

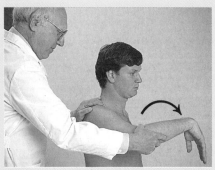

8

Figure 8
Hawkin's impingement sign

With the arm in a throwing position and flexed forward about 30°, forcibly internally rotate the humerus. Pain at a reproducible point is a positive test result that indicates impingement of the supraspinatus tendon against the coracoacromial ligament.

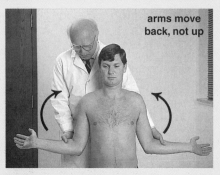

arms move back, not up

9

Figure 9
External rotation

Ask the patient to externally rotate both arms at the same time, while seated. Look for restricted motion, most commonly seen in adhesive capsulitis.

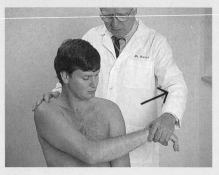

10

Figure 10
Crossed arm adduction

With the patient seated, bring the arm across the chest as far as is comfortably possible. Measure the distance between the antecubital fossa and the acromion, then repeat the test with the opposite arm. An increased distance in one arm indicates a tight posterior capsule.

shoulder

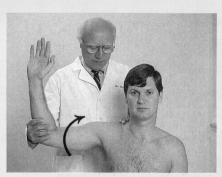

11

Figure 11
External rotation

Ask the patient to abduct the arm to 90° and externally rotate it as far as possible. Record the degree of rotation (the angle of the forearm to the horizontal plane). Next, ask the patient to internally rotate the arm, then record this measurement.

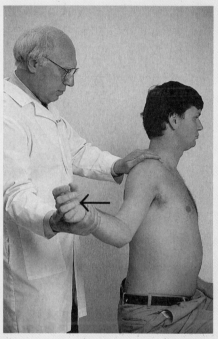

12

Figure 12
Apprehension sign

With the patient's arm in a throwing position, pull the hand backward into more external rotation and extension. If the shoulder is unstable, the patient will quickly drop the arm to the side to avoid a subluxation or dislocation.

shoulder

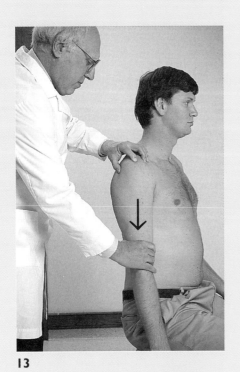

13

Figure 13
Sulcus sign

With the patient seated with arm dangling at the side, pull inferiorly and watch for deepening of the acromiohumeral sulcus, indicating inferior glenohumeral instability.

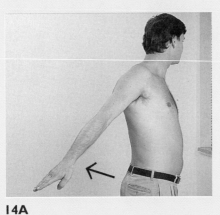

14A

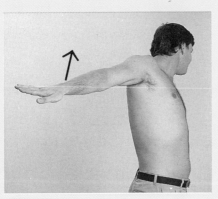

14B

Figure 14
Abduction in the coronal plane

Ask the patient to extend the shoulder posterior to the coronal plane (A) while standing, and then to abduct the shoulder (B). Shoulder pain indicates impingement of the greater tuberosity of the humerus against the lateral border of the acromion. This is another test for impingement syndrome.

Acromioclavicular Injuries

Synonyms

Shoulder separation

Acromioclavicular (AC) separation

Definition

Acromioclavicular (AC) injuries commonly result from a fall onto the tip of the shoulder (acromion) with the arm tucked into the side. These injuries are classified on the basis of the degree of separation of the end of the clavicle relative to the acromion (Figure 1). A "separation" is actually a subluxation in types I and II and a complete dislocation in types III and higher (Table 1).

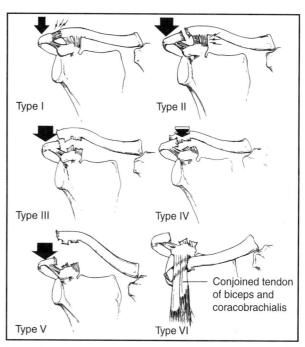

Type I   Type II

Type III   Type IV

Type V   Type VI

Conjoined tendon of biceps and coracobrachialis

Figure 1
Classification of AC separations

Table 1
Classification of AC separations

Fracture type	Injury pattern*
I	AC joint capsule, partially disrupted
II	AC joint capsule and CC ligaments partially disrupted
III	AC joint capsule and CC ligaments completely disrupted
IV	Type III + avulsion of CC ligament from clavicle, penetration of clavicle through periosteal sleeve, or major soft-tissue injury
V	Type III + posterior dislocation of clavicle behind acromion
VI	Type II + inferolateral dislocation of lateral end of clavicle

*   AC = acromioclavicular; CC = coracoclavicular

Clinical symptoms

Patients report pain over the AC joint, and lifting the arm is painful if not impossible. With type III and higher injuries, there is an obvious and cosmetically displeasing deformity.

Tests

Exam

Gentle pressure applied over the AC joint will elicit pain. With type III or higher injuries, there is an obvious deformity, especially if the patient's shirt is removed to examine the shoulder adequately. Most often, patients cannot lift the injured arm, even when asked by the physician.

Diagnostic

Obtain an AP radiograph of both shoulders to confirm a diagnosis of grade II or higher AC separations. A weighted radiographic view in which the patient holds a 10-lb weight in both hands may increase the separation in the injured shoulder, making the subluxation or dislocation more apparent on the radiograph.

Type I injuries are characterized by tenderness on palpation over the AC joint, but AP radiographs of the shoulder are normal. The ligaments holding the AC joint together are stretched slightly enough to cause pain but not sufficiently enough to allow displacement of the bone ends.

Type II injuries show widening of the AC joint compared with the opposite side. However, the distance between the clavicle and the coracoid process is normal.

With type III injuries, the distance between the acromion and clavicle and between the clavicle and the coracoid process is increased by at least 25% when compared with the uninjured side.

Type IV injuries show the clavicle penetrating through the trapezius muscle so that it tents the skin over the lateral aspect of the shoulder.

With type V injuries, the distance between the clavicle and the coracoid process is increased over 100% of that seen in the opposite shoulder.

Differential diagnosis

Fracture of the end of the clavicle (can only differentiate on radiographs)

Fracture of the acromion (can differentiate on radiographs)

Rotator cuff tear (most tenderness over the AC joint, not the rotator cuff)

Adverse outcomes of the disease

Cosmetic deformity, weakness on lifting the arm, chronic shoulder pain, and numbness in the arm are all possible. Arthritis of the AC joint may eventually develop.

Treatment

Nonsurgical treatment of type I and II injuries can consist of wearing a sling for a few days until the pain subsides. Injecting the AC joint provides rapid pain relief (See Acromioclavicular Joint Injection). Ice is helpful for the first 48 hours, and analgesics may be needed to control severe pain. When the pain allows, the patient can resume everyday activities.

Treatment of type III injuries is controversial. Many type III injuries can be treated conservatively with good functional results. However, a good argument can be made for surgical repair in a young manual laborer who does heavy overhead work.

Type IV and V injuries require surgical repair.

Adverse outcomes of treatment

Following prolonged immobilization in a sling, patients may have stiffness in the shoulder; use of a sling or tape may also cause skin breakdown.

Referral decisions/Red flags

Type IV and V injuries require early surgical repair. Heavy laborers who have type III injuries that remain painful with use need further evaluation.

shoulder

Acknowledgements

Figure 1 is reproduced with permission from Rockwood CA Jr: Subluxation and dislocations about the shoulder, in Rockwood CA Jr, Green DP (eds): *Rockwood and Green's Fractures in Adults,* ed 3. Philadelphia, PA, JB Lippincott, 1988, pp 722–985.

Table 1 is adapted with permission from Miller ME, Ada JR: Injuries to the shoulder girdle, in Browner BD, Jupiter J, Levine AM, et al (eds): *Skeletal Trauma.* Philadelphia, PA, WB Saunders, 1992, vol 2, pp 1291–1310.

procedure

Acromioclavicular Joint Injection

The acromioclavicular (AC) joint is formed by the anterior and medial face of the acromion, and the distal end of the clavicle. The sulcus formed by the rounded distal end of the clavicle and the dorsomedial surface of the acromion is most easily palpable posteriorly, especially when the patient places one hand on the opposite shoulder.

Materials

- Sterile gloves

- Skin preparation solution (iodinated soap or similar antiseptic solution)

- 1-mL syringe with a 25-gauge 1½-inch needle

- 1 mL of a 1% local anesthetic

- 3-mL syringe with a 22-gauge 1½-inch needle

- Mixture of 1 mL of a corticosteroid preparation and 0.5 mL of 1% local anesthetic

- Adhesive bandage

Step 1

Wear protective gloves at all times during the procedure and use sterile technique.

Step 2

Cleanse the skin with an iodinated soap or similar antiseptic solution.

Step 3

Use a 25-gauge needle to infiltrate the skin over the joint with 1 mL of a 1% local anesthetic.

Step 4

Through the same needle tract, insert a 22-gauge needle into the posterior aspect of the AC joint (Figure 1). If the needle is in the joint, you will feel a give as the plunger is slightly depressed, followed by a rebound as the intra-articular pressure

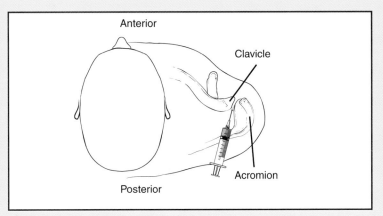

Figure 1
Location for needle insertion

Acromioclavicular Joint Injection (continued)

is transmitted back into the syringe. If the joint capsule is not intact, the solution will enter the joint easily. In either circumstance, the needle is in the joint. If the needle is in the joint capsule, you will feel no sensation of distention or rebound.

Step 5

Inject 1.5 mL of the corticosteroid/anesthetic mixture into the AC joint.

Step 6

Dress the puncture wound with a sterile adhesive bandage.

Adverse outcomes

Although rare, infection is possible. Subcutaneous fat atrophy may occur if the corticosteroid preparation is injected external to the joint. This may produce a depressed area of thin, tender, and unsightly skin.

Aftercare/patient instructions

Advise the patient to use the shoulder as dictated by comfort. Often, the shoulder will be most painful for the first 24 to 48 hours, after which the symptoms should improve. Instruct the patient to return to your office if redness, fever, or immobilizing pain occur.

Arthritis of the Shoulder

shoulder

Synonyms

Glenohumeral arthritis

Definition

Arthritis of the shoulder is characterized by destruction of joint cartilage with loss of joint space (Figure 1). Pain and stiffness develop as a result. Destruction of joint cartilage in the shoulder occurs as a result of many conditions, including osteoarthritis, rheumatoid arthritis, posttraumatic arthritis, and cuff-tear arthropathy—the arthritis that results from large, long-standing rotator cuff tears. Less common causes include osteonecrosis, infection, seronegative spondyloarthropathies, hyperparathyroidism, villonodular synovitis, and Lyme disease. Glenohumeral arthritis is more common in patients older than age 50 years.

Clinical symptoms

Patients principally report pain that is localized to the area around the shoulder and upper arm. Occasionally, patients report pain in the forearm and around the top of the shoulder. The pain is aggravated by activity and relieved by rest. Note that patients may report night pain that interferes with sleep.

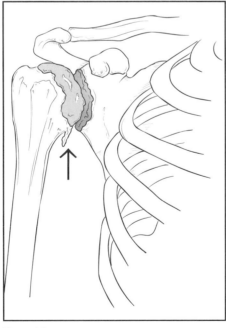

Figure 1A
Osteoarthritis of the shoulder

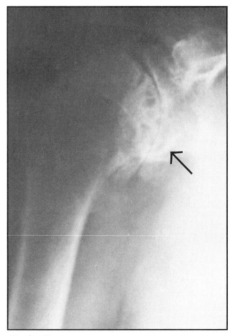

Figure 1B
Radiograph showing osteoarthritis of the shoulder

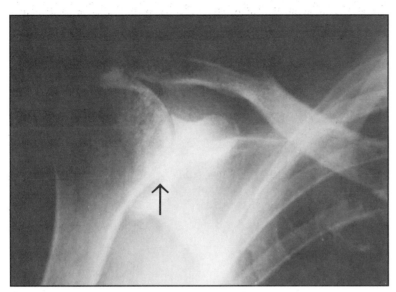

Figure 2
Radiograph of cuff tear arthropathy (arrow shows loss of space)

Progressive loss of motion in the shoulder is the second most common complaint. Patients have increasing difficulty carrying out everyday activities, such as dressing, combing their hair, and reaching overhead.

Obtaining a complete medical history is important in identifying which of the many possible conditions is responsible for destruction of the glenohumeral cartilage. A history of fracture or dislocation suggests a posttraumatic etiology. Multiple joint involvement in a patient with a positive rheumatoid factor suggests rheumatoid arthritis. Single joint involvement in an elderly patient who has no other significant medical history suggests osteoarthritis.

Patients with long-standing complete rotator cuff tears also develop glenohumeral arthritis, which is referred to as a cuff tear arthropathy (Figure 2). The humeral head touches the under surface of the acromion in this condition.

Tests

Exam
Examination may reveal wasting of the muscles about the shoulder. Swelling within the shoulder joint is not common and is difficult to detect. Palpation elicits tenderness over the front and back of the shoulder. Crepitus is commonly present with gentle rotation or flexion of the shoulder. Active and passive range of motion is usually decreased when compared with the opposite side. Patients with rotator cuff tears have normal passive range of motion.

Diagnostic
AP radiographs of the scapula and an axillary view of the shoulder are sufficient to identify joint space narrowing, which indicates cartilage destruction. Other radiographic changes include flattening of the humeral head, marginal osteophytes, and subchondral cysts in the humeral head and glenoid. If the arthritis is due to a chronic large tear of the rotator cuff, then the humeral head will be seen pressing against the undersurface of the acromion. The AP view also commonly shows arthritis of the acromioclavicular joint.

Based on the patient's history and radiographs, certain laboratory studies may be indicated, such as a rheumatoid factor, a CBC, and even a chemical screen.

Differential diagnosis

Charcot joint

Fracture of the humerus

Herniated cervical disk

Infection

Rotator cuff tear

Tumor of the shoulder girdle

Adverse outcomes of the disease

Patients may have chronic shoulder pain and loss of strength and motion. Addiction to narcotic pain medication is also a possibility.

Treatment

Conservative treatment is recommended initially, including use of NSAIDs in the early stages of the arthritis, application of heat and/or ice for relief of symptoms, gentle stretching exercises to preserve motion, and/or injection of corticosteroids for temporary relief. Note that corticosteroids do not seem to alter the natural course of the disease. Corticosteroid injections into the joint are associated with a risk of infection. Should an infection develop, the possibility of later joint replacement surgery is precluded. Activity modifications can also reduce pain. For advanced arthritis, total shoulder replacement offers a very satisfactory solution.

Adverse outcomes of treatment

NSAIDs can cause gastric, renal, and hepatic complications. Infection can develop following corticosteroid injection. Patients who undergo shoulder replacement surgery may also experience postoperative complications.

Referral decisions/Red flags

Patients with intolerable shoulder pain and loss of motion who fail to respond to at least 3 months of conservative treatment need further evaluation.

Fracture of the Clavicle

Synonyms

Collar bone fracture

Definition

Clavicle fractures are among the most common bone injuries (Figure 1). Surgery is rarely needed, and closed treatment is the gold standard.

Clinical symptoms

Patients typically report a history of significant injury, such as falling on the tip of the shoulder or being struck over the clavicle with a heavy object. Patients cannot lift their arm due to pain at the fracture site.

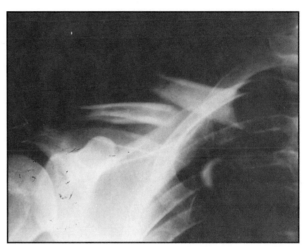

Figure 1
Radiograph of a fracture of the middle third of the clavicle

Tests

Exam
Examination typically reveals an obvious deformity, or "bump," at the fracture site. Gentle pressure over the fracture site will elicit pain, and a grinding sensation can be felt when the patient attempts to raise the arm. The skin may be "tented" over a fracture fragment, but the fragment rarely penetrates through the skin to create an open fracture.

Diagnostic
Plain radiographs are adequate for making a diagnosis of lateral and middle third clavicle fractures. Fractures or dislocations at the inner (medial) end of the clavicle are often difficult to see on plain radiographs, but are well visualized on CT scans.

Differential diagnosis

Acromioclavicular separation (deformity near the tip of the shoulder)

Sternoclavicular dislocation (deformity at the sternoclavicular junction)

Adverse outcomes of the disease

Nonunion is rare, occurring in fewer than 1% of patients. Malunion resulting in cosmetic deformity can occur, as can neurovascular complications.

Treatment

Most clavicle fractures can be treated conservatively with a simple arm sling; a commercially available figure-of-8 clavicle strap can also be used but is more uncomfortable. Immobilization for 3 to 4 weeks is adequate for a child younger than age 12 years, and 4 to 6 weeks is adequate for an adult. Usually, after three weeks of immobilization, the patient can begin shoulder exercises, as pain allows.

Adverse outcomes of treatment

Pressure over the nerves and vessels in the armpit from a tight clavicle strap may cause numbness in the arm. Malunion produces an unsightly bump. Nonunion is rare.

Referral decisions/Red flags

Painful nonunion after 4 months of conservative treatment indicates the need for further evaluation. Widely displaced (> 1 cm) lateral clavicle fractures at the acromioclavicular joint are best treated surgically.

Fracture of the Proximal Humerus

Synonyms

Humeral head fracture

Surgical neck fracture

Greater tuberosity fracture

Definition

Fractures of the proximal humerus commonly occur in elderly women with osteo-porosis. Most of these fractures (80%) are minimally displaced and can be treated with a sling and early motion. A four-part classification system is generally used to identify these fractures (Figure 1). The four parts include the greater and lesser tuberosities (the bony nodules to which the supraspinatus and infraspinatus muscles attach), the head, and the shaft. This classification system has useful implications for treatment of these fractures.

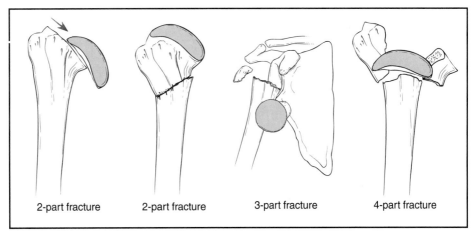

| 2-part fracture | 2-part fracture | 3-part fracture | 4-part fracture |

Figure 1
Fracture patterns for displaced fractures of the proximal humerus (Neer classification)

Clinical symptoms

Patients typically have severe pain, swelling, and discoloration around the upper arm and shoulder following a significant injury, such as a fall. The pain is worse with even the slightest movement of the arm. If the patient reports a loss of feeling in the arm, an associated brachial plexus nerve injury (the network of nerves leading from the neck to the shoulder) should be considered. If the forearm and hand appear pale, check for the presence of a radial pulse, as the axillary artery may have been injured as well.

Tests

Exam

Examination reveals swelling and discoloration around the shoulder and upper arm. A careful neurologic evaluation is also necessary as the axillary and brachial plexus nerves may have also been injured. Check for the presence of the radial pulse as well.

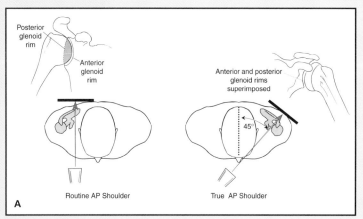
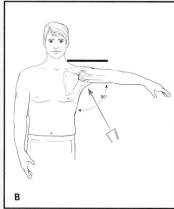

Figure 2
Plain radiographs of the shoulder should be obtained in these views: routine AP (lateral), true AP (A), and axillary lateral (B)

Diagnostic

A trauma series of plain radiographs should be ordered, consisting of AP, lateral, and axillary views of the scapula (Figure 2). If the patient has too much pain with movement of the arm, the axillary view may be impossible to obtain. If the patient cannot tolerate positioning for the axillary view, obtain a lateral view (Figure 3). A CT scan is useful in preoperative planning, but is not needed to confirm the diagnosis of a fracture of the proximal humerus.

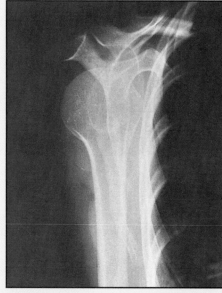

Figure 3
Radiograph showing lateral scapula view of the shoulder

Differential diagnosis

Acromioclavicular separation

Rotator cuff tear

Rupture of the long head of the biceps tendon

Shoulder dislocation (glenohumeral joint)

Adverse outcomes of the disease

Chronic pain, along with loss of motion, and nerve and vessel injury are all possible. Nonunion is also a possibility. Patients may also have posttraumatic arthritis and/or osteonecrosis of the humeral head.

Treatment

Patients with minimally displaced (less than 1 cm) fractures can be treated safely with a sling and after the first week can often begin an exercise program consisting of pendulum and circumduction exercises. After 3 weeks, the sling can be worn part time or it can be removed if pain is minimal.

Displaced two-part fractures in which the greater tuberosity is separated more than 1 cm require surgical repair. If the patient has a two-part fracture in which the lesser tuberosity is fractured, an associated posterior dislocation is also quite possible. Displaced two-part fractures through the humeral neck and displaced four-part fractures usually require surgical stabilization. Patients older than age 40 years with displaced four-part fractures need prosthetic replacement of the proximal humerus.

Adverse outcomes of treatment

Nonunion and malunion are both possible. Missed shoulder location is also a possibility. Patients sometimes report persistent stiffness in the shoulder.

Referral decisions/Red flags

Patients with displaced two-part and all three- and four-part fractures need further evaluation. In addition, patients with associated neurovascular symptoms require further evaluation as soon as possible.

Acknowledgements

Figure 1 is adapted with permission from Neer CS II: Displaced proximal humeral fractures: I. Classification and evaluation. *J Bone Joint Surg* 1970;52A:1077–1089.

Figure 2 is adapted with permission from Rockwood CA, Szalay EA, Curtis RJ, et al: X-ray evaluation of shoulder problems, in Rockwood CA, Matsen FA III (eds): *The Shoulder.* Philadelphia, PA, WB Saunders, 1990.

Figure 3 is reproduced with permission from Frymoyer JW (ed): *Orthopaedic Knowledge Update 4.* Rosemont, IL, American Academy of Orthopaedic Surgeons, 1993, p 286.

Fracture of the Scapula

Definition

Fractures of the scapula invariably result from high-energy trauma, such as that from a motorcycle accident. These fractures may involve the body of the scapula, the glenoid, or the acromion process (Figure 1). Ninety percent of patients with scapula fractures have associated injuries, many of which are life threatening. These injuries may include rib fractures, which are most common, lung injuries, head and spinal cord injuries.

Clinical symptoms

Because scapula fractures commonly occur in association with other serious injuries, the diagnosis is easily missed on initial examination. Pain about the back of the shoulder is the most common complaint of patients.

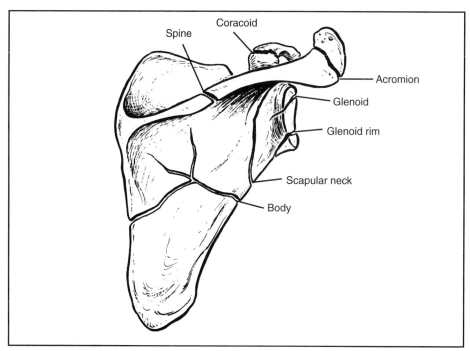

Figure 1
Fracture patterns in the scapula

Tests

Exam

Skin abrasions and swelling over the back of the shoulder are commonly seen with an underlying fracture of the scapula. Tenderness on gentle palpation over the back of the shoulder or over the acromion suggests a possible scapula fracture.

Diagnostic

In most instances, a plain AP radiograph of the shoulder and a radiograph of the chest confirm a diagnosis of a scapula fracture (Figure 2). A displaced fracture of the glenoid should be evaluated further with a CT scan of the glenoid.

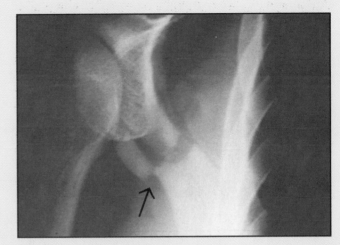

Figure 2
Radiograph showing a fracture of the body of the scapula

An axillary view best demonstrates a fracture of the acromion. Comparison axillary views help differentiate a fracture from a congenital pseudarthrosis of the acromion called os acromion.

Differential diagnosis

Acromioclavicular (AC) separation (maximum tenderness over the AC joint)

Fracture of the proximal humerus (radiograph needed to confirm diagnosis)

Fracture of the rib (evident on plain radiographs of the chest)

Shoulder dislocation (obvious deformity)

Adverse outcomes of the disease

Patients may continue to experience loss of motion and chronic pain. Malunion is possible and is usually asymptomatic. Suprascapular nerve injury and impingement syndrome are also possible.

Treatment

Conservative treatment using a sling is adequate for most patients, followed by early range of motion exercises as tolerated, usually within 1 to 2 weeks after injury. Fractures of the glenoid that are displaced more than 2 mm require open reduction and internal fixation, especially if there is a step off at the fracture site. Patients with isolated scapular body fractures should be considered for hospital admission due to the risk of pulmonary contusion.

shoulder

Adverse outcomes of treatment

Prolonged immobilization may result in shoulder stiffness.

Referral decisions/Red flags

Patients with displaced fractures of the glenoid, fractures of the acromion process with impingement syndrome, and fractures of the neck of the scapula with severe angular deformity (greater than 30°) need further evaluation.

Acknowledgements

Figure 1 is reproduced with permission from Zuckerman JD, Koval KJ, Cuomo F: Fractures of the scapula, in Heckman JD (ed): *Instructional Course Lectures 42.* Rosemont, IL, American Academy of Orthopaedic Surgeons, 1993, pp 271–281.

Frozen Shoulder

Synonyms

Adhesive capsulitis

Stiff shoulder

Definition

Frozen shoulder is characterized by insidious onset and decreased active and passive range of motion; while the cause is unknown, there is most likely an underlying inflammatory process (Figures 1 and 2).

Clinical symptoms

Pain and progressive loss of motion without any known injury are the classic signs and symptoms; in most patients, the nondominant arm is affected. At risk are women between ages 40 and 65 years, and patients with clinical depression. Insulin-dependent diabetes mellitus is associated with a recalcitrant form of frozen shoulder that is difficult to treat.

Other associated medical conditions include hypothyroidism, a recent neurosurgical procedure, Parkinson's disease, or a recent myocardial infarction. This process is typically slow to improve, often taking more than 2 years to resolve.

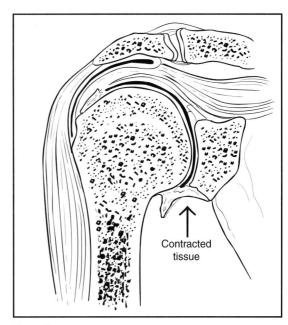

Figure 1
Contracted soft tissue about the glenohumeral joint in frozen shoulder

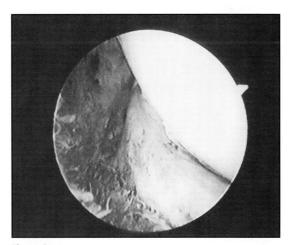

Figure 2
Arthroscopic view of a frozen shoulder. The capsule is snug against the humeral head

shoulder

Tests

Exam

Examination reveals significant reductions in both active and passive range of motion of the shoulder, at least 50%, when compared with the opposite normal shoulder. Conversely, in a rotator cuff tear, active motion is poor and passive motion is essentially normal. Motion is painful especially at the extremes; the shoulder is tender to palpation about the rotator cuff.

In addition, patients often report stiff fingers, which can be demonstrated by having the patient put his or her hands together fingertip to fingertip, in a "praying" position. Patients with a frozen shoulder often cannot touch their palms together in this position.

Diagnostic

Plain radiographs of the shoulder are indicated to rule out an underlying tumor or calcium deposit; these radiographs are usually normal. Arthrography or an MRI scan is indicated in patients whose pain and motion do not improve after 3 months of treatment. Arthrography will reveal decreased volume in the joint, which can be realized by the small amount of contrast (less than 5 mL) that can be injected.

Differential diagnosis

Fracture of the proximal humerus (evident on plain radiographs)

Impingement syndrome (good range of motion)

Osteoarthritis (evident on plain radiographs)

Rotator cuff tear (normal passive range of motion)

Shoulder instability (normal passive range of motion)

Tumor (rare, but evident on AP radiographs)

Adverse outcomes of the disease

Loss of shoulder motion and chronic pain are common.

Treatment

NSAIDs, non-narcotic analgesics, and physical therapy with a home exercise program are indicated. The Jackins exercise program is a useful home program (Figure 3). Instruct the patient in how to do these exercises and advise the patient to do the exercises four times a day (with the hope that they will be done twice a day). Some patients benefit from the supervision of a physical therapist on a once-a-week basis for 6 weeks. The key to initial treatment, however, is compliance with the home exercise program.

If there is no substantial progress after 12 weeks, consider manipulation of the shoulder under anesthesia. An exception to this would be a patient who has insulin-dependent diabetes mellitus, as these patients are not helped by manipulation. They require more aggressive surgical treatment in the form of arthroscopic capsular releases.

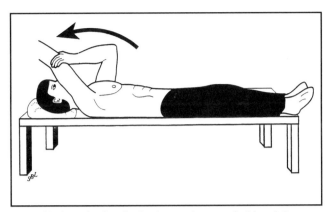

A *Stretching in overhead reach using the opposite arm as the "therapist"*

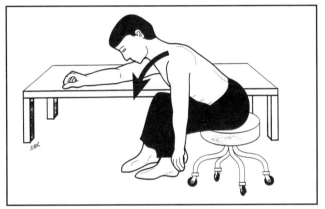

B *Stretching in overhead reach using the progressive forward lean to apply a gentle elevation force to the arm*

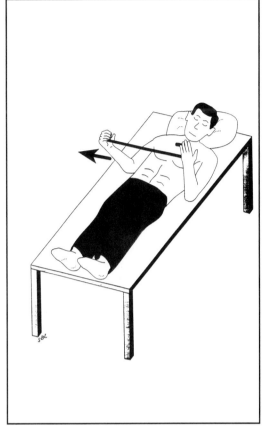

C *Stretching in external rotation using the opposite hand as the "therapist"*

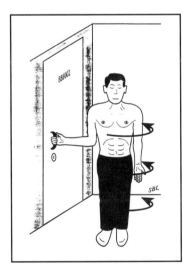

D *Stretching in external rotation by turning the body away from a fixed object to apply a gentle stretching force*

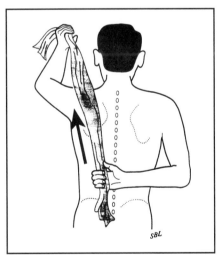

E *Stretching in internal rotation using a towel to apply a gentle stretching force*

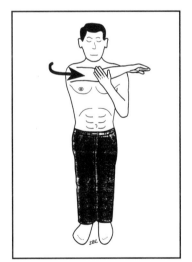

F *Stretching in cross-body reach using the opposite arm as the "therapist"*

Figure 3
University of Washington (Jackins) exercises for stiff shoulders

shoulder

Advise patients that on average, a recovery period of 2 to 2½ years is to be expected before motion is restored and pain relieved.

Although a rotator cuff tear rarely occurs in conjunction with a frozen shoulder, it does occasionally happen. Treatment should focus initially on the frozen shoulder and diagnostic studies, such as arthrogram, should be delayed until full motion is restored.

Adverse outcomes of treatment

NSAIDs may cause gastric, renal, or hepatic complications. Fracture of the humerus following manipulation is possible. Narcotic addiction from pain medication is also possible.

Referral decisions/Red flags

Patients who fail to regain motion after 3 months of conservative treatment need further evaluation.

Acknowledgements

Figure 2 is reproduced with permission from Harryman DT II: Shoulders: Frozen and stiff, in Heckman JD (ed): *Instructional Course Lectures 42.* Rosemont, IL, American Academy of Orthopaedic Surgeons, 1993, pp 247–257.

Figure 3 is reproduced with permission from University of Washington Shoulder and Elbow Service, Home exercise program for the stiff shoulder, in Matsen F III (ed): *Practical Evaluation and Management of the Shoulder.* Philadelphia, PA, WB Saunders, 1994, pp 45–49.

Glenohumeral Instability

Synonyms

Dislocation

Subluxation

Multi-directional
instability

Recurrent dislocation

Definition

Patients with glenohumeral instability are unable to keep the humeral head centered in the glenoid socket. Unlike the hip joint, the shoulder joint is not an inherently stable joint. With a subluxation, the humeral head partially slips out of the fossa, while with a dislocation, the head slips completely off the glenoid so that the articular surfaces are not touching (Figure 1). Unstable shoulders can be categorized using two acronyms—TUBS and AMBRI—based on whether an injury was the cause of the instability. TUBS stands for Traumatic, Unidirectional, Bankart lesions (a tear of the glenoid labrum), and Surgery. AMBRI refers to Atraumatic, Multi-directional, Bilateral symptoms, Rehabilitation as the preferred treatment, and Inferior capsular repair should surgery become necessary.

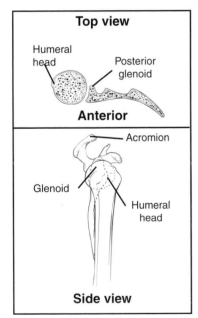

Figure 1A
Humeral head reduced

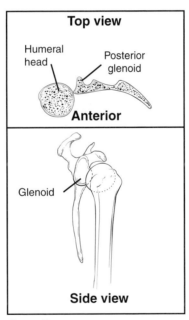

Figure 1B
Humeral head subluxated anteriorly

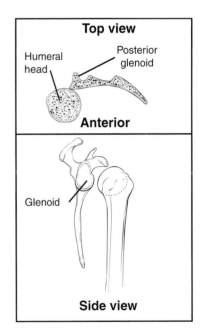

Figure 1C
Humeral head dislocated anteriorly

Clinical symptoms

Patients who have shoulder instability commonly report that their shoulder slips out of joint. Therefore, it is important to determine whether the first episode of slipping occurred after an injury. Patients should also be asked if they can voluntarily make the shoulder dislocate. The most common complaint in patients with anterior instability (shoulder slips forward) is that the shoulder slips out when the arm is in a throwing position: abducted and externally rotated. The underlying pathology with TUBS is a tear of the glenoid labrum off the front of the glenoid; this allows the humeral head to slip anteriorly (Figure 2).

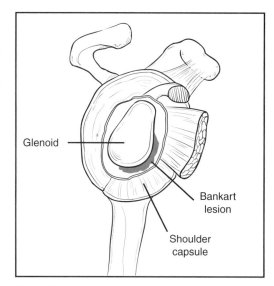

Figure 2
Detachment of labrum—Bankart lesion

Tests

Exam

The following tests or examinations should be performed:

Apprehension test: Move the patient's relaxed arm into abduction and external rotation, as in a throwing position. If this maneuver elicits pain and a sense of apprehension that the shoulder will slip out of joint, the test is positive and indicates the presence of anterior shoulder instability (Figure 3).

Sulcus sign: Pull down on the patient's arm with one hand as you stabilize the scapula with the other. If an indentation develops between the acromion and the humeral head, the test is positive (Figure 4). This suggests increased laxity in the glenohumeral joint.

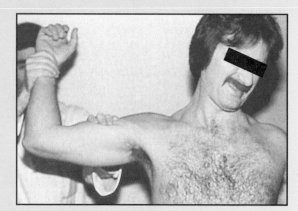

Figure 3
Apprehension test

Laxity exam: Some patients are "double jointed," which is a lay term for capsular laxity. Ask the patient to touch the thumb against the volar (flexor) surface of the forearm (Figure 5). Bend the fingers back at the knuckle to determine how far they extend past neutral with the finger and hand in a straight line. Patients with lax tissues are more likely than others to be able to voluntarily dislocate the shoulder.

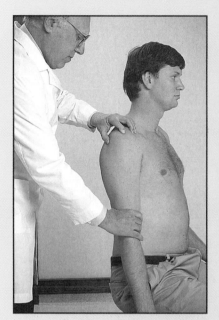

Figure 4
Sulcus sign

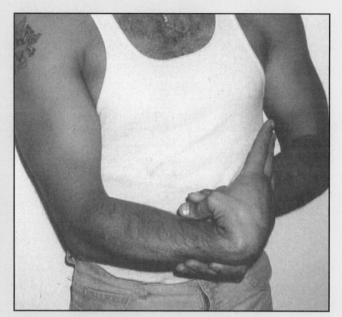

Figure 5
Patient who has generalized laxity

Sensation: Check sensation around the lateral deltoid area (tip of the shoulder) to determine if the axillary nerve has been injured.

Diagnostic

An AP radiograph of the gleno-humeral joint and an axillary view are most useful. Look for a bony irregularity at the anterior edge of the glenoid rim in the axillary view—either a bone spur or a bone defect (Figure 6). Patients older than age 35 years with a history of traumatic dislocation are also prone to tear the rotator cuff at the time of the dislocation. A shoulder arthrogram is indicated in this instance.

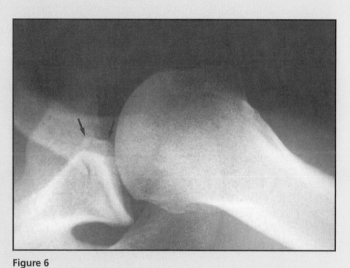

Figure 6
Axillary view showing erosion of the glenoid rim associated with anterior gleno-humeral instability

shoulder

shoulder

Differential diagnosis

Glenohumeral arthritis (confirm with radiographs)

Impingement syndrome (positive impingement sign)

Rotator cuff tear (evident on arthrogram or MRI scan)

Adverse outcomes of the disease

Pain and weakness about the shoulder are common. Patients may also have persistent instability, and/or intermittent paresthesias in the arm.

Treatment

Most acute shoulder dislocations can be reduced in the emergency department. While there are several ways to reduce a shoulder, only two are described herein (see Reduction of Anterior Shoulder Dislocation). Patients with one or more traumatic dislocations can be treated with a physical therapy program that emphasizes strengthening the muscles used for internal rotation of the shoulder—especially the subscapularis muscles (see Jackins exercises, in Frozen Shoulder). Note that after a third dislocation, the risk for another approaches 100%.

Patients with atraumatic or voluntary instability (AMBRI) should be treated conservatively if possible. The shoulder exercise program described above can be tried. Educating patients to avoid voluntarily dislocating the shoulder and to avoid positions of known instability should be part of the treatment plan.

Adverse outcomes of treatment

Axillary nerve injury, osteoarthritis of the glenohumeral joint, and/or persistent dislocation are possible. Failure to recognize a posterior dislocation can also occur.

Referral decisions/Red flags

Failure to reduce an acute dislocation by closed manipulation is an indication for further evaluation on an emergency basis. Also, recurrent dislocations (two or more), despite a 3-month trial of shoulder rehabilitation exercises, are cause for further evaluation. Patients with AMBRI whose symptoms are intolerable and who do not respond to a rehabilitation program may benefit from further evaluation.

Acknowledgements

Figure 2 is adapted with permission from Matsen F III (ed): *Practical Evaluation and Management of the Shoulder.* Philadelphia, PA, WB Saunders, 1994, p 103.

Figures 3, 5, and 6 are reproduced with permission from Bigliani LU (ed): *The Unstable Shoulder.* Rosemont, IL, American Academy of Orthopaedic Surgeons, 1996.

procedure

Reduction of Anterior Shoulder Dislocation

Two reduction techniques will be described here. For both, obtain prereduction AP and axillary radiographs demonstrating the dislocation. If the patient will not allow an axillary view, obtain a transscapular "Y" view to confirm the dislocation. Please note that reducing a first time shoulder dislocation is most safely done in the emergency department with a resuscitation cart available.

Establish an intravenous line, make sure that naloxone (Narcan) is available, and employ pulse oximetry and cardiac monitoring if you use fentanyl for anesthesia. Apply oxygen by mask or nasal cannula throughout the procedure. Fentanyl in a dose of 100 µg is given IV over 1 minute, and this dose is repeated every 3 to 5 minutes until adequate sedation is achieved. The usual total dose of fentanyl is 3 µg/kg. Patients with recurrent dislocations may not require any anesthesia.

Evaluate axillary nerve function by checking sensation over the tip of the shoulder.

First technique

Step 1

Materials

- Fentanyl
- Naloxone
- 2 sheets (for first technique)

Place the patient supine on the stretcher with one sheet folded into a band 4" to 5" wide around the patient's chest (Figure 1). Stand next to the patient on the same side as the injured shoulder, at or below the patient's waist.

Step 2

With the patient's elbow bent 90°, a second sheet is placed around the forearm just below the elbow and wrapped around your waist. Your assistant applies traction to the sheet that is wrapped around the patient's thorax while you apply a steady traction to the arm.

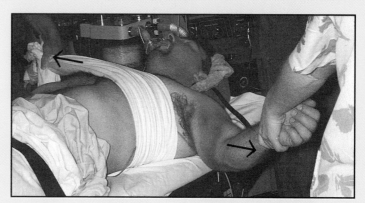

Figure 1
First technique for shoulder reduction

The Essentials of Musculoskeletal Care   105

Reduction of Anterior Shoulder Dislocation (continued)

Step 3

Reduction may be aided if you gently rotate the arm internally and externally while the longitudinal traction is applied. You usually can feel and see the shoulder reduce. Occasionally, especially in large individuals, the reduction may be subtle and you may neither feel nor see it.

Second (Stimson) technique

Step 1

Place the patient prone on a stretcher with the dislocated arm hanging off the cart (Figure 2). Secure the patient to the stretcher with a sheet.

Step 2

Either have an assistant sit on the floor and provide downward traction, or attach 10 to 15 lb of weight to the patient's arm. The weights should not touch the floor.

Step 3

For reduction of the left shoulder, place your left thumb on the patient's acromion and the fingers of your left hand over the front of the humeral head.

Step 4

As the muscles relax, gently push the humeral head caudally until it reduces.

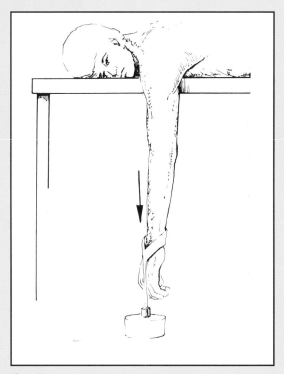

Figure 2
Stimson technique (gravity-assisted reduction with patient lying on stomach)

Reduction of Anterior Shoulder Dislocation (continued)

Adverse outcomes

Axillary nerve palsy may occur with reduction. Be sure to test axillary nerve sensation (lateral deltoid area) before and after reduction.

Inability to reduce the shoulder without general anesthesia is possible.

Fentanyl overdose is treated by administering naloxone. Initially, give 0.2 to 0.4 mg IV, and if there is no response, administer repeated doses of up to 4 to 5 mg; doses as high as 15 to 20 mg may be administered with resistant opioids such as fentanyl.

Aftercare/patient instructions

Obtain postreduction AP and axillary radiographs to confirm the reduction. A transscapular lateral radiograph also will suffice. Immobilize the arm in a sling but have the patient remove the sling and extend the elbow several times daily to prevent elbow stiffness. Begin isometric exercises for the rotator cuff. Have the patient rotate the arm externally to 0° with the hand pointing straight ahead, and flex the shoulder to 90° with the humerus parallel with the floor, at least 10 times each day for 2 weeks.

Begin strengthening exercises for the subscapularis and infraspinatus muscles at 2 to 3 weeks in the older patient and at 6 weeks in the younger patient.

Increase shoulder external rotation to 30° or 40° and shoulder flexion to 140°. This occurs at 6 weeks in patients younger than age 30 years and at 3 weeks in patients older than age 30 years.

Begin vigorous shoulder motion at 6 weeks in patients older than age 30 years, but delay this step to 3 months for patients younger than age 30 years.

The propensity for shoulder stiffness in patients older than age 30 years is an advantage in regaining stability; however, these patients are more likely to suffer a rotator cuff tear at the time of the dislocation.

In the athlete, allow return to sports activities once near full flexion and rotation of the arm and normal strength of the cuff have been regained.

Acknowledgements

Figure 2 is reproduced with permission from Rockwood CA Jr: Subluxations and dislocations about the shoulder, in Rockwood CA Jr, Green DP (eds): *Rockwood and Green's Fractures in Adults,* ed 2. Philadelphia, PA, JB Lippincott, 1984, pp 722–805.

Impingement Syndrome

Synonyms

Tendinitis

Biceps tendinitis

Shoulder bursitis

Definition

Shoulder pain is commonly caused by an impingement of the acromion, coracoacromial ligament, acromioclavicular joint, and the coracoid process on the underlying bursa, biceps, tendon, and rotator cuff (Figure 1). This syndrome may be caused by inflammation of any of these structures; commonly, all are affected. One potential outcome of impingement syndrome is a rotator cuff tear.

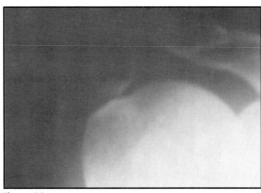

Figure 1A
Radiograph of a shoulder showing calcific tendinitis in a person with impingement syndrome

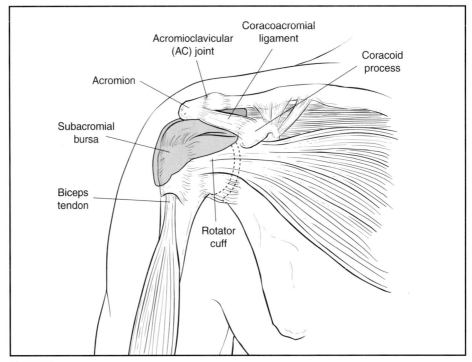

Figure 1B
Anatomy of the front of the shoulder

Clinical symptoms

Gradual onset of anterior and lateral shoulder pain is characteristic. Night pain and difficulty sleeping on the affected side are also common. Atrophy of the muscles about the top and back of the shoulder may be apparent if the patient has been symptomatic for several months. Localized areas of tenderness about the anterior and lateral shoulder are not common. However, crepitus and/or pain when the patient actively lifts the arm from 60° to 100° (the painful arc of motion) is common. The shoulder may "catch" while in this arc of motion.

Tests

Exam

Raise the patient's arm overhead while the scapula is stabilized (Figure 2). If this maneuver elicits pain, the test (Neer impingement sign) is positive for impingement. Next, flex the arm forward 90° and then rotate internally (Figure 3). Pain with this maneuver (Hawkins impingement sign) is due to impingement of the greater tuberosity against the coracoacromial ligament. After completing these tests, inject 10 mL of 1% plain local anesthetic into the subacromial space (see Shoulder Bursa Injection). Then repeat the Neer and Hawkins tests. Complete relief of pain supports a diagnosis of impingement syndrome.

To demonstrate weakness of the supraspinatus tendon, test the strength of both arms at the same time. Ask the patient to extend both arms forward, thumbs pointed to the floor, then push down on the wrists against the patient's resistance and compare shoulder strength (Figure 4).

Atrophy of the muscles about the top and back of the shoulder usually indicates a rotator cuff tear (Figure 5). Entrapment of

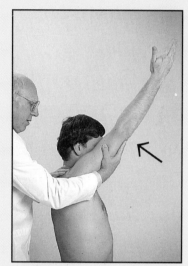

Figure 2
Neer impingement sign

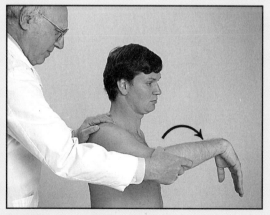

Figure 3
Hawkins impingement sign

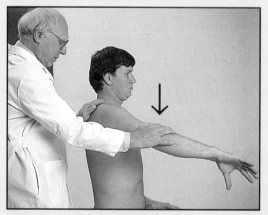

Figure 4
Testing strength of the supraspinatus

shoulder

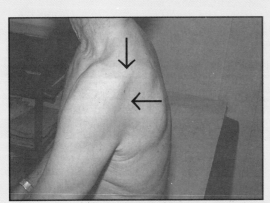

Figure 5
Atrophy of the supraspinatus and infraspinatus muscles

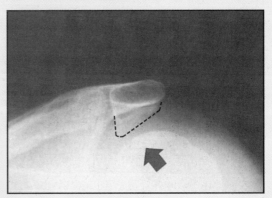

Figure 6
Radiograph of a subacromial bone spur

the suprascapular nerve is another, but less common, cause of atrophy of the supraspinatus and infraspinatus muscles.

Diagnostic
Plain radiographs of the shoulder are usually normal. As an alternative to the traditional inlet view, a special angled view is useful to demonstrate a subacromial bone spur, which can be the offending agent (Figure 6). Narrowing of the space between the head of the humerus and the under surface of the acromion, which is normally equal to or less than 8 mm, suggests a complete rotator cuff tear.

Differential diagnosis

Acromioclavicular arthritis (tenderness over the acromioclavicular joint)

Frozen shoulder (severe loss of motion)

Glenohumeral arthritis (pain on most motions; evident on plain radiographs)

Herniated cervical disk (associated neck stiffness, deltoid weakness with absent biceps reflex, possible sensory loss)

Rotator cuff tear (weakness of supraspinatus)

Suprascapular nerve entrapment (atrophy of supraspinatus and infraspinatus muscles)

Adverse outcomes of the disease

Conditions can range from edema and bursitis to a complete rotator cuff tear. Bursitis alone produces pain. The functional impairment is due primarily to pain, but when a rotator cuff tear occurs, there is weakness and loss of function.

Treatment

NSAIDs taken for 10 to 14 days, combined with rest from the offending activity, will likely be effective. Exercises to stretch the posterior capsule of the shoulder should be done four times a day (Figure 7). If this exercise program is not effective, consider a subacromial injection of a corticosteroid.

Although controversial, judicious use of corticosteroid injections in the acute shoulder (2 months) can provide significant, long-lasting pain relief (See Shoulder Bursa Injection). Pain relief should last 2 months or longer. If pain returns in less than 2 weeks after an injection, the injection should not be repeated. Do not give more than three injections due to the increased risk of rupture of the rotator cuff following repeated injections. If no improvement is noted within 6 weeks, consider ordering a shoulder arthrogram.

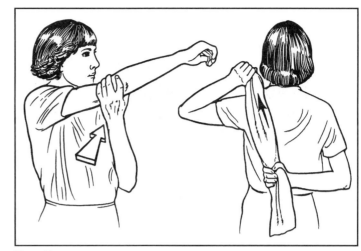

Figure 7
Exercises to stretch the posterior capsule

Adverse outcomes of treatment

NSAIDs may cause gastric, renal, or hepatic complications. Tearing of the rotator cuff and rupture of the long head of the biceps tendon can occur after repeated corticosteroid injections.

Referral decisions/Red flags

Atrophy of the supraspinatus and infraspinatus tendons or a positive arthrogram for a cuff tear are indications for surgery.

Acknowledgements

Figure 7 is reproduced with permission from Rockwood CA, Matsen FA (eds): *The Shoulder.* Philadelphia, PA, WB Saunders, 1990, p 637.

shoulder

procedure

Shoulder Bursa Injection

Materials

- Sterile gloves

- Skin preparation solution (iodinated soap or similar antiseptic solution)

- 10 mL of a 1% local anesthetic

- 2 mL of a 40 mg/mL of a corticosteroid preparation

- 10-mL syringe with a 27-gauge, 1 1/4-inch needle

- 3-mL syringe with a 25-gauge, 1 1/4-inch needle

- Spot adhesive bandage

Step 1

Wear protective gloves at all times during this procedure and use sterile technique.

Step 2

Seat the patient with the arm hanging down to distract the sub-acromial space. The injection site can be anterior or anterolateral (Figure 1).

Step 3

Palpate the acromion both anteriorly and laterally until the anterolateral corner is located.

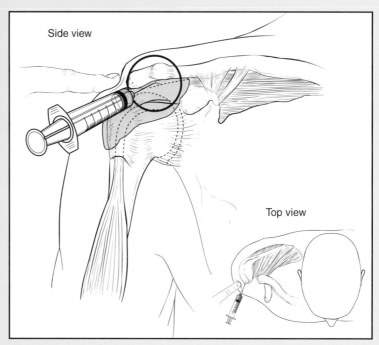

Side view

Top view

Figure 1
Location for needle insertion

Shoulder Bursa Injection (continued)

Step 4

Cleanse this area with an iodinated soap preparation or similar antiseptic solution.

Step 5

With an index finger on the lateral acromion, insert the needle about 1 cm below the palpating finger, raise a wheal with the local anesthetic, and angle the needle superiorly approximately 20° to 30° to access the subacromial space.

Step 6

Two syringes are used. Make the first injection with 5 mL of a 1% local anesthetic (plain) and a 27-gauge 1¼-inch needle. If the purpose of the injection is to evaluate the Neer impingement test, use 10 mL of local anesthetic.

Step 7

Make the second injection at the site of the previous wheal with a 25-gauge 1¼-inch needle, inserting 2 mL of a corticosteroid preparation. If there is resistance while attempting to inject the solution, partially withdraw the needle and reinsert. If the needle is in the proper place, there is little resistance to injection. If you feel the needle hit bone, redirect it superiorly if the bony obstruction is thought to be the humerus, or inferiorly if the bone is thought to be the acromion.

Step 8

Dress the puncture wound with a sterile adhesive bandage.

Adverse outcomes

Increased temporary pain is possible, and although rare, infection can occur.

Aftercare/patient instructions

Advise the patient to apply ice bags to the shoulder and take NSAIDs if increased pain occurs the night following the injection. Instruct the patient to resume usual activities as soon as tolerated, but no later than 24 to 48 hours after the injection. Advise the patient that 33% of patients will experience a temporary increase in pain for 24 to 48 hours from the corticosteroid injection.

Rotator Cuff Tear

Synonyms

Rotator cuff rupture

Rotator cuff tendinitis

Musculotendinous cuff rupture

Definition

The rotator cuff is composed of four muscles: the supraspinatus, the infraspinatus, the subscapularis, and the teres minor muscles (Figure 1). These muscles form a cover around the head of the humerus whose function is to rotate the arm and stabilize the humeral head against the glenoid.

Rotator cuff tears occur primarily in the supraspinatus tendon, which is weakened as a result of many factors, including injury, age, poor blood supply to the tendon, and subacromial impingement. This injury rarely affects people younger than age 40 years.

Clinical symptoms

Patients often report recurrent shoulder pain for several months and can often pinpoint a specific injury that triggered the onset of the pain. Night pain is characteristic, and patients often report having difficulty sleeping on the affected side. Weakness, catching, and grating are common symptoms, especially when lifting the arm overhead.

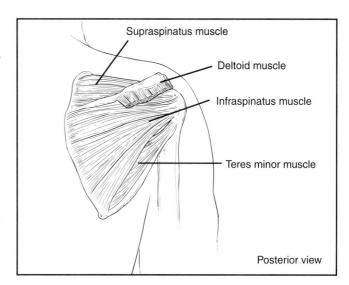

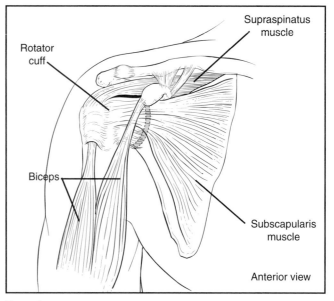

Figure 1

Muscles of the rotator cuff; posterior view (top), anterior view (bottom)

Tests

Exam

Examination reveals that the top and the back of the shoulder may appear sunken, indicating atrophy of the supraspinatus and infraspinatus muscles (Figure 2). This clinical appearance is common in long-standing rotator cuff tears. Passive range of motion of the shoulder is normal, but active range of motion is limited. Conversely, patients with a frozen shoulder have limited active and passive range of motion. As the patient lifts the arm, a "grating" sensation about the tip of the shoulder can be felt. Mild tenderness on palpation over the greater tuberosity is usually present as well.

With large tears, the patient can only shrug, or "hike," the shoulder when asked to lift the arm (Figure 2). There is a complete tear of the rotator cuff if the patient cannot hold the arm elevated when it is lifted parallel to the floor.

Diagnostic

With complete tears, AP radiographs reveal a high-riding humerus relative to the glenoid. A 20° angled view will often show a spur projecting down from the inferior surface of the acromion (Figure 3).

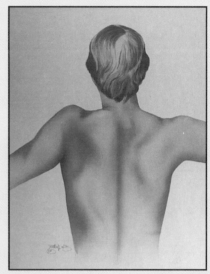

Figure 2
Atrophy of supraspinatus and infraspinatus muscles and shoulder shrug with attempted abduction

A shoulder arthrogram remains the gold standard for confirming rotator cuff tears. This test is more cost-effective and as accurate as an MRI scan; however, an MRI scan should be ordered for patients who are allergic to iodine or seafood.

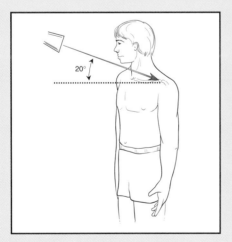

Figure 3A
Proper angle for a radiograph of the rotator cuff

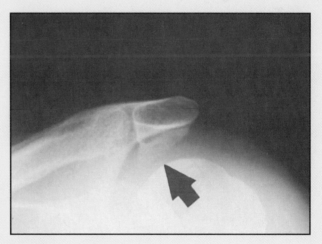

Figure 3B
Radiograph showing a bone spur of the inferior acromion

Differential diagnosis

Bursitis-tendinitis with impingement

Cervical spondylosis (neck stiffness, absent biceps reflex, sensory changes)

Frozen shoulder or adhesive capsulitis (restricted active and passive motion)

Osteoarthritis of the acromioclavicular joint (radiographic evidence of arthritis)

Osteoarthritis of the glenohumeral joint (radiographic evidence of arthritis)

Pancoast tumor (venous distention, pulmonary changes or bony metastases)

Thoracic outlet syndrome (ulnar nerve paresthesias, worse with "military brace" position)

Adverse outcomes of the disease

Loss of shoulder motion, especially the ability to lift the arm overhead, chronic pain, and/or weakness in the affected arm are all possible.

Treatment

Conservative treatment includes NSAIDs, physical therapy with strengthening and stretching exercises, subacromial corticosteroid injections, and avoiding overhead activities. Corticosteroid injections are controversial, as there is no evidence that they help the tendon to heal. While corticosteroids do relieve pain in the short term, repeated injections tend to weaken the tendon and may ultimately lead to the need for surgical repair. Therefore, patients should not receive more than three subacromial injections.

Complete cuff tears require surgery. Just as the ends of a rubber band stretched between two sticks will retract when cut, so, too, will a torn rotator cuff retract when the tendon is ruptured. However, not all patients have severe enough symptoms that they want to undergo surgery. Patients with acute rotator cuff tears tend to do better if they undergo surgery within 6 weeks of the time of the injury.

Adverse outcomes of treatment

NSAIDs may cause gastric, renal, or hepatic complications. Corticosteroid injections can result in transient increase in pain due to the injection itself and/or degeneration of cuff tissue.

Referral decisions/Red flags

Failure of conservative treatment after 3 to 4 weeks and a positive shoulder arthrogram are indications for further evaluation.

Acknowledgements

Figures 1A and 3A are adapted with permission from Rockwood CA, Matsen FA (eds): *The Shoulder.* Philadelphia, PA, WB Saunders, 1990.

Figure 1B is adapted with permission from Hunter-Griffin LY (ed): *Athletic Training and Sports Medicine, ed 2.* Park Ridge, IL, American Academy of Orthopaedic Surgeons, 1991, p 235.

Rupture of the Biceps Tendon

Synonyms

None

Definition

Ruptures of the biceps brachii commonly occur at the proximal end and involve the long head of the biceps (Figure 1). The muscle may rupture at the distal insertion onto the radius, but this is rare. Most often, ruptures occur in adults older than age 40 years who have a long history of shoulder pain secondary to an impingement syndrome. Over time, the tendon becomes frayed and weak, and ultimately ruptures, often as a result of a trivial event. These ruptures are usually associated with a rotator cuff tear, especially among the elderly.

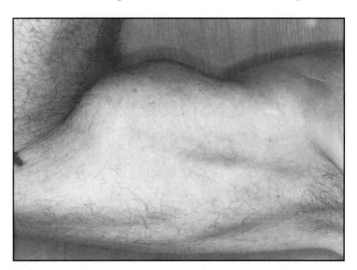

Figure 1
Rupture of the proximal biceps tendon

Clinical symptoms

Patients may report a history of sudden sharp pain in the upper arm, often accompanied by an audible snap. Subsequently, they notice a bulge in the lower arm. Tears of this tendon do occur in young adults, but usually after lifting very heavy weights.

shoulder

shoulder

Tests

Exam

Examination reveals a readily apparent bulge in the lower arm, which results from the muscle belly of the biceps retracting into the lower arm after the proximal anchor tears loose. If the bulge is not obvious, Ludington's test will accentuate the deformity (Figure 2). Ask the patient to put his or her hands behind the head and flex the biceps muscle. There is usually ecchymosis about the mid and lower arm where blood has drained down along the course of the biceps. In addition, there is an easily recognized, palpable defect in the proximal arm. Gentle pressure over the top of the arm with the arm in 10° of internal rotation will elicit pain over the bicipital groove of the humerus (Figure 3).

Diagnostic

AP and axillary radiographs of the shoulder are useful to rule out a fracture, but are not helpful in making a diagnosis of a ruptured biceps tendon. For patients with a previous history of shoulder pain, consider a shoulder arthrogram to rule out a rotator cuff tear.

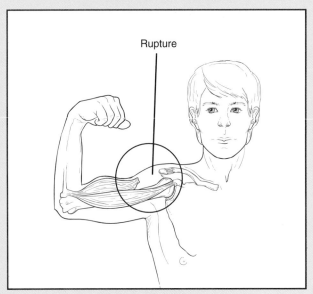

Figure 2
Rupture made more obvious by attempted contraction

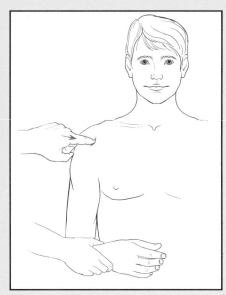

Figure 3
Palpation of the bicipital groove

Differential diagnosis

Dislocated biceps tendon

Fracture of the proximal humerus

Glenohumeral arthritis

Glenohumeral instability

Impingement syndrome

Rotator cuff tear

Rupture of the pectoralis major muscle

Adverse outcomes of the disease

Patients may lose approximately 10% of supination in turning the forearm from palm down to palm up (the screwdriver motion). Cosmetic deformity of the arm in the form of a bulge in the lower arm is also possible.

Treatment

Conservative treatment is effective for most patients, resulting in little loss of function and acceptable cosmetic deformity. Most patients regain full range of motion and normal elbow flexion strength with a therapy program, consisting of range of motion and strengthening exercises as pain allows. Patients, especially elderly patients, who have a long history of shoulder pain may need an arthrogram to confirm a diagnosis of a concomitant rotator cuff tear.

In adults younger than age 40 years who work as heavy laborers and need the extra strength for lifting, surgical repair of the tendon is a reasonable option.

Adverse outcome of treatment

No improvement in function following surgery is possible, as is postoperative stiffness and loss of motion. Note that there is an added expense with surgery while the potential for improvement may be small.

Referral decisions/Red flags

Young patients who are heavy laborers need further evaluation to explore the possibility of surgical repair of the ruptured tendon. Older patients with long-standing shoulder pain and a positive arthrogram who find their symptoms to be intolerable may also benefit from surgical treatment.

Acknowledgements

Figure 2 is adapted with permission from Netter FH: The CIBA Collection of Medical Illustrations. Summit, NJ, CIBA-GEIGY, 1987.

Thoracic Outlet Syndrome

Synonyms

None

Definition

Thoracic outlet syndrome (TOS) refers to a combination of signs and symptoms that affects the neck, shoulder, arm, and hand. TOS is a result of compression of structures, such as the brachial plexus, and subclavian artery or vein, as they exit the narrow space between the superior shoulder girdle and the first rib (Figure 1). These structures may be affected individually or in combinations, and symptoms are often worse with the arm in an overhead position. Women between the ages of 20 and 50 years are most commonly affected.

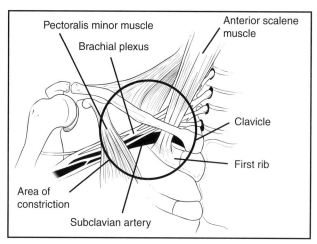

Figure 1
Anatomy of the thoracic outlet

While there is no exact etiology for TOS, congenital anomalies such as cervical rib and abnormally long transverse process of the seventh cervical vertebra, or an anomalous fibromuscular band in the thoracic outlet have been implicated. Posttraumatic fibrosis of the scalene muscles may also be a cause.

No gold standard exists for the diagnosis of thoracic outlet syndrome; thus, the diagnosis is clinical and controversial as the symptoms are often vague and variable. While the optimal treatment is uncertain, most patients are treated nonsurgically.

Clinical symptoms

Nerve compression accounts for most symptoms, and the paresthesias of the arm, forearm, and ulnar side of the hand (little and ring fingers) often mimics ulnar nerve entrapment at the elbow. Vague, aching pain in the upper arm, shoulder, and neck may also occur along with the paresthesias. Some patients also experience headaches, weakness, and loss of coordination. Psychological disturbances, including depression, are seen with TOS, but whether these changes result from TOS or contribute to its development is not clear.

In contrast, symptoms from vascular compression include intermittent swelling and discoloration of the arm. Occasionally the venous compression can result in a thrombosis of the subclavian vein. Arterial compression or an aneurysm of the artery distal to the first rib are the least common forms of TOS and may produce ulcerations in the hand and fingers, arterial emboli, a painful cool extremity, and weakness and fatigue in the arm, especially with overhead use.

Tests

Exam

Unilateral depression of the shoulder narrows the space between the rib and the clavicle, increasing the chance of neurovascular compression. Look for swelling or discoloration of the arm and palpate the supraclavicular fossa to rule out a mass lesion. Auscultation over this area may reveal the presence of a bruit, especially while doing the provocative maneuvers described below. Gentle palpation or light percussion over the brachial plexus may create paresthesias that radiate into the arm and hand. Test range of motion of the neck and look for restricted movements that might indicate a herniated disk or cervical osteoarthritis. Examine the shoulder for signs of an impingement syndrome or shoulder instability. Finally, perform a sensory and motor examination of the upper extremity.

Several provocative maneuvers have been described to diagnose TOS. For a test to be positive, the patient has to volunteer that the maneuver exactly reproduces the symptoms. Adson's test is performed by palpating the radial pulse with the arm at the side, the neck extended, and the head turned toward the affected side (Figure 2). Absence of or a diminished radial pulse is due to compression of the vessels by the scalenus anticus muscle in the neck. With the costoclavicular or "military brace" maneuver, the patient is instructed to move the shoulders back and down (Figure 3). A loss of pulse occurs from compression of the subclavian artery by the clavicle or pectoralis minor muscle. The elevated arm stress test (EAST) maneuver is performed by having the patient hold the arms overhead and then opening and closing the hands (Figure 4). People who do not have TOS should be able to perform this task up to 3 minutes while patients with TOS report pain and fatigue within 30 seconds.

Diagnostic

AP and lateral radiographs of the cervical spine identify cervical ribs or overly long transverse processes. PA and lateral views of the chest help rule out an apical lung tumor or infection. No laboratory studies currently exist to confirm the diagnosis. A cervical MRI scan may be needed if the patient has signs and symptoms of a cervical disk rupture or a cervical spondylosis. AP and axillary radiographs of the shoulder are indicated if the patient has shoulder symptoms. Somatosensory evoked potentials, nerve conduction studies, and ultrasound are not reliable in confirming the diagnosis, but can be useful in ruling out alternative diagnoses (eg, ulnar nerve entrapment).

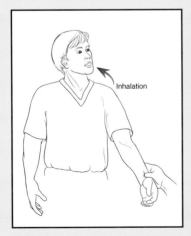

Figure 2
Adson's test

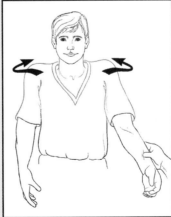

Figure 3
Costoclavicular maneuver

Figure 4
Elevated arm stress test (EAST)

shoulder

Differential diagnosis

Brachial plexus neuritis (sudden onset, severe pain, proximal muscle weakness)

Carpal tunnel syndrome (numbness on the radial side of the hand, positive Phalen's test)

Herniated cervical disk (symptoms may appear on the radial side of hand and arm, but not on the ulnar side)

Impingement syndrome (positive impingement test)

Pancoast tumor (venous congestion, lesion on apical lordotic chest radiograph)

Ulnar nerve entrapment (Tinel's sign at elbow, abnormal nerve conduction studies)

Adverse outcomes of the disease

Weakness and loss of coordination of the upper extremity, chronic headaches, and the inability to work with the arm overhead are all possible. Ulcerations on the arm and hand and Raynaud's phenomenon are also possible. Serious problems such as venous thrombosis and aneurysm of the subclavian artery can develop as well. Patients may also experience psychological changes related to TOS.

Treatment

Most patients can be treated conservatively with a physical therapy program that emphasizes muscle strengthening and postural education exercises (Figure 5). The physical therapy program should be continued for 3 months. Strenuous activities, such as carrying heavy objects, should be avoided, as should placing straps over the affected shoulder, including bras, purses, and seat belts. Prolonged overhead activities, strenuous aerobic exercises, and sleeping on the affected shoulder should be discouraged as well. Exercises that aggravate the patient's symptoms by placing the arm in a provocative position should also be avoided.

Maintaining proper posture is also important; therefore, the patient should be taught to stand up straight with the shoulders back, not slumped forward. Other conservative measures include altering the work station, when possible, to eliminate prolonged overhead activity. The use of NSAIDs, muscle relaxants, and TENS units can help decrease the severity of the symptoms. Weight reduction, when indicated, should be encouraged as well.

In the presence of a congenital anomaly such as a cervical rib, physical therapy may not be helpful, and surgery should be considered. Since the success rate from surgery is so variable and the complication rate so significant, every effort should be made to treat these patients nonsurgically.

Adverse outcomes of treatment

The following conditions may develop following or as a result of treatment: reflex sympathetic dystrophy, intercostal neuroma, frozen shoulder, brachial plexus injury, or pneumothorax.

Referral decisions/Red flags

Vascular compromise with swelling and/or ulceration necessitates early consultation. Further, patients with TOS and a cervical rib or extra-long transverse process need early specialty evaluation, especially when these findings are associated with loss of sensation, muscle atrophy, and weakness. Finally, failure of a well-supervised exercise program in a patient with disabling symptoms is an indication for a surgical consultation.

The following exercises are designed to stretch the soft-tissue structures that may be compressing the neurovascular bundle. Do the exercises two times daily, 10 repetitions each.

1. Corner stretch
Stand in a corner with your hands at shoulder height. Lean into corner until you feel a gentle stretch. Hold for 5 seconds.

2. Neck stretches
Put your left hand on your head with your right hand behind your back. Pull your head towards your left shoulder until you feel a gentle stretch. Hold for 5 seconds. Switch hand positions and repeat the exercise in the opposite direction.

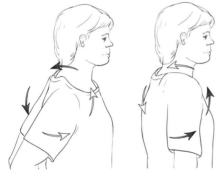

3. Shoulder rolls
Shrug your shoulders up, back, and then down in a circular motion.

4. Neck retraction
Pull your head straight back, keeping your jaw level. Hold for 5 seconds.

If any of these exercises causes an increase in your symptoms, discontinue exercises and consult your physician.

Figure 5
Thoracic outlet syndrome exercise protocol

Acknowledgements

Figure 5 is adapted with permission from Visual Health Information, Tacoma, WA.

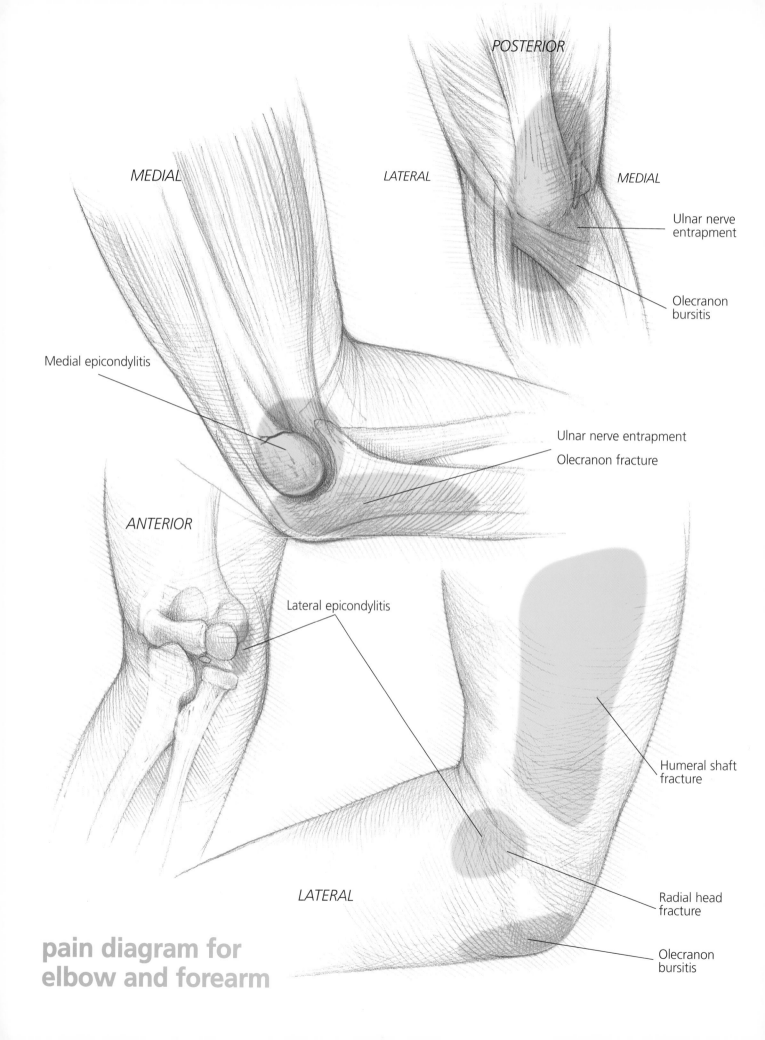

MEDIAL

LATERAL

POSTERIOR

MEDIAL

Ulnar nerve
entrapment

Olecranon
bursitis

Medial epicondylitis

Ulnar nerve entrapment

Olecranon fracture

ANTERIOR

Lateral epicondylitis

Humeral shaft
fracture

Radial head
fracture

LATERAL

Olecranon
bursitis

**pain diagram for
elbow and forearm**

section three

elbow and forearm

Section Editor

Thomas R. Johnson, MD
Orthopedic Surgeons, P.S.C.
Billings, Montana

elbow and forearm—
an overview

elbow and forearm

Diagnosis of elbow and forearm problems is straightforward, as detailed in this section of *Essentials of Musculoskeletal Care.* Patients typically have one of three complaints: pain (which is the most common), stiffness, or swelling.

Most structures about the elbow are superficial and easily palpable. Therefore, both physician and patient can usually pinpoint the source of the pain with one finger. If the patient indicates a more diffuse area of origin, making a correct diagnosis becomes more difficult, if not impossible. With diffuse pain, a work-related or cumulative trauma disorder should be considered.

Type of pain

Pain and swelling that develop suddenly following an injury may be due to fracture or dislocation at the elbow joint or a tendon rupture. If the elbow looks grossly deformed as well, suspect a dislocation. AP and lateral radiographs will confirm a fracture or dislocation. If the radiographs are normal, a tendon rupture, such as the distal biceps in the antecubital fossa, is likely.

Swelling over the tip of the elbow usually indicates an olecranon bursitis, but sudden pain and swelling in the absence of trauma may indicate infection or other inflammatory conditions like crystalline deposition disease (gout) or rheumatoid arthritis.

Location of pain

Anterior elbow pain
Anterior elbow pain is difficult to diagnose. Plain AP and lateral radiographs will confirm a diagnosis of arthritis in the elbow joint. Tenderness to palpation, along with swelling and ecchymosis over the distal biceps tendon, suggests a rupture at its insertion on the bicipital tuberosity of the radius. This diagnosis is reinforced if the patient is a middle-aged man who demonstrates weakness in resisted supination (screwdriver motion) and a history of pain that developed after heavy lifting or a sudden jerking movement.

Medial elbow pain
Pain over the medial or inner aspect of the elbow is likely one of two conditions: ulnar nerve entrapment or medial epicondylitis (medial tennis elbow). Pain from ulnar nerve entrapment will be associated with numbness into the little and ring fingers. Nerve conduction studies of the ulnar nerve at the elbow will help confirm this diagnosis. Check for excessive ulnar nerve mobility (subluxation) by flexing the patient's elbow while palpating the nerve in the ulnar groove. The nerve slips out of the groove in as many as 10% of patients. Next, tap over the nerve in the cubital tunnel to see if the patient experiences a "pins and needles"

sensation in the little and ring fingers. Finally, examine the hand for any wasting and atrophy in the intrinsic muscles between the metacarpals. Patients with pain due to medial tennis elbow will have tenderness to palpation over and just distal to the medial epicondyle. They will also have increased pain when they flex the wrist or pronate the forearm against resistance.

Lateral elbow pain

Pain over the lateral epicondyle is referred to as lateral epicondylitis (tennis elbow). Patients with tennis elbow will have increased pain when the wrist is extended and supinated against resistance. They may also experience pain over the course of the radial nerve in the proximal forearm. Unfortunately, this pattern of referred pain often leads to a mistaken diagnosis of radial nerve entrapment. While this condition does occur, it is rare and overdiagnosed. Lateral elbow pain associated with crepitation over the radio-capitellar joint during forearm rotation is likely from an arthritis between the radius and humerus.

Posterior elbow pain

Posterior elbow pain is usually associated with olecranon bursitis, either acute or chronic. Patients with acute pain typically have mildly painful swelling, which can be treated by aspiration and compression. With chronic bursitis, patients may report recurrent swelling, despite injections, or they may have a "gravelly" sensation of small, tender nodules, which represent scarring in the olecranon bursa. Occasionally, this bursa may become infected and require surgical drainage.

Stiffness

Stiffness or loss of motion may occur with an effusion, arthritis, an osteochondral loose body, or heterotopic bone formation. Joint effusions can result from trauma or an inflammatory condition such as rheumatoid arthritis. A history of sudden, painful intermittent locking suggests the presence of a bony or cartilaginous loose body in the joint. Tomogram radiographs or a CT scan with contrast material will help confirm the diagnosis. An MRI scan is seldom indicated or needed to confirm diagnosis in the elbow. However, it is appropriate when evaluating the extent of an osteochondritis of the capitellum, or for a suspected tumor where the MRI scan can better define the pathology.

Measure the range of motion in flexion, extension, pronation, and supination. Normal flexion is from 0° (arm out straight) to 130° of flexion. Normal pronation and supination is 80° each way. Finally, it is important to evaluate the "carrying angle." Ask the patient to place his or her arm at the side with the palms facing forward. Normally, the angle the forearm makes with the arm is 5° to 8°. If the angle is negative (ie, the forearm points in and the elbow juts out) the patient has cubitus varus, which usually is due to a malunion of a supracondylar fracture.

A final cause of stiffness is attributed to the formation of heterotopic bone, or bone that exists where it should not, such as in a muscle. Patients with head injuries or burns about the elbow are prone to this problem.

Unlike the shoulder, the elbow is a stable joint, and instability is difficult to diagnose. Since the elbow is not a weightbearing joint, pain and limited motion from osteoarthritis is rare.

elbow and forearm—physical exam

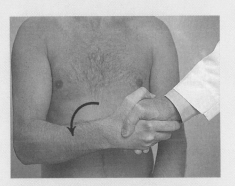

1

Figure 1
Resisted supination

To test the strength of forearm supinators, the most powerful of which is the biceps muscle, shake hands with the patient while resisting the patient's maximum effort to turn the palm up. There is weakness with proximal or distal rupture or tendinitis of the biceps tendon, proximal subluxation of the biceps tendon at the shoulder, or a lesion of the musculocutaneous nerve or the C5, C6 nerve root. Patients with lateral tennis elbow will also experience pain with this maneuver.

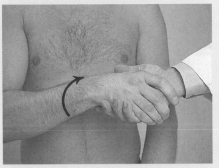

2

Figure 2
Resisted pronation

To test the strength of forearm pronators, the most powerful of which is the pronator teres, which originates from the medial epicondyle, shake hands with the patient while resisting the patient's maximum effort to turn the palm down. There is weakness with rupture of the pronator origin from the medial epicondyle, fracture of the medial elbow, and lesions involving the median nerve or the C6, C7 nerve root. Patients with medial tennis elbow will also experience pain with this maneuver.

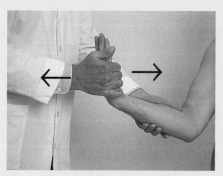

3

Figure 3
Resisted wrist flexion

To test the strength of the wrist flexors, the most powerful of which is the flexor carpi ulnaris, resist the patient's effort to flex the wrist with the elbow flexed. There is weakness with rupture of the muscle origin, fracture of the medial elbow, tendinitis of the medial elbow, or lesions involving the ulnar nerve (C8, T1) or median nerve (C6, C7).

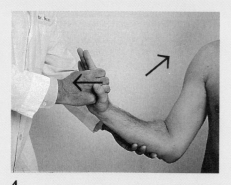

4

Figure 4
Resisted wrist extension

To test the strength of the wrist extensors, the most powerful of which are the extensor carpi ulnaris and extensor carpi radialis brevis, resist the patient's effort to dorsiflex the wrist with the elbow flexed. There is weakness with rupture of the extensor origin, fracture of the lateral elbow, lateral tennis elbow, or lesions involving the radial nerve or C6–C8 nerve root.

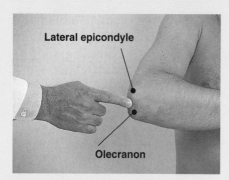

Lateral epicondyle

Olecranon

5

Figure 5
Elbow effusion

Compare the patient's elbows. A bulge between the ulna and the lateral epicondyle is associated with reduced elbow flexion. The dots show where the lateral epicondyle and the olecranon form the boundaries of the fossa, which distends when there is a joint effusion.

elbow and forearm

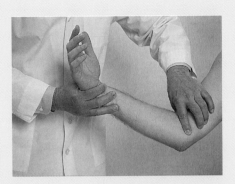

6

Figure 6
Medial epicondylitis

Palpate the medial epicondyle of the distal humerus. The pointing finger indicates the area of tenderness in the common origin of the wrist and hand flexor muscles at or slightly distal to the medial epicondyle.

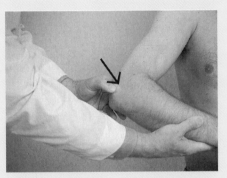

7

Figure 7
Lateral epicondylitis (tennis elbow)

Palpate the lateral epicondyle of the distal humerus. The pointing thumb indicates the area of tenderness in the common origin of the wrist and hand extensor muscles at or slightly distal to the lateral epicondyle.

8

Figure 8
Ulnar nerve palpation

Palpate the ulnar nerve in the ulnar groove of the distal humerus. Palpation and light tapping here may induce paresthesias in the forearm or ulnar two fingers of the hand, suggesting entrapment of the ulnar nerve.

elbow and forearm

Dislocation of the Elbow

Synonyms

None

Definition

Dislocation of the elbow is the most common type of dislocation in children and the second most common type in adults, second only to shoulder dislocation. Young adults between age 25 and 30 years are most often affected, and sports activities account for almost 50% of these injuries.

The most common mechanism of injury is a fall on the outstretched hand. The dislocation can be anterior or posterior, with posterior dislocations most common, occurring 98% of the time (Figure 1). Elbow dislocations seldom occur in isolation. Associated injuries include fracture of the radial head, which occurs in up to 10% of elbow dislocations. Neurovascular structures, including the brachial artery and median nerve, may be injured as well.

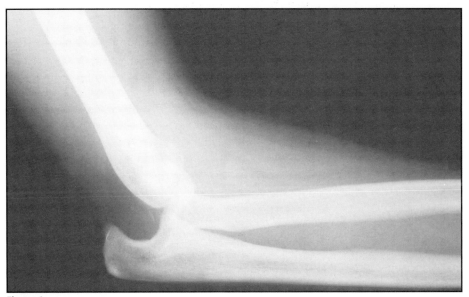

Figure 1
Posterior dislocation of the elbow

Clinical symptoms

Extreme pain, swelling, and inability to bend the elbow following a fall on the outstretched hand are characteristic. In addition, patients often report pain in the shoulder and wrist if there are associated injuries of these structures.

Tests

Exam

The most important part of the examination is the neurovascular evaluation. Check the radial pulse as well as the sensation over the median, radial, and ulnar nerve distribution. Motor function of these nerves must be assessed as well.

Diagnostic

Plain AP and lateral radiographs are adequate to make a diagnosis. CT and MRI scans are seldom indicated.

Differential diagnosis

Fracture-dislocation of the elbow (same deformity, fracture appears on radiographs)

Fracture of the distal humerus

Fracture of the olecranon process of the ulna

Hemarthrosis (positive fat pad sign)

Synovitis (normal radiographs)

Adverse outcomes of the disease

Loss of motion in the elbow, especially extension, and instability are possible. Associated neurovascular injury may also be present. In addition, arthritis of the elbow joint may eventually develop. Formation of ectopic bone is also possible.

Treatment

Reduce an elbow dislocation as soon after injury as possible. Use of anesthetic for reduction in the emergency department may be necessary (see Injections and Corticosteroids). The elbow may also be reduced by injecting 10 mL of 1% local anesthetic into the hematoma located in the dislocated elbow joint. Insert the needle just above the olecranon process (Figure 2). If the dislocation is several hours old, general anesthesia is often required because of swelling and muscle spasm.

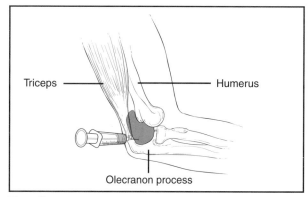

Figure 2
Injecting hematoma with a local anesthetic

Once adequate anesthesia is obtained, reduce the elbow by having an assistant hold the patient's upper arm to apply counter traction during the reduction. Next, hold the patient's hand and the forearm and apply a slow, steady pull in line with the long axis of the arm (humerus). Usually, the reduction is easily felt. Flex the elbow to a right angle, then rotate (pronate and supinate) the forearm to make certain that this motion is smooth. Finally, gently extend and flex the elbow to determine its stability. Failure to obtain full motion suggests the possibility of an intra-articular loose body such as a fragment of bone and cartilage. A postreduction radiograph is required to check the adequacy of the reduction and to confirm that a fracture has not occurred during the reduction. Apply a well-padded posterior splint, extending from just below the wrist to just below the axilla, and leave it in place for 10 days.

Once the splint is removed, initiate early range of motion exercises. NSAIDs are useful during this period and may also decrease the incidence of ectopic bone formation. If the elbow is quite unstable at the time of reduction, extend splint immobilization to 3 weeks. However, the splint cannot remain on longer than 3 weeks because of the potential for loss of elbow motion.

Adverse outcomes of treatment

NSAIDs may cause gastric, renal, or hepatic complications. Reduction can cause neurovascular injury and even a fracture about the elbow. Prolonged immobilization can result in elbow contracture.

Referral decisions/Red flags

Patients with associated neurovascular or bony injuries need further evaluation, as do patients with a contracture of 45° or greater, 3 weeks following the injury.

Acknowledgements

Figure 1 is reproduced with permission from Crosby LA, Lewallen DG (eds): *Emergency Care and Transportation of the Sick and Injured,* ed. 6 revised. Rosemont, IL, American Academy of Orthopaedic Surgeons, 1997, p 544.

elbow and forearm

Entrapment of the Ulnar Nerve

Synonyms

Cubital tunnel syndrome

Tardy ulnar palsy

Ulnar nerve neuritis

Definition

Pressure on the ulnar nerve as it passes through a tunnel along the inside of the elbow causes sensory and motor changes in the hand. This condition may develop as a result of a direct blow to the elbow or from repetitive motion, including flexing and extending the elbow. The nerve is quite superficial as it passes through the tunnel and is bumped easily. Bending the elbow stretches the nerve (Figure 1) When the ulnar nerve is irritated, it creates the sensation of "hitting the funny bone." Patients with diabetes or alcoholism are at increased risk for irritation of the ulnar nerve at the elbow.

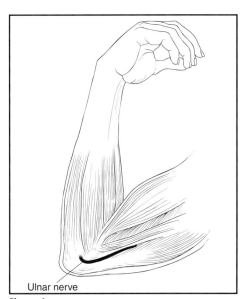

Ulnar nerve

Figure 1
Ulnar nerve compressed during elbow flexion

Clinical symptoms

Patients commonly report tingling and numbness of the little and ring fingers, loss of hand coordination, decreased grip and pinch strength, and tenderness along the inside of the elbow. The ring and little fingers become numb at night.

Conversely, patients with carpal tunnel syndrome commonly awaken with numbness of the thumb, index, and middle fingers. Patients with ulnar nerve entrapment may have pain along the inside border of the shoulder blade on the same side as the nerve irritation, but this is less common.

Tests

Exam

Palpating the ulnar nerve or gently tapping over the nerve in the cubital tunnel causes pain about the inner elbow and tingling in the ring and little finger. Because the finger-spreader (intrinsic) muscles of the hand are innervated predominantly by the ulnar nerve, constant long-term pressure on this nerve causes atrophy or muscle wasting. Muscle wasting can be best visualized by looking at the back of the hand with the palm facing down. A hollowed out appearance of the web space between the thumb and index finger and between the metacarpal bones indicates intrinsic muscle wasting. Lack of sweating, as indicated by dry, pale skin over the little finger and half of the ring finger, suggests chronic nerve irritation.

Ask the patient to bend the elbow as far as it will flex comfortably and report any tingling or numbness in the hand as soon as it is felt (elbow flexion test). Record how quickly these symptoms appear; if symptoms do not develop within 30 seconds, the test is considered negative.

With the patient looking away, move the two points of an ECG caliper (with the points initially separated by 5 mm) lightly across the fingertip. Repeat the test with only one of the points next. The patient should report feeling either one or two points. If the calipers have to be separated by 7 mm or more before the patient can distinguish two separate points, the patient's sensation is considered impaired.

Ask the patient to spread all the fingers as widely as possible; the inability to separate the index from the middle finger, and the little from the ring finger, indicates intrinsic muscle weakness, most likely due to an ulnar nerve lesion (intrinsic muscle test).

Diagnostic

The ability of a nerve to conduct an electrical impulse is impaired by pressure on the nerve. Nerve conduction studies provide an objective measure of nerve compression. A reduction in velocity of 30% or more suggests significant compression of the ulnar nerve.

Plain radiographs of the elbow are indicated if previous elbow trauma has occurred.

Differential diagnosis

Carpal tunnel syndrome (numbness in thumb, index, and middle fingers)

Herniated cervical disk (rare to involve C8 and T1 nerve roots)

Medial tennis elbow (no distal weakness, paresthesias, or numbness; tenderness over the medial epicondyle)

Thoracic outlet syndrome (normal nerve conductions at the elbow, rare wasting in the hand)

Ulnar nerve entrapment at the wrist (strong wrist flexors and ulnar deviators)

Adverse outcomes of the disease

Loss of grip and pinch strength and loss of sensation in the ring and little fingers are possible. Pain and tenderness at the elbow may also persist.

elbow and forearm

elbow and forearm

Treatment

Advise patients to keep the elbow straight as much as is practical. At night, an elbow splint that keeps the elbow from flexing to a right angle can be worn. A towel wrapped around the elbow is sufficient if a commercial splint is not available. A sports elbow protector can be used at work to keep from bumping the elbow. A 1-week course of a decreasing dose of an oral corticosteroid is generally more effective than NSAIDs for an acute severe episode. Altering the work station for typists or computer operators should be tried before surgical treatment is considered.

Adverse outcomes of treatment

An elbow splint may not be tolerated. Fluid retention, flushing of the face, agitation, and nausea may occur as a result of taking the oral corticosteroid. Loss of motion and a tender scar may remain following surgery.

Referral decisions/Red flags

Progressive wasting and atrophy of the intrinsic muscles, accompanied by decreased strength and increasing numbness despite conservative treatment, indicates the need for further evaluation.

Acknowledgements

Figure 1 is adapted with permission from Mackinnon SE, Dellon AL: *Surgery of the Peripheral Nerve*. New York, NY, Thieme Medical Publishers, 1988.

Epicondylitis

Synonyms

Tennis elbow

Elbow tendinitis

Lateral epicondylitis

Lateral tendinosis of the elbow

Radial tunnel syndrome

Definition

Epicondylitis (tennis elbow) is the term used to describe a condition that is characterized by pain about the lateral aspect of the elbow (the lateral epicondyle) (Figure 1). The pain is caused either by an injury to the elbow or by repetitive overuse activities of the upper extremity. Although historically the condition has been called a tendinitis, implying an inflammatory origin, histologic evaluation of surgical specimens does not support this diagnosis. The usual signs of inflammation are absent. Rather, a histologic pattern of tissue degeneration is seen. Therefore, a more accurate term for this condition is lateral tendinosis of the elbow.

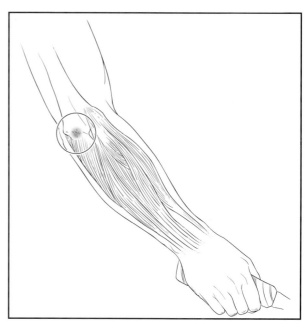

Figure 1
Location of pain with lateral epicondylitis

Tennis elbow is the common cause of pain in the upper forearm and elbow and most commonly affects patients between the age of 30 and 60 years, with peak incidence in patients in their 40s. It most commonly develops as a result of overuse of the forearm muscles in motions that require rotation of the arm and extension of the wrist. Typical precipitating activities include turning a screwdriver, painting a wall, or hitting a ball in racquet sports with improper technique. A direct blow to the outside of the elbow can also trigger tennis elbow. A similar condition exists on the inside of the elbow in handball players and golfers.

Clinical symptoms

Patients report pain, often severe, over the outside of the elbow and the back of the upper forearm. Lifting, especially if the palm is facing down, increases the pain. Holding lightweight objects such as a cup may be difficult. Swelling may be present, but it is difficult to detect.

elbow and forearm

Tests

Exam

The most consistent physical finding is pain to pressure with one finger over the wrist extensor muscles a fingerbreadth below the lateral epicondyle. The Losee elbow position helps to "uncover" the origin of the extensor carpi radialis brevis so that this structure, not any of the other extensor muscle origins, can be palpated (Figure 2). Ask the patient to bend the elbow to 90° and place it across the abdomen with the palm facing up. Next, apply pressure one fingerbreadth below the lateral epicondyle, as this will reproduce the pain of tennis elbow.

Diagnostic

Order plain radiographs of the elbow to rule out an underlying arthritis between the radial head and the capitellum. Also, an osteochondral loose body may be seen, especially if the patient reports a history of locking in addition to pain. An area of calcification may be seen at the attachment of the extensor muscles to the lateral epicondyle of the humerus, but this is rare.

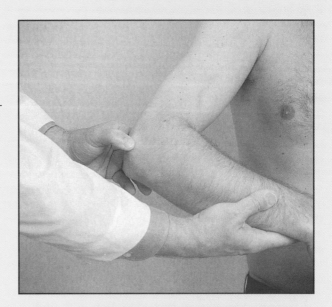

Figure 2

Differential diagnosis

Fracture of the radial head

Osteoarthritis of the elbow joint

Osteochondral loose body

Radial nerve irritation in the upper forearm

Synovitis of the elbow

Triceps tendinitis

Adverse outcomes of the disease

Weakness at the elbow and arm, especially with motions that involve lifting with the palm down and with motions that involve turning the forearm, can develop. Persistent pain may also be a problem.

Treatment

Activities that cause the pain should be eliminated and a short course of NSAIDs (10 to 14 days) prescribed. Alternatively, a 6- or 7-day course of corticosteroids may be effective. Use of a commercial tennis elbow strap (or one made with two pieces of Velcro) worn just below the elbow during heavy lifting activities is helpful as well. Application of heat or ice (whichever works best) may relieve pain and inflammation. Once the pain has decreased, general stretching and forearm strengthening exercises can be initiated.

If conservative treatment fails to relieve the pain, injection of a steroid into the area of maximum tenderness is indicated (see Tennis Elbow Injection). Advise patients that there is a one in three chance they will experience a significant increase in pain for 1 to 2 days after the injection. No more than three injections should be given. If the pain recurs, surgical fasciotomy should be considered.

Adverse outcomes of treatment

Fluid retention, flushing of the skin, and shakiness can occur as a result of taking oral corticosteroids. NSAIDs may cause gastric, renal, or hepatic complications. A cyst or draining sinus can develop following fasciotomy.

Referral decisions/Red flags

Failure of conservative treatment indicates the need for further evaluation.

elbow and forearm

procedure

Tennis Elbow Injection

The classic tender spot in lateral epicondylitis of the elbow (tennis elbow) is at or just distal to the lateral epicondyle of the humerus with the elbow in flexion.

Materials

- Sterile gloves

- Skin preparation solution (iodinated soap or similar antiseptic solution)

- 4-mL syringe with a 27-gauge needle

- 3 to 4 mL of a 1% local anesthetic and epinephrine solution

- 2-mL syringe with a 25-gauge needle

- 2 mL of a corticosteroid preparation

- Adhesive bandage

Step 1

Wear protective gloves at all times during this procedure and use sterile technique.

Step 2

Place the patient's arm in the Losee position, with the arm against the chest or

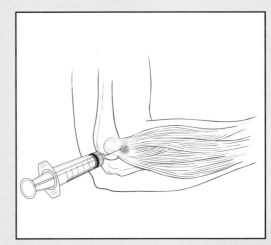

Figure 1
Location for needle insertion

abdomen, the elbow flexed at least 90°, and the forearm fully supinated (palm up). In this position, the extensor carpi radialis brevis (ECRB) muscle origin is easily palpated just distal to the lateral epicondyle.

Step 3

Palpate this prominence and mark a circle with a pen, 1″ in diameter, just distal to the lateral epicondyle.

Step 4

Cleanse the skin with an iodinated soap or similar antiseptic solution.

Step 5

Insert the 27-gauge needle, make a subcutaneous skin wheal with the local anesthetic and advance through the tendon of the ECRB to inject the remaining 2 to 3 mL of local anesthetic (Figure 1).

elbow and forearm

Tennis Elbow Injection (continued)

Step 6

Wait 3 to 4 minutes, then insert the 25-gauge needle through the skin and tendon and inject the corticosteroid preparation.

Step 7

Dress the puncture wound with a sterile adhesive bandage.

Adverse outcomes

Subcutaneous fat atrophy may follow subcutaneous infiltration of the cortisone, leading to a waxy appearing depression in the skin. Although rare, infection is possible.

Aftercare/patient instruction

Advise the patient that pain may increase 24 to 48 hours following the injection. Pain often improves with the application of ice. In some patients, administration of narcotic analgesics may be necessary for pain relief.

Fracture of the Distal Humerus

Synonyms

Supracondylar fracture

Transcondylar fracture

Intercondylar fracture

T condylar fracture

Lateral/medial condylar fracture

Definition

Fractures of the distal humerus are relatively uncommon, accounting for only 2% of fractures in adults (Figure 1). However the morbidity from these fractures is high. Improved results have been achieved with rigid internal fixation and early motion of displaced fractures.

In adults, the fracture usually involves both condyles and often extends into the joint. While there are several classification schemes for these fractures, many of which are quite complex, perhaps the most useful way to consider them is either as nondisplaced or displaced (Figure 2). A displaced fracture involves one or both condyles, and the joint surface may or may not be involved. This schema has implications for treatment. All displaced fractures, except severely comminuted fractures, require open reduction. Nondisplaced fractures can be treated by splinting and early motion.

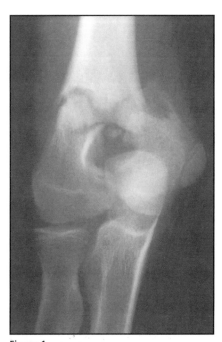

Figure 1
Displaced supracondylar fracture

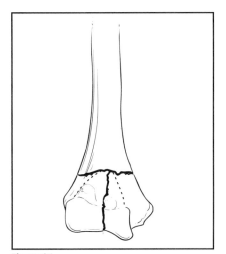

Figure 2A
Nondisplaced T condylar fracture of the distal humerus

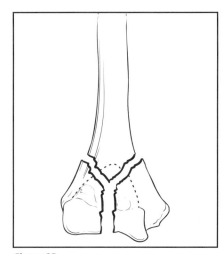

Figure 2B
Displaced intercondylar fracture of the distal humerus

Clinical symptoms

Marked swelling, ecchymosis, deformity, and pain about the elbow following an injury are common. Patients report increased pain with attempted flexion of the elbow.

Tests

Exam
Evaluate the neurovascular status of the injured extremity first. Inspect the skin around the site to identify any open wounds. Palpation about the joint may reveal an effusion, and crepitus may be felt with gentle flexion. A deformity should be visible with any displaced fracture. Check the wrist and shoulder on the affected side for associated injuries to these joints.

Diagnostic
AP and lateral plain radiographs should be adequate in most instances to make the diagnosis. When there is no radiographic evidence of fracture, look carefully for a fat pad sign, which indicates bleeding into the joint, often from an occult fracture (Figure 3). *See also: Figure 1, p 659.*

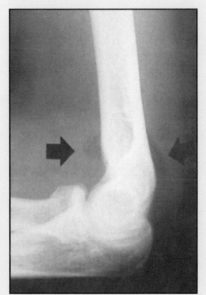

Figure 3
Anterior and posterior fat pad signs

Differential diagnosis

Rupture of the distal biceps tendon

Rupture of the triceps muscle

Adverse outcomes of the disease

The patient may continue to have pain and stiffness. Other complications include deformity with malunion, nonunion, and ulnar neuropathy.

Treatment

The goal of treatment is to achieve and to maintain a stable reduction of the fracture that allows for early motion. Therefore, all displaced distal humerus fractures are potential candidates for surgical stabilization. Conservative treatment, consisting of splinting for 10 days followed by early gentle range of motion, is appropriate for stable nondisplaced fractures and fractures in patients with severe osteopenia.

elbow and forearm

Adverse outcomes of treatment

Pain and stiffness can persist following treatment. Deformity, nonunion, and ulnar neuropathy can develop, as can symptoms related to the hardware used during surgery.

Referral decisions/Red flags

Patients with displaced fractures of the distal humerus need further evaluation to determine whether surgical stabilization of the fracture is necessary. Likewise, patients with associated neurovascular injury need further evaluation.

Acknowledgements

Figure 1 is reproduced with permission from Beaty JH, Kasser JR: Fractures about the elbow, in Jackson DW (ed): *Instructional Course Lectures 44.* Rosemont, IL, American Academy of Orthopaedic Surgeons, 1995, pp 199–215.

Figure 2 is adapted from Mehne DK, Jupiter JB: Fractures of the distal humerus, in Browner BD, Jupiter JB, Levine AM, et al (eds): *Skeletal Trauma: Fractures, Dislocations, Ligamentous Injuries.* Philadelphia, PA, WB Saunders, 1992, pp 1146–1176.

Fracture of the Humeral Shaft

Synonyms

None

Definition

Fractures of the humeral shaft are fairly common, constituting up to 5% of all fractures (Figure 1). These fractures may result from a direct blow to the arm such as occurs in a motor vehicle accident or from a fall on the out-stretched arm. The vast majority of these fractures can be treated conservatively with a high rate of union: close to 100%. No one standard classification system for humeral shaft fractures exists that is universally accepted.

Clinical symptoms

Severe arm pain, swelling, and deformity are characteristic of a displaced fracture of the humerus. With gentle palpation and movement of the arm, it is often possible to detect motion at the fracture site. If the radial nerve has been injured, patients are unable to extend the wrist or fingers and may have loss of sensation over the back of the hand (Figure 2). Humeral fractures may occur as a result of sports activities involving a vigorous throwing motion, but these instances are rare.

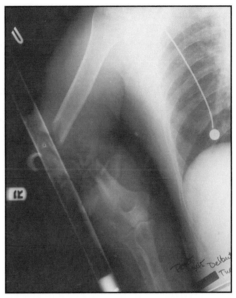

Figure 1
Humeral shaft fracture

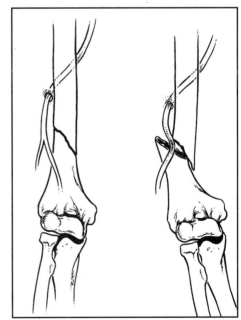

Figure 2
Entrapment of the radial nerve at the fracture site

elbow and forearm

elbow and forearm

Tests

Exam

Examination reveals red or blue discoloration and may reveal crepitus with a displaced fracture. Ask the patient to flex and extend the wrist and fingers to test function of the median, radial, and ulnar nerves. Assess sensation over the tip of the thumb, the tip of the little finger, and the back of the first web space near the thumb. Check the radial pulse and record the color and temperature of the hand. Look for puncture wounds in the skin near the fracture site, as these indicate an open (compound) fracture. Because these fractures commonly result from high-energy trauma, look for injuries to other areas of the body.

Diagnostic

AP and lateral radiographs confirm the diagnosis. These views should show the shoulder and elbow joints as well. CT and MRI scans are not indicated unless the fracture is pathologic, occurring through bone weakened by tumor or osteoporosis. In this instance, a bone scan can be useful to identify other possible skeletal metastases.

Differential diagnosis

Ruptured biceps tendon

Ruptured pectoralis muscle

Subcutaneous or intramuscular hematoma

Adverse outcomes of the disease

Radial nerve injury, indicated by weakness in the wrist or finger extensors and numbness in the thenar web space, is possible. Injury to the brachial plexus or vascular system may also occur. While nonunion is rare, malunion resulting in little functional impairment, 30° of varus, or an anterior bow can occur. Shortening of the limb is also possible, as is persistent stiffness in the shoulder and elbow.

Treatment

Most humeral shaft fractures can be treated conservatively. Fractures with minimal shortening (2 cm or less) can be treated with a U-shaped coaptation splint for 2 weeks, followed by a Sarmiento-type functional humeral fracture brace fitted by an orthotist.

A coaptation splint is applied as follows (Figure 3): place 12 thicknesses of plaster or a commercially available prepackaged splint around the elbow and extend it over the top of the shoulder over the acromion process. Use a collar and cuff made from stockinette to support the forearm and wrist. *See also:* Splinting the upper extremity, *pp 61-63.*

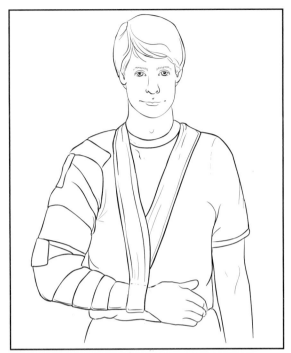

Figure 3
Coaptation splint

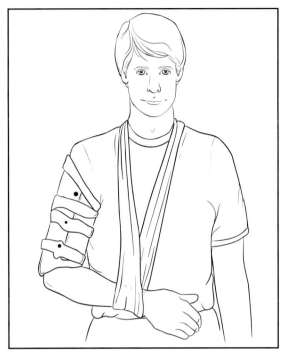

Figure 4
Humeral fracture brace

Instruct the patient to exercise the wrist, fingers, and shoulder at least three times a day. Allow the patient to flex the elbow as tolerated, and to extend the elbow to 100° as pain allows. The coaptation splint may need to be reapplied during these first 2 weeks. After 2 weeks, the patient is fitted with a humeral fracture brace (Figure 4), which is worn for at least the next 6 weeks or until there is radiographic evidence of healing. During this time, encourage range of motion exercises for the shoulder, elbow, wrist, and hand.

A special problem associated with humeral fractures is injury to the radial nerve. Even if the nerve is not functioning immediately following the injury, observation is appropriate since 95% of patients will regain nerve function within 6 months. During this period of observation, the patient should be fitted with a wrist/hand splint and work with a therapist if available. Electromyelograms (EMGs) are indicated after 4 to 6 months if radial nerve function does not return. However, if the patient has normal radial nerve function initially and loses function after a manipulation or during treatment, then surgical exploration of the nerve and the fracture is indicated.

Adverse outcomes of treatment

Radial nerve injury after manipulation, stiffness of the shoulder and elbow, and discomfort and/or skin irritation from the splint are all possible. The patient may need to sleep sitting in a chair.

Referral decisions/Red flags

Patients who have one of the following conditions need further evaluation: associated vascular injury; a nerve injury that develops after manipulation; an open fracture; a segmental fracture; a "floating elbow" in which the radius and ulna are fractured along with the humerus; nonunion following 3 months of treatment; an associated head injury or seizure disorder; a pathologic fracture; or skin breakdown under the fracture brace. In addition, obese patients may need additional evaluation for proper fit of a splint.

Acknowledgements

Figure 1 is reproduced with permission from Crosby LA, Lewallen DG (eds): *Emergency Care and Transportation of the Sick and Injured,* ed. 6 revised. Rosemont, IL, American Academy of Orthopaedic Surgeons, 1997, p 542.

Figure 2 is reproduced with permission from Rockwood CA, Green DP, Bucholz RW, et al: *Rockwood & Green's Fractures in Adults 4th edition.* Philadelphia, PA, Lippincott/Raven, 1996, p 1044.

Fracture of the Olecranon

Synonyms

Elbow fracture

Definition

The olecranon is the portion of the ulna that constitutes the bony prominence of the elbow. Because of its subcutaneous position, the olecranon is easily fractured as a result of any one of the following: a direct blow to the elbow, a fall on the elbow with the elbow flexed, or a fall on an outstretched arm in association with a dislocation. Like many types of fractures, olecranon fractures can be either nondisplaced or displaced. Displaced fractures may be further classified as two-part fractures with transverse or oblique fracture lines, comminuted fractures, or fracture-dislocations (Figure 1).

Clinical symptoms

Swelling, often quite marked, and ecchymosis following a history of trauma are typical findings. If the patient has an associated dislocation, then the elbow will appear deformed as well. Due to pressure from the swelling on nerves about the elbow, especially the ulnar nerve, patients may report numbness in some or all of the fingers.

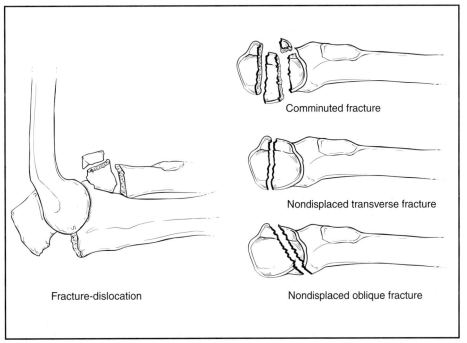

Comminuted fracture

Nondisplaced transverse fracture

Fracture-dislocation

Nondisplaced oblique fracture

Figure 1
Types of olecranon fractures

Tests

Exam

Examination reveals marked swelling of the entire elbow joint. The skin over the tip of the elbow may be abraded if the mechanism of injury is a direct blow to the elbow. Gentle palpation may reveal a defect if the olecranon fragments have separated. Movement of the elbow joint produces pain and is met with resistance. Associated nerve injury is possible; therefore, assess sensation over the tip of the little finger, the tip of the thumb, and the back of the first web space. The adequacy of circulation to the hand must be ascertained as well.

Diagnostic

AP and lateral radiographs are usually adequate to confirm the diagnosis.

Differential diagnosis

Dislocation of the elbow (more grotesque deformity)

Fracture of the coronoid process of the olecranon

Fracture of the distal humerus

Fracture of the radial head (plain radiographs make the distinction)

Adverse outcomes of the disease

Loss of motion and/or stability in the elbow is possible, as is loss of strength in the arm and forearm. Elbow pain may persist, and arthritis of the joint may eventually develop.

Treatment

Nondisplaced fractures of the olecranon can be treated with a posterior splint, which holds the elbow in 90° of flexion. Obtain follow-up radiographs 10 days after the injury to ensure that the fracture has not become displaced. Instruct patients to do hand exercises with a rubber ball or commercially available hand exerciser for 10 minutes at least twice each day. Wrist flexion and extension exercises can also be done at these sessions.

Most displaced fractures are best treated surgically by internal fixation with plates or a combination of wire and pins or screws. Displaced fractures in debilitated elderly patients who are poor risks for surgery can be treated with a sling and early range of motion as pain allows. Many of these patients do surprisingly well with an acceptable and functional range of motion.

Adverse outcomes of treatment

Elbow stiffness and loss of motion can develop despite treatment. Nonunion or displacement of the fracture are also possible. Following surgical treatment, irritation from the hardware used to fix the fracture can occur.

Referral decisions/Red flags

Patients with displaced fractures or open (compound) fractures need further evaluation for possible surgical treatment.

Acknowledgements

Figure 1 is adapted with permission from Jupiter JB, Mehne DK: Trauma to the adult elbow and fractures of the distal humerus, in Browner BD, Jupiter JB, Levine AM, et al (eds): *Skeletal Trauma: Fractures, Dislocations, Ligamentous Injuries.* Philadelphia, PA, WB Saunders, 1992, vol 2, pp 1125–1175.

Fracture of the Radial Head

elbow and forearm

Synonyms

None

Definition

Fractures of the radial head result from a fall on the outstretched hand with the arm and forearm turned in. Dislocations of the elbow are often associated with radial head fractures. A commonly used classification separates these fractures into three types, as follows: type I fractures are nondisplaced; type II fractures are marginal radial fractures that are displaced; and type III fractures are comminuted fractures involving the entire radial head (Figure 1).

Clinical symptoms

Patients with a history of a fall on the outstretched arm often have pain about the outside of the elbow and swelling of the joint. Loss of flexion and extension of the elbow as well as rotation of the forearm are also common.

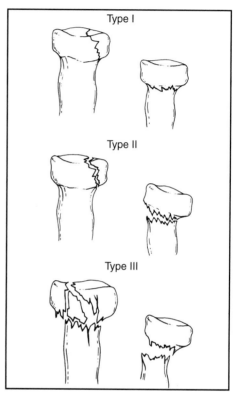

Figure 1
Classification of radial head fractures

Tests

Exam
Patients are unable to pronate or supinate the forearm. Flexion and extension of the elbow is painful and limited. Gentle palpation over the outside of the elbow produces pain, and swelling over the joint is frequently apparent.

Diagnostic
AP and lateral radiographs are adequate to confirm diagnoses of types II and III fractures. Since type I fractures are nondisplaced, they may be missed on initial radiographs, but may be seen on radiographs obtained 3 weeks after the injury.

Differential diagnosis

Dislocation of the elbow

Fracture of the olecranon process (the tip of the elbow)

Hemarthrosis of the elbow

Supracondylar fracture of the humerus

Synovitis of the elbow

Adverse outcomes of the disease

Loss of elbow motion, instability, and nonunion are all possible. Posttraumatic arthritis may also develop.

Treatment

Treatment of type I fractures consists of use of a splint or a sling for 3 days, followed by early motion within the patient's pain tolerance. Early motion is the key to a successful outcome with these fractures. Most type II fractures can also be treated by early range of motion exercises. The exception is for type II fractures with a large (ie, greater than 30% of the head) single fragment that is displaced more than 2 mm. Surgery is preferred in this instance. Type III fractures are best treated by early surgical excision of the bone fragments.

Adverse outcomes of treatment

Loss of motion, instability, and wrist pain from excision of the radial head can all occur.

Referral decisions/Red flags

Type III fractures require early surgical excision, so further evaluation is needed as soon as possible after diagnosis. Failure of conservative treatment, manifested by persistent pain and limited motion after 2 months, also indicates the need for further evaluation.

Acknowledgements

Figure 1 is adapted with permission from Jupiter JB, Mehne DK: Trauma to the adult elbow and fractures of the distal humerus, in Browner BD, Jupiter JB, Levine AM, et al (eds): *Skeletal Trauma: Fractures, Dislocations, Ligamentous Injuries*. Philadelphia, PA, WB Saunders, 1992, vol 2, pp 1125–1175.

elbow and forearm

Olecranon Bursitis

elbow and forearm

Synonyms

None

Definition

Olecranon bursitis is an inflammation of the bursa, which is located between the skin and the tip of the ulna at the elbow (Figure 1). Because of its superficial location, the bursa is bruised and irritated easily.

Clinical symptoms

The swelling associated with bursitis develops either gradually (chronic) or suddenly (infection or trauma). Pain is variable, but more intense, and may limit motion after an acute injury or with infection. Because of the size of the mass, patients often report difficulty putting on long-sleeved shirts. As the mass recedes, patients may feel "lumps" that are tender when the elbow is bumped. These lumps or nodules are scar tissue left when the fluid recedes and are often referred to as bits of "gravel."

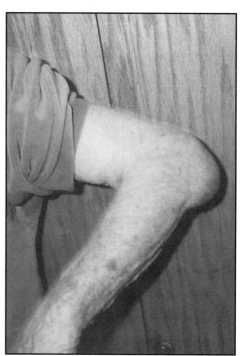

Figure 1
Swollen olecranon bursa

Tests

Exam

Examination reveals a large mass, up to 6 cm in diameter, over the tip of the elbow. The skin may be abraded or even lacerated. Redness and heat are not uncommon with acute bursitis and may indicate infection. Exquisite tenderness usually means an infectious or traumatic origin. Chronic, recurrent swelling is usually not tender.

Diagnostic

If the mass is large and symptomatic, aspiration can be both diagnostic and therapeutic (see Olecranon Bursa Aspiration). Following an acute injury, either blood or serum can be found. Culture any fluid that is cloudy or has a foul odor. If the origin is traumatic, obtain radiographs to rule out a fracture of the olecranon process of the ulna.

Differential diagnosis

Fracture of the olecranon process of the ulna

Gout

Rheumatoid arthritis

Synovial cyst of the elbow joint

Adverse outcomes of the disease

Infection, chronic recurrence or drainage, and swelling or limited motion can develop.

Treatment

Aspirate the cyst and culture any suspicious fluid. Apply a compression bandage consisting of a circular-shaped piece of foam, 8 cm in diameter, and an elastic wrap. Reassess the patient in 1 week. Recurrence of the fluid is common, and the sac may have to be aspirated two or more times. If swelling recurs following aspiration, and if the fluid does not look infected, inject 2 mL of a corticosteroid preparation into the sac. If the patient's elbow is at risk for repeated trauma, then recommend an elbow protector to prevent further trauma.

Adverse outcomes of treatment

Infection, chronic drainage, or recurrence are all possible.

Referral decisions/Red flags

Recurrence of fluid despite repeated (three or more) aspirations or an infection that is unresponsive to antibiotic treatment usually requires surgical treatment.

procedure

elbow and forearm

Olecranon Bursa Aspiration

Anatomy

The olecranon bursa lies on the extensor aspect of the elbow, over the olecranon process of the ulna. The ulnar nerve lies adjacent to the medial face of the olecranon, behind the ulnar groove of the distal humerus. For this reason, aspiration is best done from the lateral side.

Materials

- Sterile gloves

- Skin preparation solution (iodinated soap or similar antiseptic solution)

- 1-mL syringe with a 27-gauge, 3/4-inch needle

- 1-mL syringe with a 25-gauge, 3/4-inch needle

- 1 mL of a 1% local anesthetic

- 10-mL syringe

- 18-gauge needle

- 1 mL of a 40 mg/mL corticosteroid preparation (optional)

- Adhesive bandage

Step 1

Wear protective gloves at all times during this procedure and use sterile technique.

Step 2

With the patient in a prone position, cleanse the skin with an iodinated soap or similar antiseptic solution.

Step 3

Use a 27-gauge needle to infiltrate the skin over the lateral aspect of the bursa with 1 mL of a 1% local anesthetic.

Step 4

Through the same needle tract, insert the 18-gauge needle attached to the 10-mL syringe into the enlarged and fluid-filled bursa, and aspirate the contents slowly until the bursa is flat (Figure 1). If there is any concern about infection, send the fluid for culture and sensitivity and do not inject the corticosteroid preparation into the cavity.

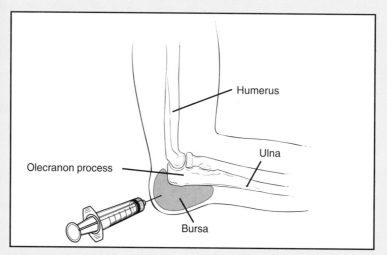

Figure 1
Location for needle insertion

Olecranon Bursa Aspiration (continued)

Step 5

If infection does not seem probable, remove the aspirating syringe and attach a 1-mL syringe containing 1 mL of a 40 mg/mL corticosteroid preparation. Inject this into the bursal cavity.

Step 6

Dress the puncture wound with a sterile adhesive bandage.

Step 7

Lightly wrap the elbow with an elastic dressing.

Adverse outcomes

Infection is possible, though not likely. Recurrence of the bursal effusion is possible if the patient rests the elbow on table tops or similar surfaces in the first few days following aspiration.

Aftercare/patient instructions

Advise the patient to limit elbow motion for the first day or two following aspiration. If the patient has a recurrent bursitis, use a posterior plaster splint to limit elbow motion for a week or two following the aspiration.

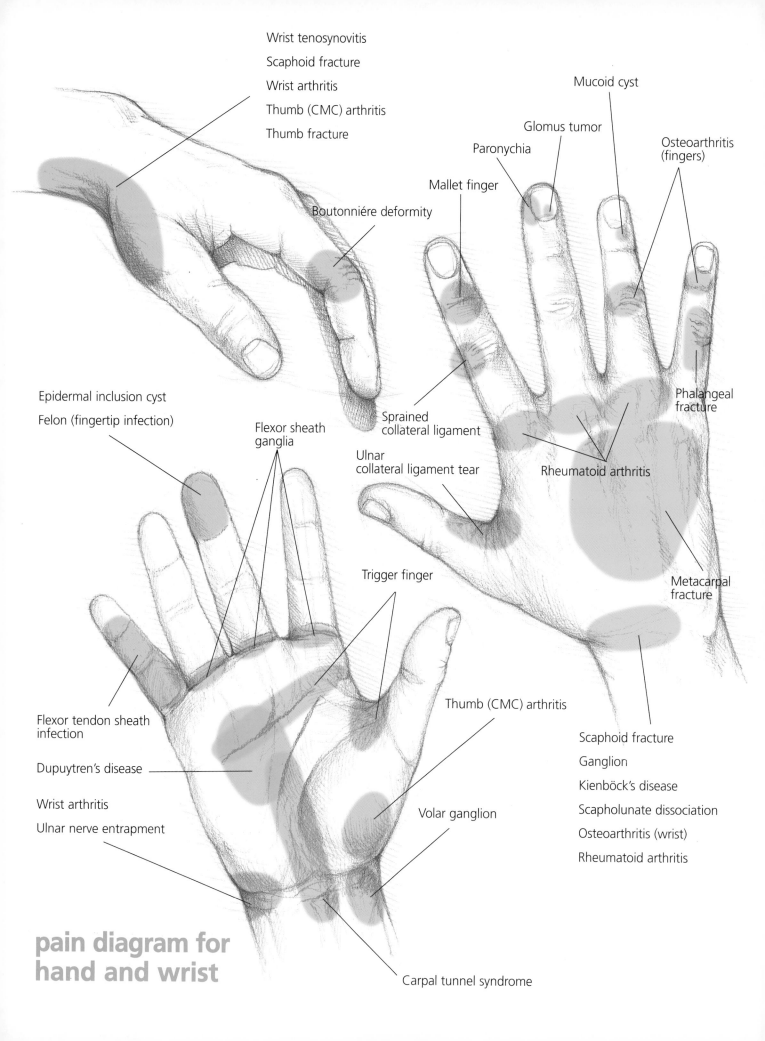

Wrist tenosynovitis

Scaphoid fracture

Wrist arthritis

Thumb (CMC) arthritis

Thumb fracture

Mucoid cyst

Glomus tumor

Paronychia

Osteoarthritis (fingers)

Mallet finger

Boutonniére deformity

Epidermal inclusion cyst

Felon (fingertip infection)

Flexor sheath ganglia

Sprained collateral ligament

Phalangeal fracture

Ulnar collateral ligament tear

Rheumatoid arthritis

Trigger finger

Metacarpal fracture

Flexor tendon sheath infection

Dupuytren's disease

Thumb (CMC) arthritis

Wrist arthritis

Ulnar nerve entrapment

Volar ganglion

Scaphoid fracture

Ganglion

Kienböck's disease

Scapholunate dissociation

Osteoarthritis (wrist)

Rheumatoid arthritis

pain diagram for hand and wrist

Carpal tunnel syndrome

Section Editor

Thomas R. Johnson, MD
Orthopedic Surgeons, P.S.C.
Billings, Montana

William Anderson, MD
St. Vincent Hospital and Health Center
Billings, Montana

W. L. Gorsuch, MD
Great Falls Orthopaedic Associates
Great Falls, Montana

Charles D. Jennings, MD
Great Falls Orthopaedic Associates
Great Falls, Montana

hand and wrist— an overview

Patients with hand and wrist problems will typically have one (or more) of the following seven complaints: 1) pain, 2) instability, 3) stiffness, 4) swelling, 5) weakness, 6) numbness, or 7) a mass. This section of *Essentials of Musculoskeletal Care* focuses on hand and wrist problems related to these complaints. Carpal tunnel syndrome, trigger finger, ganglia, carpometacarpal (CMC) arthritis of the thumb, and radiocarpal arthritis constitute more than 90% of the musculoskeletal problems in the hand and wrist.

Obtaining a complete history, including patient age, exact location of the problem, and whether the problem is acute or chronic, accompanied by a thorough physical examination should result in making a correct diagnosis in more than 90% of patients. Because all the structures in the hand and wrist are within 1.5 cm of your hand and there is little in the way of overlying soft tissue, palpation for swelling and tenderness is relatively easy (Figure 1). Diagnostic testing should be limited to plain radiographs in most patients, as CT and MRI scans are rarely indicated for these conditions.

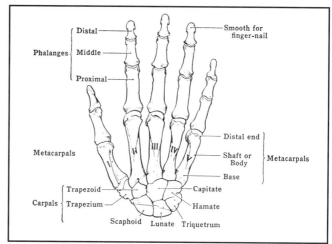

Figure 1
Bones of the hand

Location of pain

Figure 2 shows the regions of the hand and wrist: radial, ulnar, volar, or dorsal. By localizing the pain to one of these four areas first, the list of possible diagnoses is reduced considerably.

Radial pain

Wrist pain in patients younger than age 30 years most commonly results from trauma. Posttraumatic tenderness and pain over the radial aspect of the wrist suggest a possible fracture of the scaphoid, which is the most commonly missed diagnosis associated with trauma. In the absence of trauma, pain associated with tenderness over the radial styloid is most likely deQuervain's (wrist) tenosynovitis. Pain that occurs without numbness in patients older than age 40 years is likely due to posttraumatic arthritis or osteoarthritis. Pain in women in this age group is likely to be CMC arthritis at the base of the thumb. Performing Watson's stress test easily confirms the diagnosis.

Dorsal pain

Pain in this region associated with a well-defined mass over the dorsoradial aspect of the wrist is typically a ganglion. Pain and loss of motion in the wrist in young adults is commonly Kienböck's disease (osteonecrosis of the lunate). Plain radiographs easily confirm the diagnosis in the later stages of the latter condition.

Ulnar pain

Pain in this region following trauma may be due to a tear of the triangular fibrocartilage complex, which is located distal to the ulnar styloid. Swelling and tenderness over the dorsoradial or volar aspects of the wrist are likely caused by a tendinitis of the ulnar wrist extensor or flexor tendons.

Volar pain

Volar wrist pain may occur due to arthritis between the pisiform and the triquetrum bones. Carpal tunnel syndrome is by far the most common cause of volar wrist pain. Other causes include a volar ganglion, which should be easily palpable on the volar radial aspect of the wrist. Swelling over the volar region suggests inflammation of the finger flexor tendons. Patients with radiocarpal arthritis may have pain over both dorsal and volar aspects of the wrist.

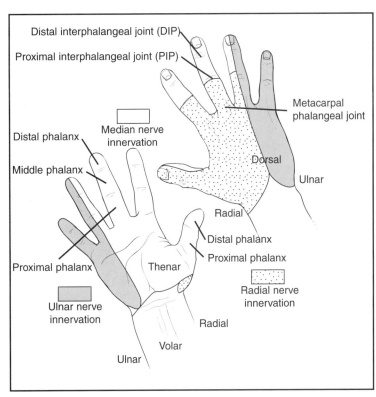

Figure 2
Regions and sensory distribution of the hand

Instability

Diagnosis of wrist instability is often quite complex, requiring considerable experience in interpreting radiographs, localizing signs and symptoms, and performing the physical examination. Patients may experience sensations of slipping, snapping, or clunking with certain wrist motions following an injury. The unstable structure may be a joint (following a tear of the supporting and stabilizing ligaments that hold together the carpal bones) or may be a subluxating tendon (following a tear of the restraining ligaments that guide the tendon). Plain PA radiographs of the wrist may show separation between the scaphoid and the lunate (Terry Thomas sign), indicating a tear of the ligaments binding the scaphoid and the lunate (scapholunate dissociation).

hand and wrist

Stiffness

Morning stiffness is a common complaint in association with carpal tunnel syndrome or trigger finger, two conditions that are often mistaken for arthritis. Patients with trigger finger often have pain as well and localize the problem to the proximal interphalangeal (PIP) joint when it locks or "jumps" as the finger is flexed. In fact, the problem lies in the palm where the thickened flexor tendon hangs up under the proximal tendon pulley at the distal palmar crease. Tenderness just distal to this crease confirms the diagnosis of trigger finger.

Swelling

Swelling in the joints of the hand and wrist is caused by synovitis, which may be secondary to infection or a systemic disease such as rheumatoid arthritis or gout. A history of penetrating trauma will help distinguish infection from systemic disease. Plain radiographs are useful in diagnosing osteoarthritis and rheumatoid arthritis.

Swelling about the tendons can occur in association with rheumatoid arthritis and/or overuse syndromes such as deQuervain's tenosynovitis. Pain and swelling about the wrist flexor or extensor tendons suggests calcific tendinitis. Plain radiographs will show a calcific deposit in close proximity to the involved tendon.

Weakness

Weakness in the hand may be secondary to pain, as with CMC or radiocarpal arthritis. Weakness without pain suggests possible peripheral nerve entrapment. Ulnar nerve entrapment at the elbow will result in decreased grip and pinch strength in addition to loss of sensation in the little and ring fingers. Wasting of the intrinsic muscles is easily seen in advanced cases. Weakness in which patients report dropping things also commonly occurs in association with carpal tunnel syndrome.

Numbness

Figure 2 also shows the typical sensory distribution of the median, ulnar, and radial nerves. Loss of sensation in some or all of the fingers should be considered indicative of carpal tunnel syndrome, until proven otherwise. With this condition, the numbness characteristically occurs in the thumb, index, middle, and radical half of the ring finger; Phalen's test is also positive. Some patients report that their entire hand is numb. Hand and wrist pain are common with carpal tunnel syndrome, but note that this condition rarely affects individuals younger than age 20 years.

Loss of sensation in the little and ring fingers is usually caused by entrapment of the ulnar nerve at the elbow (also possibly at the wrist, but this is rare). Patients with thoracic outlet syndrome often report symptoms at the ulnar side of the hand and forearm, but this condition is much less common than ulnar nerve entrapment at the elbow. Tinel's sign and elbow flexion tests will both be negative with thoracic outlet syndrome.

Neck pain associated with loss of sensation in the thumb and index finger suggests a C6 radiculopathy; Phalen's test will be negative.

Mass

The most common mass in the hand and wrist is a ganglion cyst; these cysts most often occur in four locations: dorsoradial and volar radial aspects of the wrist; the proximal finger flexor crease, and the distal interphalangeal (DIP) joint. Multilobulated masses along the sides of the finger are most likely giant cell tumors. Nontender nodules in the palm and cords that cross both the DIP and PIP joints are consistent with Dupuytren's disease. A hard mass at the dorsal base of the index metacarpal is a carpal boss. This is a bony mass consisting of spurs from the index and middle metacarpals, the trapezoid, and the capitate.

Acknowledgements

Figure 1 is reproduced with permission from Anderson JE: *Grant's Atlas of Anatomy*, ed. 8. Baltimore, MD, Williams & Wilkins, 1983.

hand and wrist

hand and wrist—
physical exam

1

Figure 1
Wrist extension (dorsiflexion)

Ask the patient to maximally dorsiflex the wrist to test for range of motion. The patient's ability to maximally dorsiflex the wrist is reduced when flexion deformities of the wrist, joint effusions, or arthritis are present.

2

Figure 2
Wrist flexion (palmar flexion)

Ask the patient to maximally flex the wrist to test for range of motion. The patient's ability to maximally flex the wrist is reduced when extension deformities of the wrist, joint effusions, or arthritis are present.

hand and wrist

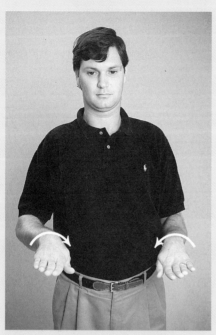

3

Figure 3
Forearm pronation

Ask the patient to maximally pronate the forearm (turn the palm down) while the elbow is at the side. The patient's ability to maximally pronate the forearm is reduced when wrist fractures, congenital conditions, or elbow arthritis are present.

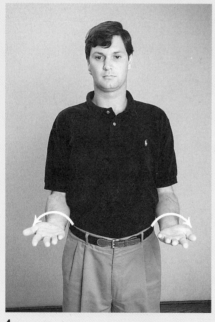

4

Figure 4
Forearm supination

Ask the patient to maximally supinate the forearm (turn the palm up) while the elbow is at the side. The patient's ability to maximally supinate the forearm is reduced when wrist fractures, congenital conditions, or elbow arthritis are present.

5

Figure 5
Phalen's test

Ask the patient to sit comfortably with wrist and elbow flexed. When the test is positive, the patient will experience numbness or tingling within 45 seconds. In some patients, performing this test will recreate their wrist, thumb, or forearm ache.

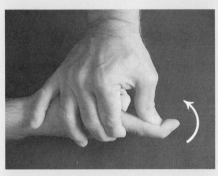

6

Figure 6
Profundus test

Extend the patient's proximal interphalangeal joint while the patient flexes the tip of the finger. Inability to flex the distal interphalangeal joint indicates an injury to the profundus tendon to that finger.

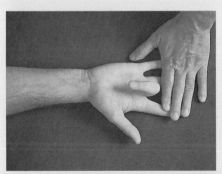

7

Figure 7
Sublimus test

With the patient's hand palm up on the examining surface, maintain the metacarpophalangeal, proximal interphalangeal, and distal interphalangeal joints in extension in all fingers but the one to be tested. Ask the patient to flex the unblocked finger. Inability to flex this finger indicates an injury to the sublimus tendon to that finger.

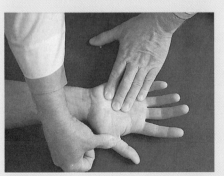

8

Figure 8
Carpometacarpal (CMC) stress test

With the patient's hand palm up on the examining surface, press down on the thumb metacarpophalangeal joint. Pain indicates arthritis or instability of the CMC joint of the thumb.

hand and wrist

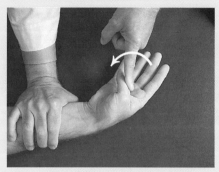

9

Figure 9
Finkelstein's test

With the patient's wrist deviated to the ulnar side, push the proximal phalanx of the thumb into flexion. Pain at the wrist indicates deQuervain's disease or tenosynovitis of the abductor pollicus longus and extensor brevis tendons.

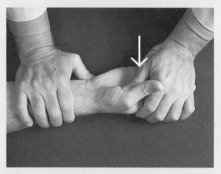

10

Figure 10
Opponens strength

With the patient's hand palm up on the examining surface, ask the patient to place the thumb straight up and resist your attempt to push it down onto the table (abduction and extension). Weakness indicates damage to the motor branch of the median nerve, most commonly related to carpal tunnel syndrome.

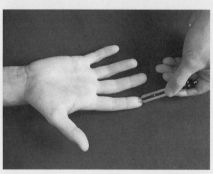

11

Figure 11
Sensory testing

Ask the patient to close both eyes as you lightly move both points of an ECG caliper across the fingertip. Determine whether the patient is able to distinguish two points 5 mm apart as separate points or a single point. Inability to discriminate the two points indicates diminished sensation. The median nerve supplies the tip of the index finger; the ulnar nerve supplies the tip of the small finger; the radial nerve supplies the dorsal web space between the thumb and second metacarpal.

12

Figure 12
Thenar atrophy

With the patient's palms facing each other, compare the two thenar masses. Asymmetry indicates damage to the median nerve or the C8, T1 root, or arthritis of the carpometacarpal joint.

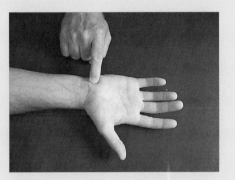

13

Figure 13
Pisiform tenderness

Palpate the pisiform bone just lateral to the proximal palmar crease and slightly distal to the wrist crease. Tenderness indicates pisotriquetral arthritis or inflammation of the flexor carpi ulnaris tendon, which envelopes the pisiform.

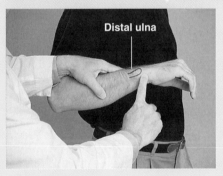

Distal ulna

14

Figure 14
TFCC palpation

Palpate the area distal to the ulnar styloid and proximal to the carpal bones. Tenderness indicates injury to the triangular fibrocartilage complex, usually as a result of trauma.

hand and wrist

Animal Bites

Synonyms

None

Definition

As many as 3 million people in the United States sustain animal bites each year. Animal bites most commonly occur on the fingers in the dominant hand of children. Because rabies is a possibility as a result of the bite, it is important to be familiar with the status of rabies in your area.

In the United States, more than 90% of rabies comes from wild animals, especially bats, skunks, raccoons, and foxes. However, outside the United States the dog is the most common source of rabies. Approximately 50% of rabies cases reported to the Centers for Disease Control and Prevention (CDC) in 1985 were due to dog bites that occurred in foreign countries. Dog bites account for up to 90% of animal bites, while cat bites are second most common, constituting 5% of animal bites.

Animal bite wounds often become infected with a variety of organisms, including *Pasteurella multocida,* which is a bacterium associated with dog and cat bite wounds. Aerobic organisms, including alpha-hemolytic streptococci and *Staphylococcus aureus,* and anaerobic organisms such as *Bacteroides* and *Fusobacterium* are also isolated in dog bite wounds. The risk of infection from a dog bite is 5% to 10%; the risk of infection from a cat bite is much higher (30% to 50%) because a cat's sharp teeth create deeper puncture wounds.

Clinical symptoms

Pain, swelling, and redness around the puncture wound indicate an infection secondary to the bite. The patient may also report loss of sensation and motion distal to the bite, suggesting that a nerve or tendon is severed. It is important to determine whether the bite was provoked by a sudden movement toward the animal. An animal that initiates an unprovoked attack is more likely to be rabid. The animal should be found, if possible, and observed for 10 days to ensure that it does not become ill.

hand and wrist

Tests

Exam

Examination may reveal an irregular, jagged wound with devitalized tissue at the margins and swelling and redness around the wound. Try to determine the *depth of the wound* and the *time at which the patient was bitten.* Purulent drainage may be present if the wound is more than 10 to 12 hours old. Test for sensation and tendon function in the affected hand or finger. Inspect the forearm for any sign of lymphangitis, which is indicated by the presence of red streaks, and palpate the inner aspect of the elbow and the axilla for the presence of enlarged lymph nodes. Ensure that the patient's temperature is checked to rule out a fever.

Diagnostic

Obtain plain AP and lateral radiographs of the affected part to rule out a fracture or presence of a foreign body. These views may also reveal gas in the soft tissues. Routine laboratory studies are not needed for wounds seen shortly after injury. However, if an infection is suspected, a swab of the wound should be sent for a Gram stain and aerobic and anaerobic cultures.

Differential diagnosis

Foreign body with secondary infection

Adverse outcomes of the disease

Any of the following conditions may develop as a result of an untreated animal bite: sepsis in the joint, deep space infection, septic tenosynovitis, osteomyelitis, and/or rabies. Patients may also lose sensation and motion, and possibly their fingers, following an animal bite. Lymphedema with hand and finger stiffness is also possible.

Treatment

For superficial wounds that do not involve a fracture, or nerve or tendon injury, debridement, wound irrigation with 500 to 1,000 mL of saline, and outpatient antibiotics are appropriate. Use of an anesthetic block will facilitate debridement of bite wounds on the finger. If the bite wound is on the back of the hand, 3 to 10 mL of local anesthetic should be infiltrated about the wound.

Primary suturing of animal bite wounds is somewhat controversial. Because of a higher rate of infection associated with cat bites, these wounds should never be sutured primarily. Most dog bite wounds can be sutured primarily as long as the closure is loose enough to allow for drainage, or the wound is closed over a penrose drain. Wound closure can be delayed for at least 3 days if there is no sign of infection.

Empiric therapy can be started with IV ampicillin-sulbactam (Unasyn), 1.5 to 3.0 g every 6 hours, or oral amoxicillin-clavulanate acid (Augmentin), 500 mg 2 times a day for 5 days. These antibiotics may be changed based on culture results and sensitivities. Tetracycline can be used in patients with a penicillin allergy. Tetanus prophylaxis should be given as outlined in Table 1. If the animal is believed to be possibly rabid, contact local or state public health officials regarding the need for rabies prophylaxis.

Table 1
Guide to tetanus prophylaxis in wound management

History of tetanus toxoid (doses)	Clean, minor wound		Contaminated wound	
	Td	**TIG**	**Td**	**TIG**
Unknown	Yes	No	Yes	Yes*
Fewer than 3 doses	Yes	No	Yes	Yes*
Three or more	Yes, if more than 10 years since last dose	No	Yes, if more than 5 years since last dose	No

Td = combined tetanus and diphtheria toxoid adsorbed (dose = 0.5 mL IM). TIG = tetanus immune globulin (dose = 250 IU IM).
* Td and TIG should be administered at different sites.

Adverse outcomes of treatment

Infection secondary to primary wound closure can occur. Patients may also have allergic reactions to antibiotics.

Referral decisions/Red flags

Any patient with an animal bite that involves the tendon, nerve, joint capsule, or an underlying fracture needs further evaluation.

Acknowledgements

Table 1 is adapted with permission from the Centers for Disease Control and Prevention.

hand and wrist

Arthritis of the Hand

Synonyms

Osteoarthritis

Rheumatoid arthritis (RA)

Degenerative joint disease

Definition

Rheumatoid arthritis is a systemic condition affecting synovial tissue (Figure 1). All deformities, joint destruction, and pathologic anatomy that occur in patients with rheumatoid arthritis are a result of the alteration of the synovial tissue. The articular cartilage is destroyed by the rheumatoid synovium and subchondral bone may be invaded. The rheumatoid synovium may surround the flexor and extensor tendons, disrupting the delicate balance of the bones, joints, and ligaments.

The most common types of arthritic disease are osteoarthritis and degenerative joint disease. These conditions are characterized by progressive loss of articular cartilage, possible reactive bony changes at the margins, and subchondral cyst formation. The etiology of primary arthritis is unknown. Secondary arthritis arises in joints affected by trauma, mechanical problems, or preexisting lesions.

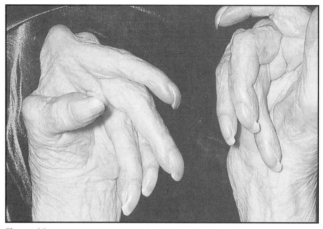

Figure 1A
Clinical appearance of rheumatoid hands

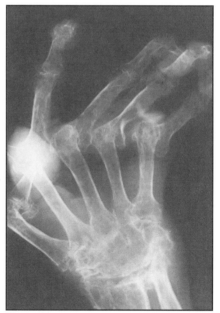

Figure 1B
Radiograph of a rheumatoid hand

Clinical symptoms

Patients with rheumatoid arthritis may have swelling in the joints, especially in the metacarpophalangeal (MP) joints and the wrist. Extensor or flexor tenosynovitis is also common. In osteoarthritis, the distal interphalangeal (DIP) and proximal interphalangeal (PIP) joints are most often involved. Patients may also report stiffness and loss of motion in the fingers.

Tests

Exam

Patients with rheumatoid arthritis often have fusiform swelling of multiple joints, with some joints swollen more than others. A boggy mass over the dorsum of the hand and crepitation with movement is quite common. Volar flexion and flexor tenosynovitis with crepitus at the wrist or in the fingers is also characteristic. There may also be ulnar drift of the fingers at the level of the MP joint. Other findings include contractures of the fingers at the PIP joints or hyperextension at the PIP joints with flexion at the DIP joints (swan neck deformity).

Patients with osteoarthritis may have bony nodules at the DIP joint (Heberden's nodes), and while these nodules may be painful at first, the pain usually resolves. Nodules may also occur at the PIP joints (Bouchard's nodes). Involvement of the MP joints is much less common and is often the result of previous trauma.

Diagnostic

PA, oblique, and split finger lateral views should be obtained. If a single digit is involved, isolated PA and true lateral views of the digit are also necessary. Blood studies should be ordered for patients who have the characteristic changes of inflammatory arthritis, but for whom the diagnosis has not been established.

Differential diagnosis

Pyogenic arthritis (may resemble the swollen joint of rheumatoid arthritis)

Adverse outcomes of the disease

Rheumatoid arthritis may be slowly progressive, resulting in the typical rheumatoid hand deformity. Progressive osteoarthritis may cause joint destruction, particularly at the DIP, PIP, and wrist joints. Fusion and surgery for pain relief or stabilization may be necessary.

Treatment

There is no cure for rheumatoid arthritis. However, cortisone injections may be extremely helpful to the involved joint or structure, such as flexor tenosynovitis in the finger, extensor tenosynovitis on the dorsum of the hand, or isolated PIP or MP joint involvement (see MP or PIP Joint Injection). Note that these injections are given in addition to the systemic rheumatoid treatment. Splinting for ulnar drift can slow the deformity but not prevent it from occurring.

Treatment of osteoarthritis typically includes NSAIDs and occasionally temporary splinting of an involved joint for pain relief.

Adverse outcomes of treatment

NSAIDs may cause gastric, hepatic, and renal complications. Cortisone injections should be used very judiciously as tendon rupture, particularly of the extensor tendons, may occur. Infection is also a risk, but can be largely avoided by use of careful sterile technique.

Referral decisions/Red flags

If a cortisone injection for extensor tenosynovitis is not effective, surgery is necessary to prevent rupture of the extensor tendon. Patients who cannot extend their fingers, especially the small finger, ring finger, or the thumb, may have ruptured the extensor tendon, and further evaluation is needed immediately. Patients with rheumatoid arthritis who report increasing deformity and increasing pain in the hand may need reconstructive surgery. Further evaluation is also needed for patients with osteoarthritis whose pain is no longer controlled with splinting and NSAIDs and who have radiographic evidence of joint destruction.

procedure

MP or PIP Joint Injection

Injections are performed on the extensor aspect of either the metacarpophalangeal or interphalangeal joint.

The metacarpophalangeal (MP) joint is most often defined by the head of the metacarpal. The actual joint is below the prominent metacarpal head, and is palpable as a small sulcus with the finger flexed 20°. Identify this sulcus while moving the joint through a small amount of flexion and extension.

Identify the proximal interphalangeal (PIP) joint in the same manner, although be aware that the dorsal rim of the proximal phalanx is often more easily palpated with this joint in extension.

Materials

Sterile gloves

Skin preparation solution (iodinated soap or similar antiseptic solution)

1-mL syringe with a 25-gauge needle

1-mL syringe with a 23-gauge needle

0.2 mL of a 1% local anesthetic solution

0.5 mL of a 40 mg/mL corticosteroid preparation

Adhesive dressing

Step 1

Wear protective gloves at all times during this procedure and use sterile technique.

Step 2

Cleanse the skin with an iodinated soap or similar antiseptic solution.

Step 3

Use a 25-gauge needle to infiltrate the skin over the joint with 0.2 mL of a 1% local anesthetic (Figure 1). This anesthetizes an area for the needle tract, but does not distort the tissues for further palpation.

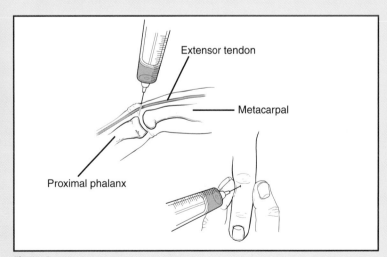

Figure 1
Location for needle insertion

MP or PIP Joint Injection (continued)

Step 4

Insert the 23-gauge needle through the previous needle tract and into the joint. Slight pressure against the syringe plunger should make the joint bulge slightly on either side. Inject 0.5 mL of a 40 mg/mL corticosteroid preparation.

Step 5

Dress the puncture wound with a sterile adhesive bandage.

Adverse outcomes

Although rare, infection is possible. Subcutaneous fat atrophy may occur if the corticosteroid preparation is injected external to the joint. This may produce a depressed area of thin, tender, and unsightly skin.

Aftercare/patient instructions

The joint may often be sore for 24 to 48 hours following injection. Instruct the patient to return to your office if undue swelling, pain, or redness occur.

Arthritis of the Thumb

Synonyms

Carpometacarpal (CMC) degenerative arthritis

Degenerative arthritis of the basal joint

Metacarpotrapezial degenerative arthritis

Definition

Idiopathic basal joint degenerative arthritis of the thumb most commonly occurs in women between the ages of 30 and 60 years (Figure 1). However, in men, this condition usually results from previous trauma, either fracture-dislocation or dislocation. The idiopathic variety is thought to be caused by anatomic factors (joint configuration and ligamentous laxity) that predispose the joint to instability, shear forces, and subsequent degenerative change.

Clinical symptoms

The most common symptom is pain at the base of the thumb that occurs with grip and pinch activities and may radiate proximally into the wrist and forearm. Decreased pinch strength is a common complaint. Patients may also note instability, "catching," or "clicking" with certain movements. Late manifestations include stiffness of the CMC joint in adduction with secondary metacarpophalangeal (MP) tenosynovitis hyperextension (swan neck deformity).

Note that carpal tunnel syndrome may coexist with or mimic the symptoms of degenerative arthritis of the basal joint of the thumb.

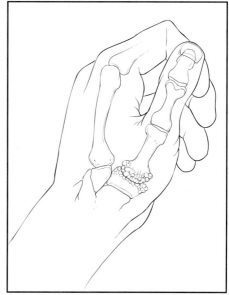

Figure 1
CMC arthritis of the thumb

hand and wrist

Tests

Exam

The hallmark of this condition is tenderness over the volar and radial aspects of the joint in the region of the base of the thumb metacarpal. Manipulation of the joint with simultaneous longitudinal pressure typically causes pain and often some crepitation or instability. Watson's stress test readily reproduces the pain of CMC arthritis (Figure 2). With the palm facing up and the back of the hand resting on the table, the thumb is pushed down toward the table with the MP and IP joints extended.

Diagnostic

PA and lateral radiographs of the thumb show joint space narrowing, subchondral sclerosis, and varying degrees of subluxation or dislocation at the CMC joint (Figure 3).

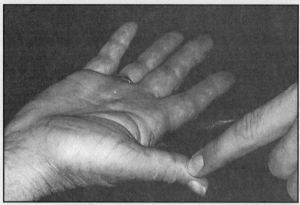

Figure 2
Watson's stress test for CMC arthritis of the thumb

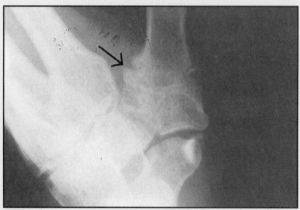

Figure 3
Radiographic appearance of CMC arthritis of the thumb

Differential diagnosis

Arthritis of the wrist (radioscaphoid arthritis)

Carpal tunnel syndrome

deQuervain's tenosynovitis

Flexor carpi radialis tendinitis

Fracture of the scaphoid

Radioscaphoid arthritis

Volar radial ganglion

Adverse outcomes of the disease

Chronic pain, loss of pinch and grip strength, and flexion contracture of the thumb (thumb held against the index finger) are all possible.

Treatment

Initial treatment should consist of placing the thumb in a thumb spica splint for 3 weeks and NSAIDs. If symptoms recur, continue intermittent splinting that immobilizes the entire thumb. If splinting fails, a corticosteroid preparation should be injected into the joint (see Thumb Arthritis Injection). Although injections do not alter the natural history of the disease, many patients report pain relief lasting a few months with each injection. At least three injections can be given.

Adverse outcomes of treatment

NSAIDs may cause gastric, renal, or hepatic complications. Infection can develop following corticosteroid injection. Also possible is the rapid progression of arthritis from multiple corticosteroid injections.

Referral decisions/Red flags

Failure of conservative treatment indicates the need for further evaluation.

hand and wrist

procedure

Thumb Arthritis Injection

Materials

Sterile gloves

Alcohol

Skin preparation solution (iodinated soap or similar antiseptic solution)

3-mL syringe with a 27-gauge, 3/4-inch needle

3-mL syringe with a 25-gauge, 3/4-inch needle

0.5 mL of 40 mg/mL corticosteroid preparation

2 mL of a 1% local anesthetic (without epinephrine)

Adhesive bandage

Step 1

Wear protective gloves at all times during this procedure and use sterile technique.

Step 2

Cleanse the skin with alcohol, allow it to dry, then cleanse the area with iodinated soap or similar antiseptic solution.

Step 3

The joint can be injected from the dorsum (back of the hand) or from the volar radial side. Insert a 27-gauge needle on the volar (palmar) and radial side of the thumb, as there is less chance of penetrating a sensory nerve in this area. Inject 1% of the anesthetic subcutaneously. To locate the joint space, manipulate the metacarpal and then palpate the base of the metacarpal. Next pull on the end of the thumb to open the joint space.

Step 4

Advance the needle into the joint and inject 0.5 to 1 mL of a 1% local anesthetic (Figure 1).

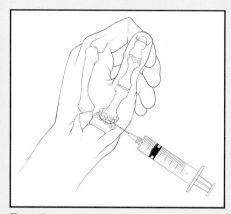

Figure 1
Injecting CMC joint of the thumb

Thumb Arthritis Injection (continued)

Step 5

Using the 25-gauge needle, inject 0.5 mL of 40 mg/mL corticosteroid preparation through the same needle tract. The injection of fluid should meet little resistance if the needle is in the CMC joint.

Step 6

Dress the puncture wound with a sterile adhesive bandage.

Adverse outcomes

Injury to sensory branches of the radial nerve, joint space infection, and depigmentation and fat atrophy at the site of injection are all possible.

Aftercare/patient instructions

Advise the patient that 33% of patients may experience a "flare" manifested by increased joint pain for 1 to 2 days. NSAIDs or an analgesic may be given to alleviate this pain. Also, ice may be helpful in the first 24 hours of pain. If available, the patient may wear a thumb sprain splint for 2 to 3 days after the injection.

Arthritis of the Wrist

hand and wrist

Synonyms

Gout

Synovitis

Pseudogout

Osteoarthritis

Infectious arthritis

Rheumatoid arthritis

Definition

Arthritis in the wrist is associated primarily with either osteoarthritis (primary or secondary to trauma) or rheumatoid arthritis. Patients often have associated deformity, which significantly alters hand tendon function.

Clinical symptoms

Patients with rheumatoid arthritis typically report generalized swelling, tenderness, and limited motion. Hand function is often impaired by the supination (rotation) of the carpus within the wrist joint since the tendons are no longer working on stable joints. In addition, there is often significant metacarpophalangeal (MP) joint involvement with ulnar drifting of the fingers. Decreased grip strength and pain are common complaints.

Osteoarthritis is usually associated with swelling, pain, and limited motion, but involvement of other joints is less common.

Tests

Exam

Examination reveals swelling, heat, redness, and limited motion. In patients with rheumatoid arthritis, involvement of the MP joints and deformity at the wrist are also common. The ulna appears prominent in these patients.

In primary osteoarthritis, the distal interphalangeal (DIP) joints may be involved, but they usually appear normal with posttraumatic osteoarthritis.

Diagnostic

PA and lateral radiographs are very helpful in distinguishing among the various types of arthritis. Generalized thinning of bone structure (osteopenia) with erosions in the area of the joint surface is characteristic of rheumatoid arthritis. Subchondral sclerosis, joint space narrowing, spur formation, and in some cases, erosion, characterize osteoarthritis (Figure 1). Early calcification may indicate

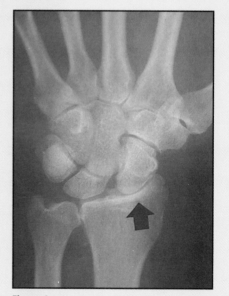

Figure 1
Radiograph showing osteoarthritis of the wrist

pseudogout, which, along with gout, can be confirmed by the presence of crystals in synovial fluid aspirate.

Laboratory studies, including erythrocyte sedimentation rate, rheumatoid factor (RF), antinuclear antibodies (ANA), and uric acid, confirm the diagnosis.

Differential diagnosis

Synovitis (normal radiographs)

Wrist joint infection (erythrocyte sedimentation rate, leukocytosis, culture of aspirate)

Adverse outcomes of the disease

Pain, loss of motion and/or strength, and impaired function in the fingers are all possible.

Treatment

Immobilization in a wrist splint may help relieve pain and swelling. Further treatment depends on the etiology, but is usually nonsurgical, typically intermittent use of NSAIDs. In the absence of infection, injection of a corticosteroid may provide temporary pain relief (see Wrist Aspiration/Injection). Surgical treatment is usually necessary if hand function decreases, if the joint becomes unstable, or if nonsurgical means fail to relieve pain.

Adverse outcomes of treatment

NSAIDs may cause gastric, renal, or hepatic complications. Loss of motion, nonunion, and persistent pain are other possibilities.

Referral decisions/Red flags

Patients with a possible wrist infection require early evaluation. Those with radiographic evidence of advanced disease and who do not respond to splinting and NSAIDs also are candidates for further evaluation.

procedure

Wrist Aspiration/Injection

Materials

Sterile gloves

Alcohol

Skin preparation solution (iodinated soap or similar antiseptic solution)

3-mL syringe with a 25-gauge needle

3-mL syringe with an 18-gauge needle

1-mL syringe with a 25-gauge needle (optional)

0.5 to 1 mL of 2% local anesthetic in breakable ampules (no preservatives)

0.5 to 1 mL of 40 mg/mL corticosteroid preparation (optional)

Adhesive bandage

Step 1

Wear protective gloves at all times during this procedure and use sterile technique.

Step 2

Cleanse the skin with alcohol, allow it to dry, then cleanse the area with iodinated soap or similar antiseptic solution.

Step 3

Palpate the distal edge of the radius between the extensor carpi radialis brevis and extensor digitorum communis tendons. The depression just distal to the distal edge of the radius indicates the radial carpal joint.

Step 4

Use a 25-gauge needle to infiltrate 2% local anesthetic in the subcutaneous tissue and joint capsule.

Step 5

Aspirate joint fluid with an 18-gauge needle. The joint fluid should be submitted for crystal analysis, cell count, smear Gram stain, and culture.

Step 6

If infection is not present, use the second 25-gauge needle to inject 0.5 to 1 mL of 40 mg/mL corticosteroid preparation, using the same technique. The corticosteroid preparation may be mixed with an equal amount of 2% local anesthetic for pain relief.

Step 7

Dress the puncture wound with a sterile adhesive bandage.

Wrist Aspiration/Injection (continued)

Adverse outcomes

Infection is possible. Corticosteroids can mask the usual signs of infection.

Aftercare/patient instructions

Instruct the patient to watch for signs of infection, such as increasing pain, swelling, heat, or redness, and to call your office if any of these signs occur.

Boutonnière Deformity

hand and wrist

Synonyms

Button hole deformity

Extensor tendon injury

Definition

Boutonnière deformity is characterized by a tear of the central portion of the extensor tendon at the level of the proximal interphalangeal (PIP) joint (Figure 1). The PIP joint flexes from the unopposed pull of the flexor (sublimus) tendon, and the distal interphalangeal (DIP) joint extends because the remaining part of the extensor tendon is intact.

Clinical symptoms

Patients typically report a history of trauma and subsequent inability to extend the PIP joint. The PIP joint is painful and tender if the injury is recent.

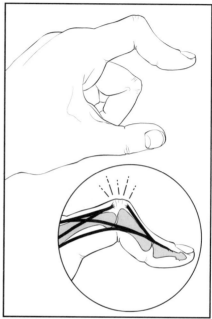

Figure 1
Boutonnière deformity

Tests

Exam
Hold the metacarpophalangeal (MP) and wrist joints in flexion and ask the patient to extend the PIP joint. Patients who lack 15° to 20° of extension at the PIP joint probably have a rupture of the central slip of the extensor tendon.

Diagnostic
AP and lateral radiographs will rule out a fracture.

Differential diagnosis

Fracture about the PIP joint (radiographs required to confirm)

Sprain of the PIP joint (may be difficult to differentiate on initial exam)

Adverse outcomes of the disease

Flexion contracture of the PIP joint and extension contracture of the DIP joint are both possible.

Treatment

Splint the PIP joint in extension for 6 weeks in a young patient and for 3 weeks in an elderly patient (Figure 2). The DIP joint is left free. Active and passive motion should be initiated at the DIP joint.

If the injury is more than 1 or 2 weeks old, it may not be possible to achieve full extension at the first visit. Use a dynamic extension splint until full extension is achieved and then begin the static splinting program (Figure 3).

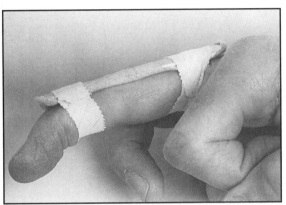

Figure 2
Static extension splint for PIP joint

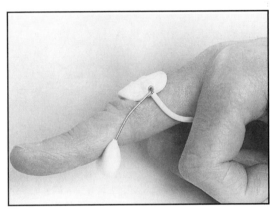

Figure 3
Dynamic extension splint for the PIP joint

Adverse outcomes of treatment

Failure to achieve full extension and/or residual deformity are possible.

Referral decisions/Red flags

Failure to achieve full extension and/or residual deformity are possible indications for further evaluation.

Acknowledgements

Figure 1 is adapted with permission from Steinberg GG, Akins CM, Baran DT: *Ramamurti's Orthopaedics in Primary Care,* ed 2. Philadelphia, PA, Williams & Wilkins, 1992, p 112.

Figures 2 and 3 are reproduced with permission from Culver JE: Office management of athletic injuries of the hand and wrist, in Barr JS (ed): *Instructional Course Lectures XXXVIII.* Park Ridge, IL, American Academy of Orthopaedic Surgeons, 1989, pp 473–482.

hand and wrist

Carpal Tunnel Syndrome

Synonyms

Median nerve entrapment at the wrist

Median nerve compression

Definition

Carpal tunnel syndrome (entrapment of the median nerve at the wrist) is the most common compression neuropathy in the upper extremity and produces paresthesias, pain, and sometimes paralysis (Figure 1).

Clinical symptoms

The pain associated with carpal tunnel syndrome is typically described as a vague aching that radiates into the thenar area, the proximal forearm, and occasionally to the elbow. The more acute the onset of the nerve entrapment, the more proximal the radiation of pain. The pain may extend to the shoulder or neck and even cause headaches in young people or patients with sudden synovitis of the wrist, such as in rheumatoid arthritis. The pain is also typically accompanied by paresthesias and numbness in the median distribution (thumb, index, long, and radial ring fingers, or some combination thereof).

Patients usually awaken at night with pain or numbness; during the day they have trouble with fixed flexed wrist activities when the hand is stationary and the wrist is slightly flexed, such as when driving or reading. They frequently awaken in the morning with stiffness in the hand and typically report needing to rub or shake the hand to "get the circulation back." They often report being unable to open jars or twist off lids. If the compression is severe and long-standing, persistent numbness and thenar muscle atrophy may occur (Figure 2).

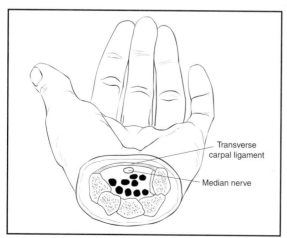

Transverse
carpal ligament

Median nerve

Figure 1
Cross-section of the carpal tunnel

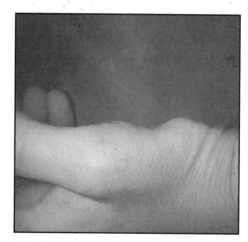

Figure 2
Thenar atrophy

Tests

Exam

Patients with carpal tunnel syndrome may not exhibit all the signs and symptoms described below; on rare occasions, they may have no physical findings, yet have abnormal results on nerve conduction studies. If placing the patient's wrist in maximal flexion or extension causes numbness or aching within 45 seconds (often 15 seconds or less), then Phalen's test is positive for carpal tunnel syndrome (Figure 3). Tapping over the median nerve at the wrist may produce tingling in some or all of the digits in the median nerve distribution (Tinel's sign) (Figure 4). Thumb pressure over the median nerve at the wrist for up to 30 seconds may elicit pain or paresthesias in the median distribution. Thenar atrophy may be apparent, and testing thumb opposition against resistance may reveal weakness of the thenar muscles (Figure 5). Patients with carpal tunnel syndrome will be unable to distinguish the two points of a caliper as separate points when they are closer than 5 mm together (Figure 6).

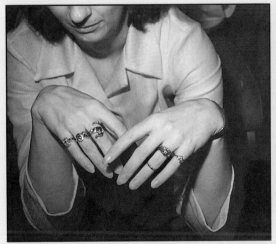

Figure 3
Phalen's test

Figure 4
Median nerve percussion

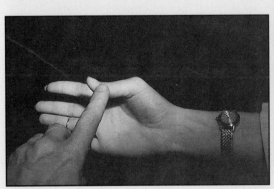

Figure 5
Testing for strength of opposition

Figure 6
Testing for two-point discrimination

Diagnostic

Obtain radiographs of the wrist if the patient has limited wrist motion. The most helpful objective diagnostic test is a median nerve conduction study. Some patients have no clinical signs or symptoms, yet have prolonged motor latency on nerve conduction studies. However, as many as 5% of patients with carpal tunnel syndrome may have normal nerve conduction studies.

Differential diagnosis

Arthritis of the carpometacarpal joint of the thumb (painful motion of the joint)

Cervical radiculitis (numbness in the thumb and index fingers only)

Diabetes mellitus with neuropathy (history and elevated fasting blood glucose)

Flexor carpi radialis tenosynovitis (tenderness near the base of the thumb)

Hypothyroidism

Kienböck's disease (painful wrist motion)

Median nerve compression at the elbow (wrist flexor weakness)

Pisotriquetral arthritis (ulnar palmar tenderness)

Triscaphoid arthritis (painful wrist motion, reduced grip)

Ulnar neuropathy (first dorsal interosseus weakness, numbness of the ring and little fingers)

Volar radial ganglion (mass near the base of the thumb, above wrist flexion crease)

Adverse outcomes of the disease

Permanent loss of sensation is possible, as is thenar atrophy and weakness of apposition.

Treatment

For mild cases, splinting the wrist, and a short-term course of NSAIDs or oral corticosteroids can be tried. The splint should be worn at night (at a minimum) and may be worn during the day if it does not interfere with the patient's work or daily activities. If these measures fail, consider injecting a corticosteroid into the carpal canal (see Carpal Tunnel Injection). Injection has diagnostic as well as therapeutic benefits, but improvement may only be temporary. Care must be taken to avoid direct injection into the median nerve.

Work-related carpal tunnel syndrome may be improved with ergonomic modifications, such as using keyboard or forearm supports, adjusting the height of computer keyboards, and avoiding holding the wrist in a flexed position (such as a dental hygienist). Occasionally, patients with acute carpal tunnel syndrome may have wrist pain rather than the more typical signs of numbness and thenar weakness. These patients respond well to corticosteroid injection.

Carpal tunnel syndrome that occurs during pregnancy usually resolves when the pregnancy terminates; therefore, treatment should consist of splinting and other conservative measures, such as injection of corticosteroid.

Surgical treatment is necessary for patients who have atrophy or weakness of the thenar muscles, decreased sensation, and for those who have intolerable symptoms despite a course of conservative treatment.

Adverse outcomes of treatment

NSAIDs may cause gastric, renal, or hepatic complications. Fluid retention, flushing of the skin, and shakiness can occur as a result of taking oral corticosteroids. Injecting corticosteroids is associated with the risk of an intraneural injection, which may have long-term adverse consequences. Prolonged conservative treatment can result in loss of sensation and thenar atrophy.

Referral decisions/Red flags

Persistent numbness, weakness, and/or atrophy of the thenar muscles are indications for surgical treatment. Failure of conservative treatment after 3 months warrants further evaluation.

Acknowledgements

Figure 1 is adapted with permission from Szabo RM, Steinberg DR: Nerve entrapment syndromes in the wrist. *J Am Acad Orthop Surg* 1994;2:116.

procedure

Carpal Tunnel Injection

Materials

Sterile gloves

Alcohol

Skin preparation solution (iodinated soap or similar antiseptic solution)

3-mL syringe with a 25-gauge, 3/4-inch needle

Mixture of 2 mL of a 1% local anesthetic (without epinephrine) and 1 mL of a 40 mg/mL depo corticosteroid preparation

Adhesive bandage

Step 1

Wear protective gloves at all times during this procedure and use sterile technique.

Step 2

Cleanse the volar aspect of the wrist with alcohol, allow it to dry, then cleanse the area with iodinated soap or similar antiseptic solution.

Step 3

Inject 1 mL of the local anesthetic/corticosteroid mixture at the proximal wrist. The needle is inserted 1 cm proximal to the wrist flexion crease and 3 to 5 mm from the ulnar side of the palmaris longus. The needle is directed toward the hand at an angle of 30° to 45° (Figure 1). If the palmaris longus is absent, the direction of the needle may be aligned with the ring finger. Have the patient flex the fingers fully into the palm, then advance the needle approximately 1 to 1.5 cm until resistance is felt.

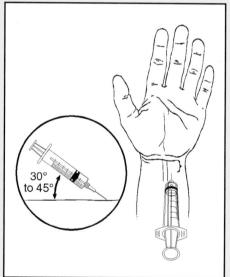

Figure 1
Location for needle insertion

hand and wrist

Carpal Tunnel Injection (continued)

Step 4

Instruct the patient to slightly wiggle the tips of the ring and little finger. If this causes slight movement of the tip of the needle, the needle is safely positioned.

Step 5

Ask the patient to extend the fingers as gentle pressure is applied to the needle. The distal excursion of the ring and little flexor tendons will carry the point of the needle into the carpal canal.

Step 6

If the patient reports any tingling, the needle has entered the nerve. If this occurs, do not continue with the injection. Otherwise, inject 1 to 2 mL of local anesthetic/corticosteroid mixture into the carpal canal.

Step 7

Dress the puncture wound with a sterile adhesive bandage.

Adverse outcomes

Infection or intraneural injection is possible.

Aftercare/patient instructions

Advise the patient that there may be occasional mild soreness and that the injection may require 24 to 48 hours to work.

Dupuytren's Disease

Synonyms

Palmar fasciitis

Viking disease

Palmar fibromatosis

Definition

Dupuytren's disease involves the thickening and contraction of the palmar fascia and has a dominant genetic component, particularly involving people of northern European descent (Figure 1). It has also been called the Viking disease. The tendon is not involved, although it may appear to be in advanced contractures. Trauma may accelerate and in some cases may initiate the process. This disease most commonly affects men older than age 40 years; it also affects individuals who have epilepsy, diabetes, pulmonary disease, or alcoholism.

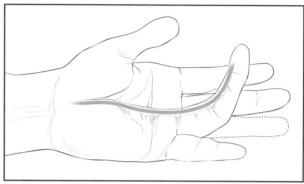

Figure 1A
Dupuytren's contracture of the ring finger

Clinical symptoms

Patients typically notice one or more painless nodules near the distal palmar crease that are initially moderately sensitive to pressure. The nodule(s) may gradually thicken and contract, drawing the finger into flexion at the metacarpophalangeal (MP) joint and occasionally, the proximal interphalangeal joint (PIP) joint as the disease progresses. The ring finger is most commonly involved, followed by the small, long, thumb, and index fingers. While extension is limited, finger flexion is usually normal. Invariably, the condition is painless in its later stages.

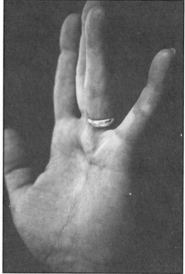

Figure 1B
Clinical appearance of Dupuytren's contracture

As the contractures increase, patients may have trouble putting on gloves, putting their hand in a pocket, or grasping large objects. Sensation in the fingers is usually normal unless the patient has concomitant carpal tunnel syndrome.

Tests

Exam

Examination reveals a painless palmar skin nodule that may look very much like a callus. There may be a pit in the skin in the area of the nodule with fascial bands that extend distally and sometimes proximally to the nodule. These bands may cross the MP joint, which is most common, and the PIP joint holding the finger in a contracted position. The bands are seldom tender to palpation unless the patient is in the early stages of the disease.

Diagnostic

Diagnosis is made upon clinical examination. Radiographs are not needed.

Differential diagnosis

Flexion contracture secondary to joint or tendon injury (no cords or bands)

Locked trigger finger (no associated nodules)

Adverse outcomes of the disease

Both progressive flexion contracture of the fingers and limited function are possible.

Treatment

There is no conservative treatment for this disease at this time. Splinting is not effective.

Adverse outcomes of treatment

Patients may experience nerve injury, skin slough, and recurrence of the disease postoperatively.

Referral decisions/Red flags

Patients with significant contractures (greater than 30°) of the MP joints who are troubled by their lack of extension are candidates for further evaluation. Likewise, patients with involvement of the PIP joint in association with Dupuytren's disease require close follow up.

Acknowledgements

Figure 1A is adapted with permission from the American Society for Surgery of the Hand: Brochure: Dupuytren's disease. Englewood, CO, 1995.

hand and wrist

Fingertip Infections

hand and wrist

Synonyms

Felon

Paronychia

Definition

Infections of the fingertip typically occur in two locations: in the digital pulp (felon) (Figure 1) and in the soft-tissue fold about the finger (paronychia). Paronychia is the most common type of hand infection. *Staphylococcus aureus* is the most common causative bacterial organism in both conditions.

Clinical symptoms

Felons are characterized by severe pain and swelling in the pad of the fingertip, usually caused by a puncture wound. They most commonly occur in the thumb and index finger. With a felon, the entire pulp of the fingertip is swollen, tense, red, and very tender. There may also be a visible puncture wound. Note that the swelling does not cross the distal finger flexion crease.

Paronychia infection is characterized by swelling of the tissues about the fingernail, usually along one side and about the base of the nail. These infections often follow a manicure or a nail deformity, such as a hangnail or an ingrown nail. Occasionally, the swelling may extend completely around the nail, referred to as a "run-around abscess." The pain associated with a paronychia is not as intense as that with a felon.

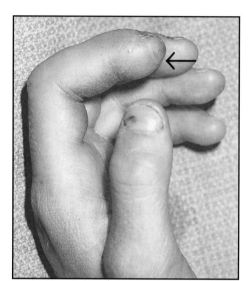

Figure 1
Felon

Tests

Exam

Passively flex and extend the finger to determine if motion increases the pain; increased pain suggests a deep infection in the flexor tendon sheath. The fingertip should also be inspected for small vesicles, as these suggest a diagnosis of herpetic whitlow (Figure 2).

This distinction between a felon and herpetic whitlow is important because surgery is contraindicated in treating a whitlow. Health care workers who are exposed to human saliva, such as dental hygienists and respiratory therapists, are at increased risk for herpetic whitlow.

Diagnostic

Plain radiographs show soft-tissue swelling early in the course of the infection.

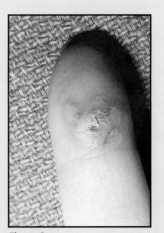

Figure 2
Herpetic whitlow

Differential diagnosis

Chronic fungal infection (not responsive to antibiotics)

Epidermal inclusion cyst (usually not that painful)

Herpetic felon (obvious vesicles)

Septic tenosynovitis (increased pain with active and passive motion of finger)

Adverse outcomes of the disease

Untreated felons may lead to osteomyelitis of the distal phalanx, which is suggested by resorption of the bony tuft. Other problems include nail deformity, ischemic necrosis of the fingertip, and skin slough. In addition, the fingertip may remain tender or lose sensation. Sepsis of the distal interphalangeal joint and/or septic tenosynovitis may develop as well.

Treatment

Most felons require surgical drainage under digital block anesthesia (see Digital Anesthetic Block [Hand]) and tourniquet control. Two different incisions can be used: a central longitudinal incision or a dorsal midaxial hockey stick incision. Use a central longitudinal incision, extending from the flexion crease to the fingertip, if a collection of pus can be seen under the skin on the pad side of the finger (Figure 3). Culture the drainage for aerobic and anaerobic organisms. Use open packing, such as a gauze packing strip, and then remove 3 days later. Allow the wound to close by secondary intention; never suture the wound.

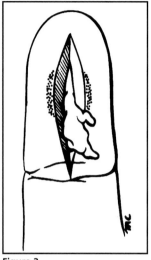

Figure 3
Volar longitudinal incision of the pulp

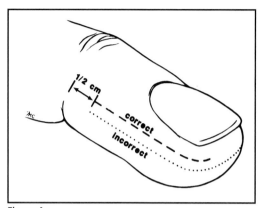

Figure 4
Dorsal midaxial incision for drainage of a felon

Use a dorsal midaxial hockey stick incision if no collection of pus is readily visible under the skin (Figure 4). For the thumb, make the incision on the radial (noncontact) side and for the fingers on the ulnar (noncontact) side. Extend the incision to the fingertip, but do not wrap it around the tip. Make the incision down to the bone where the soft-tissue attachments to the bone can be separated by blunt dissection with a small hemostat. Use open packing gauze and then remove in 3 days.

Treatment of an early stage paronychia should be nonsurgical. Application of warm, moist soaks for 10 minutes four times a day combined with an oral antibiotic for 5 days is usually adequate. Since *Staphylococcus aureus* is the most likely organism, an oral cephalosporin such as cephalexin (Keflex) 250 mg four times a day or dicloxacillin 250 mg four times a day is a good initial choice.

In later stages when pus is present, remove 25% of the lateral nail on the affected side (Figure 5). Wrap a penrose drain around the base of the finger as a tourniquet and use a digital anesthetic block (see Digital Anesthetic Block [Hand]). Elevate the outer quarter of the nail with a hemostat or metal probe, cut it with heavy scissors, and then remove it. Pack the defect with an iodine-impregnated packing strip and then remove it in 3 days.

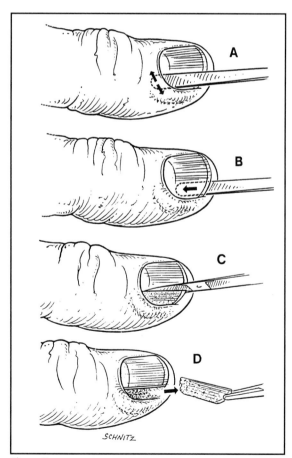

Figure 5
Portion of nail removed for paronychia

hand and wrist

Adverse outcomes of treatment

Following treatment, patients may have a painful scar from a misplaced incision or a nail deformity from injury to the germinal matrix. Inadequate drainage with persistent or recurrent infection may occur.

Referral decisions/Red flags

Persistent swelling and/or drainage of the fingertip despite antibiotic treatment and surgical decompression is an indication for further evaluation.

Acknowledgements

Figures 1-4 are reproduced with permission from Stern PJ: Selected acute infections, in Greene WB (ed): *Instructional Course Lectures XXXIX*. Park Ridge, IL, American Academy of Orthopaedic Surgeons, 1990, pp 539–546.

Figure 5 is reproduced with permission from American Society for Surgery of the Hand, Regional Review Course Manual 1992, p 75.

hand and wrist

procedure

Digital Anesthetic Block (Hand)

Materials

Sterile gloves

Skin preparation solution (iodinated soap or similar antiseptic solution)

Ethyl chloride spray

3-mL syringe with a 27-gauge needle

3 mL of 1% local anesthetic *without* epinephrine

Adhesive dressing

Step 1
Wear protective gloves at all times during this procedure and use sterile technique.

Step 2
Cleanse all surfaces of the base of the finger with iodinated soap or similar antiseptic solution.

Step 3
Draw 3 mL of the local anesthetic into the syringe.

Step 4
Freeze the skin with an ethyl chloride spray.

Step 5
Insert the 27-gauge needle into the web space alongside the extensor tendon (Figure 1). Advance the needle until it is almost at the volar skin and inject 1 mL of the local anesthetic.

Step 6
Withdraw the needle and insert it along the top of the extensor tendon (Figure 2). Inject 1 mL of the solution.

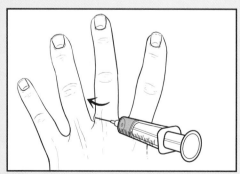

Figure 1
Location for needle insertion

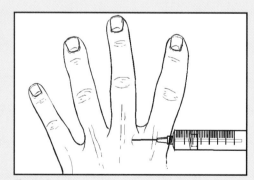

Figure 2
Location for needle insertion

Digital Anesthetic Block (Hand) (continued)

Step 7

Repeat the procedure in the web space on the opposite side of the finger (Figure 3).

Step 8

Dress the puncture wound with a sterile adhesive dressing.

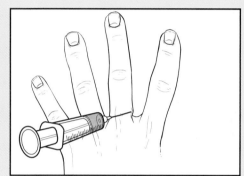

Figure 3
Location for needle insertion

Adverse outcomes

Although rare, infection is possible. Necrosis of a digit is possible if epinephrine is used in the anesthetic solution. Note that the agent should be injected only on three sides of the finger to avoid a circumferential nerve block that can cut off the circulation to the finger from hydrostatic pressure.

Aftercare/patient instructions

Advise the patient that local swelling at the site of the block should resolve in a few hours. Instruct the patient to call your office if the finger becomes dusky or completely white. Avoid hot surfaces or sharp objects until all sensation has returned.

hand and wrist

procedure

Fishhook Removal

The presence of a barb in the fishhook makes retrograde extraction of the hook difficult. Most hook injuries involve the skin and subcutaneous tissues only.

Materials

Sterile gloves

Skin preparation solution (iodinated soap or similar antiseptic solution)

24-inch length of 0 or #1 nylon or silk suture

3-mL syringe with a 27-gauge needle

3 mL of a 1% local anesthetic

Wire cutter

Although many techniques of fishhook removal have been described, only two will be discussed here (Figures 1 and 2). Remember to wear protective gloves at all times during this procedure and use sterile technique.

First technique

Step 1

Cleanse the skin with an iodinated soap or similar antiseptic solution.

Step 2

Use a 27-gauge needle to infiltrate the skin with 2 to 3 mL of 1% local anesthetic.

Step 3

Grasp the exposed end of the hook (shank) with the thumb and index finger and rotate the hook to force the barb out through the skin.

Step 4

Cut the barbed end of the hook with a wire cutter. Remove the rest of the hook retrograde. It will back out easily once the barb is gone.

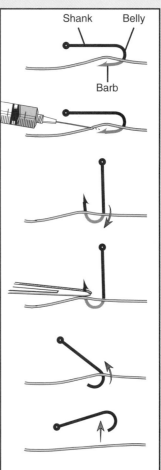

Figure 1
Technique 1 for removing a fishhook

Fishhook Removal (continued)

Second technique

Step 1

Cleanse the skin with an iodinated soap or similar antiseptic solution.

Step 2

Inject 2 to 3 mL of 1% local anesthetic about the hook.

Step 3

Loop a size 0 or #1 nylon or silk suture around the belly of the hook at the point where it penetrates the skin.

Step 4

Grasp the shank of the hook with your left thumb and middle finger and press against the skin. At the same time, press gently downward on the belly of the hook with the left index finger to disengage the barb from the surrounding tissues.

Step 5

With your right hand, grasp the suture 10″ to 12″ from the hook and pull sharply to remove the hook. The hook often disengages with considerable velocity, so care must be taken that bystanders are not impaled by the flying fishhook.

Step 6

Dress the wound with a sterile adhesive bandage.

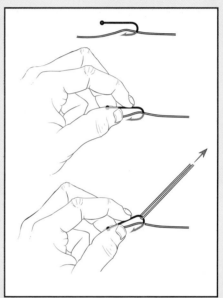

Figure 2
Technique 2 for removing a fishhook

hand and wrist

hand and wrist

Fishhook Removal (continued)

Adverse outcomes

Use of the first technique may inflict further soft-tissue damage by pushing the barb through the skin. Breakage of the hook or infection is possible.

Aftercare/patient instructions

Check if the patient is up-to-date on tetanus prophylaxis. Advise the patient to keep the wound clean until it is healed, usually in 3 to 4 days. Instruct the patient to return to your office if redness, fever, or proximal swelling occur.

Acknowledgements

Figures 1 and 2 are adapted with permission from Hospital Medicine 1980, Cahners Publishing Co.

Flexor Tendon Injuries

Synonyms

Tendon laceration or rupture

Jersey finger

Definition

The flexor tendons of the hand are vulnerable to rupture or laceration most commonly in association with rheumatoid arthritis and athletics (football, wrestling, and rugby), and usually involve the profundus tendon of the ring finger. Complete lacerations of both the flexor digitorum sublimus (FDS) and flexor digitorum profundus (FDP) cause immediate loss of flexion at the proximal interphalangeal (PIP) and distal interphalangeal (DIP) joints. Incomplete or partial lacerations may be missed, only to present 1 week later as a spontaneous rupture in the weakened tendon.

Clinical symptoms

A brief description of anatomy is useful in a discussion of diagnosing flexor tendon lacerations (Figure 1). Each finger is supplied with two extrinsic flexor tendons—the muscle is outside the hand—and the thumb has only one extrinsic flexor tendon. The FDP flexes the DIP and PIP joints and the metacarpophalangeal (MP) joint. The FDS flexes the PIP and MP joints, while the intrinsic muscles flex only the MP joint. Therefore, if the FDP is cut, patients can flex the PIP and MP joints, but not the DIP joint. If the FDS is cut, patients can still flex all three joints (Figure 2). Because of the close proximity of the digital nerves to the flexor tendons, open injuries to the flexor tendons are commonly associated with injuries to the digital

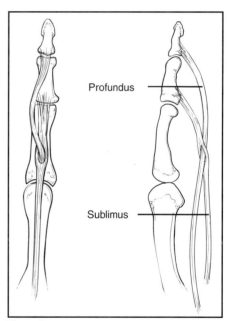

Figure 1
Extrinsic flexor tendons of the finger

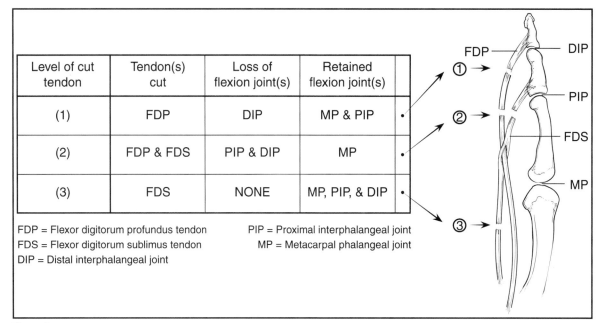

Level of cut tendon	Tendon(s) cut	Loss of flexion joint(s)	Retained flexion joint(s)	
(1)	FDP	DIP	MP & PIP	•
(2)	FDP & FDS	PIP & DIP	MP	•
(3)	FDS	NONE	MP, PIP, & DIP	•

FDP = Flexor digitorum profundus tendon   PIP = Proximal interphalangeal joint
FDS = Flexor digitorum sublimus tendon   MP = Metacarpal phalangeal joint
DIP = Distal interphalangeal joint

Figure 2
Effects of flexor tendon injuries on flexion of the finger joint

nerves as well. With these injuries, patients will report numbness on one or both sides of the finger.

A classic method of tendon rupture (a closed injury) occurs when a football player grabs another's jersey (thus the name jersey finger) and the ring finger becomes caught (Figure 3). The profundus tendon is avulsed from its insertion and possibly accompanied by a bony fragment. These finger injuries are often missed or diagnosed late, as they are often considered to be a "jammed" finger. In patients with rheumatoid arthritis, ruptures are usually "silent," meaning that the patient notices that the finger will not bend, but does not remember when the function was lost.

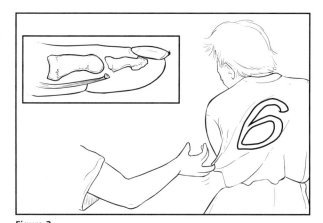

Figure 3
Mechanism of injury in rupture of the profundus tendon

Tests

Exam

First, test for active flexion, and then for strength of flexion. If the peritendinous structures are intact, the patient may retain flexion even in the presence of a complete laceration; however, flexion will be weak. Test flexion at both the DIP and PIP joints by asking the patient to flex the injured finger against your finger as you apply resistance. Check flexion of the FDP by asking the patient to flex the fingertip at the DIP joint while the PIP joint is held in extension (Figure 4).

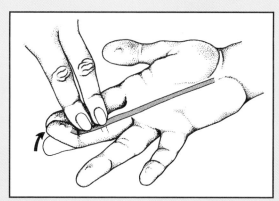

Figure 4
Test for function of the FDP

Diagnosis of an FDS laceration can be difficult, as the FDP can perform the function of the superficialis; that is, it can flex the PIP joint. To test the integrity of the FDP, hold the fingers straight, then have the patient flex each finger individually at the PIP joint (Figure 5). With a lacerated FDS but an intact FDP, PIP flexion will not occur if the FDP is made inoperative by holding down the tips of the remaining fingers. This is due to the intertendinous connection of the FDPs of the long, ring, and small fingers. In the index finger, this test cannot be performed reliably as the FDP is independent. In the little finger, one third of the population does not have an independently functioning FDS, and the FDP provides all the flexion for the DIP and PIP joints.

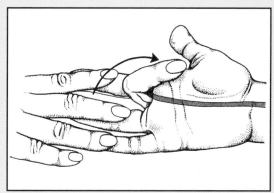

Figure 5
Test for function of the FDS

Partial lacerations can pose a diagnostic challenge. Patients typically have full range of motion, but they have more pain with active flexion than would be expected. If the diagnosis is unclear, referral for possible surgical exploration should be strongly considered to identify if there is a partial laceration.

Patients with a tendon rupture may have mild swelling over the flexion surface of the DIP joint. Test the strength of flexion at the DIP joint. If flexion is weak, rupture of the FDP must be considered a possibility. Patients with rheumatoid arthritis may not remember the point at which the tendon ruptured; they can usually only report that the finger will not flex.

Sensation in the finger should also be evaluated as open tendon injuries are often accompanied by injuries to the nearby digital nerves.

Diagnostic

PA and lateral radiographs of the involved finger may show a small avulsed fragment from the distal phalanx in an FDP rupture. These views may also identify a fracture.

Differential diagnosis

Anterior interosseous nerve paralysis (no laceration)

Partial tendon laceration (full flexion with pain and weakness)

Stenosing tenosynovitis with the finger locked in extension (no visible wound; tenderness over the proximal flexor pulley)

Adverse outcomes of the disease

Loss of flexion and grip and pinch strength in the involved and adjacent fingers is possible.

Treatment

The principal goal of initial treatment is to correctly identify the condition and ensure that the patient is evaluated for surgical repair. Flexor tendon injuries ultimately require surgical repair. Initially, clean and repair superficial wounds and splint the hand in a position for flexor tendon injuries (Figure 6). Surgical exploration and repair should be done within 2 weeks of the injury.

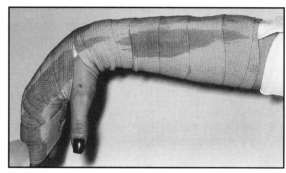

Figure 6
Splint for flexor tendon injury

Adverse outcomes of treatment

Postsurgical infection or failure of the repair is possible.

Referral decisions/Red flags

Patients with any type of suspected flexor tendon injury (rupture or laceration) require further evaluation for surgical repair.

Acknowledgements

Figures 1 and 2 illustrations by Elizabeth Roselius, © 1988. Reproduced with permission from Green DP (ed): *Operative Hand Surgery, ed 2*, New York, NY, Churchill Livingstone, 1988, p 1971.

Figure 3 is reproduced with permission from Carter PR: *Common Hand Injuries and Infections: A Practical Approach to Early Treatment.* Philadelphia, PA, WB Saunders, 1983.

Figures 4 and 5 are reproduced with permission from American Society for Surgery of the Hand: *The Hand: Examination and Diagnosis,* ed 3. New York, NY, Churchill Livingstone, pp 18–19.

Flexor Tendon Sheath Infections

Synonyms

Septic tenosynovitis

Purulent tenosynovitis

Definition

The flexor tendons of the fingers and thumb are covered by a double layer of synovium, the function of which is to allow the tendons to slide freely within a fibro-osseous tunnel (Figure 1). Infections can occur in the space between these two layers from a puncture wound or by spread from a subcutaneous abscess.

Clinical symptoms

Patients typically report a history of a puncture wound to the flexor surface of the finger or thumb. The entire finger becomes swollen, and any movement is painful. Occasionally, patients have an abrasion on the palm of the hand that preceded the onset of the tenosynovitis.

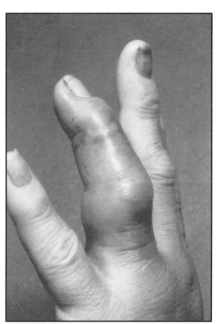

Figure 1
Septic tenosynovitis of the ring finger

Tests

Exam

Kanavel's cardinal signs help make the diagnosis of septic tenosynovitis: 1) swelling along the entire flexor surface; 2) tenderness over the course of the tendon sheath; 3) pain on passive extension; and 4) flexed position of the finger at rest (Figure 2). Patients with septic tenosynovitis will exhibit all of these signs.

Diagnostic

Plain radiographs of the hand and finger may reveal a foreign body or air in the sub-cutaneous tissues.

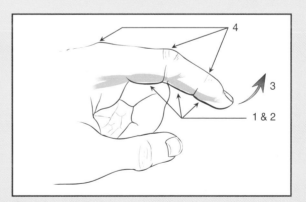

Figure 2
Kanavel's cardinal sign for septic tenosynovitis

hand and wrist

Differential diagnosis

Cellulitis (little or no pain with active motion of the finger)

Subcutaneous abscess (no pain with passive extension)

Adverse outcomes of the disease

Loss of skin, or even of the entire digit, is possible. Other problems include stiffness and the possibility of either osteomyelitis or septic arthritis.

Treatment

Established tenosynovitis infections need both surgical drainage and antibiotics, as a course of antibiotics alone is inadequate. If some, but not all, of Kanavel's signs are present and septic tenosynovitis is suspected, give parenteral antibiotics for 12 to 24 hours and reevaluate. Cefazolin sodium is a good initial choice, as it is effective for both Staphylococcus and Streptococcus infections. If the finger appears to have responded to the antibiotic, continue the course for another 24 to 36 hours. If no improvement is seen or if the finger looks worse, refer for possible surgical drainage.

Adverse outcomes of treatment

Finger stiffness can develop.

Referral decisions/Red flags

Patients with all Kanavel's signs need surgical evaluation for drainage. Patients with suspected tenosynovitis who do not respond to antibiotic treatment also need additional evaluation.

Acknowledgements

Figure 1 is reproduced with permission from Stern PJ: Selected acute infections, in Greene WB (ed): *Instructional Course Lectures XXXIX.* Park Ridge, IL, American Academy of Orthopaedic Surgeons, 1990, pp 539–546.

Figure 2 is adapted with permission from Carter PR: *Common Hand Injuries and Infections: A Practical Approach to Early Treatment.* Philadelphia, PA, WB Saunders, 1983, p 220.

Fracture of the Metacarpals and Phalanges

Synonyms

Boxer's fracture

Fighter's fracture

Definition

Phalangeal fractures are most common in children, with one third of fractures involving the epiphysis (80% of these fractures are Salter type II fractures). The epiphysis appears when children are about age 3 years; it fuses in girls at about age 15 years and in boys at about age 18 years. Growth disturbances from epiphyseal injuries are rare in pediatric fractures.

Metacarpal fractures are most common in adults (Figure 1). The metacarpal of the small finger is most commonly fractured, followed by the metacarpals in the thumb, index, long, and ring fingers, respectively. In adult phalangeal fractures, the distal phalanx is the most commonly injured, followed by the proximal and the middle phalanx (Figure 2). Approximately 20% of these fractures are intra-articular.

Clinical symptoms

Patients typically have a history of trauma. Local tenderness, swelling, deformity, or decreased range of motion are common findings.

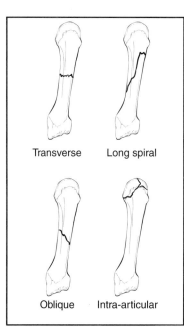

Figure 1
Types of metacarpal fractures

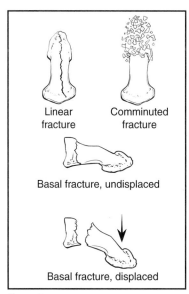

Figure 2
Types of distal phalanx fractures

Tests

Exam

Examination reveals swelling over the fracture site. If the joint is involved, there may also be decreased range of motion. The involved finger may appear shortened, or the knuckle may be depressed. Often, the distal fragment of a fractured bone is rotated in relation to the proximal fragment; this is not easy to see with the fingers extended. However, if the patient makes a partial fist, the rotated fragments cause the involved finger to overlap onto its neighbor.

Diagnostic

Radiographs are always indicated for suspected fractures. For fractures of the phalanges, PA and true lateral views of the individual digits should be obtained. For metacarpal fractures, PA, lateral, and oblique views of the hand are indicated.

Differential diagnosis

Metacarpophalangeal and interphalangeal collateral ligament injuries

Metacarpophalangeal and interphalangeal joint dislocations

Sprained finger

Adverse outcomes of the disease

Malunion, nonunion, and loss of finger motion are all possible.

Treatment

Extra-articular fracture (no joint surface involvement): For nondisplaced fractures, casting or splint immobilization for 3 weeks is indicated for phalangeal fractures and 4 weeks for metacarpal fractures. If loss of position is a possible problem, radiographs should be repeated 1 week following the injury.

Do not wait for radiographic evidence of healing to begin range of motion exercises, as full healing may not be apparent on radiographs for 2 to 5 months. If the fracture seems clinically stable after cast removal, motion should be started.

Immobilization of closed phalangeal fractures for more than 3 to 4 weeks will result in stiffness. When casting or splinting, always include the joint above and below the fracture and the adjacent digits. The hand should be positioned in a cast or splint such that the MP joints are in flexion and the PIP and DIP joints are held in extension.

Displaced extra-articular fractures: Displaced transverse fractures of the metacarpals and phalanges tend to angulate (Figure 3). Spiral fractures tend to rotate, and oblique fractures tend to shorten. Patients with these types of fractures need further evaluation upon initial presentation.

Metacarpal neck fractures (boxer's fracture, fighter's fracture): Apply an ulnar gutter splint for 2 to 3 weeks in patients with 10° to 15° of angulation at the metacarpal neck of the ring and small fingers (Figure 4). If there is more than 15° of angulation, but no extensor lag (the patient can fully extend the finger), use of an ulnar gutter cast or splint for 2 to 3 weeks is also appropriate. Advise the patient that although this fracture will not result in any functional deficit, there may be a loss of metacarpal prominence. If this is unsatisfactory to the patient, further evaluation is indicated.

If there is more than 15° of angulation and there is an extensor lag, or if there is 40° of angulation, referral for reduction is necessary. If there is less than 10° of angulation at the metacarpal neck of the index and long fingers, use of a radial gutter splint is appropriate. If there is more than 10° of angulation, functional loss may result; therefore, referral for reduction is necessary.

Intra-articular fractures: Splint nondisplaced intra-articular fractures with the MP joints in flexion and the PIP and DIP joints in extension. Repeat radiographs in 1 week to assess for continued articular congruity, then initiate active range of motion at 3 weeks. Any displacement in intra-articular fractures is unacceptable due to loss of joint congruity; therefore, patients with displaced intra-articular fractures need further evaluation.

A summary of fracture type and treatment information is presented in Table 1.

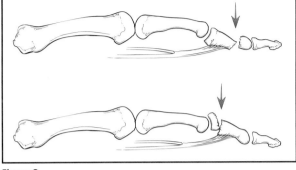

Figure 3
Phalangeal fractures angulate due to pull by the flexor tendons

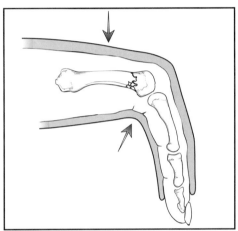

Figure 4
Fracture of the metacarpal neck (boxer's fracture)

Adverse outcomes of treatment

Joint stiffness is the most common problem seen following hand fractures and is directly related to prolonged immobilization. Malunion due to inadequate reduction is also possible.

Referral decisions/Red flags

Displaced fractures and intra-articular fractures usually require surgical pinning.

hand and wrist

Table 1
Metacarpal and phalangeal fractures

Type of fracture	Bone	Treatment	Duration	Possible adverse outcomes
Extra-articular, nondisplaced	Phalanx	Cast, splint	3 weeks	Loss of position is possible. To check, repeat radiograph 1 week following the injury.
	Metacarpal, ring and small finger; <15° angulation, or >15° angulation with no extensor lag (inability to fully extend the MP joint)	Ulnar gutter cast or splint	2 to 3 weeks	Loss of metacarpal prominence may occur. If this is not acceptable, consider surgical pinning.
	Metacarpal, ring and small finger; >15° angulation with extensor lag, or 40° angulation with no extensor lag	Consider surgical pinning	Per surgeon	
	Metacarpal, thumb, index or long finger	Cast, splint	3 weeks	
Extra-articular, displaced	All phalanges and metacarpals; transverse, oblique, or spiral	Consider surgical pinning		Rotational and angular fractures. Malunions are common without surgery.
Intra-articular, displaced	MP and IP joints	Consider surgical pinning		Joint stiffness and posttraumatic arthritis are common.

Acknowledgements

Figure 1 is adapted with permission from Dabezies EJ, Schutte JP: Fixation of metacarpal and phalangeal fractures with miniature plates and screws. *J Hand Surg* 1986;11A:283–288.

Figures 2 and 3 are adapted with permission from Rockwood CA, Green DP, Bucholz RW, et al (eds): *Rockwood and Green's Fractures in Adults*, ed 4. Philadelphia, PA, Lippincott-Raven, 1996, vol 2, pp 614, 627.

Figure 4 is adapted with permission from Green DP (ed): *Operative Hand Surgery*, ed 3. New York, NY, Churchill Livingstone, 1993.

Fracture of the Scaphoid

Synonyms

Navicular fracture

Carpal navicular fracture

Definition

The scaphoid, which in Greek means boat-shaped, is the most proximal radial wrist carpal bone and the most commonly fractured carpal bone. The blood supply to this bone is precarious as the blood vessels enter the bone primarily in the distal third of the bone. Articular cartilage covers 80% of the scaphoid. Because of these anatomical features, displaced fractures of the scaphoid (less than 1 mm) have a nonunion rate of 55% to 90%. Fractures of the middle third result in osteonecrosis in 33% of patients, and more proximal fractures almost always result in osteonecrosis. Of all scaphoid fractures, approximately 20% occur in the proximal pole, 60% in the middle or waist, and 20% in the distal pole.

Men between ages 20 and 40 years are most commonly affected following a fall on the outstretched (or dorsiflexed) wrist. These fractures rarely occur in children because the growth plate of the distal radius is weaker than the bone of the scaphoid and fails first. Thus, a fall on the outstretched wrist typically results in a Salter type I or II fracture of the distal radius in children.

Clinical symptoms

Patients report pain and tenderness about the radial (thumb) side of the wrist. Any type of wrist motion, such as gripping, is painful. Swelling about the back and radial side of the wrist is also common. If the patient reports a history of a high-energy injury, such as a car accident, additional ligamentous injuries are possible in addition to the scaphoid fracture.

Tests

Exam
Palpation over the anatomical snuffbox reveals marked tenderness (Figure 1). This area is defined by the abductor and long thumb extensor tendons just distal to the radial styloid. Likewise, pressure over the scaphoid tubercle on the underside of the wrist will produce pain with a scaphoid fracture.

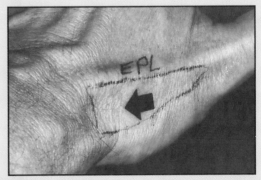

Figure 1
Anatomical snuffbox

Diagnostic

Scaphoid fractures may not be visible on initial plain PA and lateral radiographs of the wrist. Thus, if these radiographs appear normal, obtain a PA view with the wrist in ulnar deviation and an oblique view to help to visualize the fracture (Figure 2). A comparison view of the opposite wrist may be useful in difficult diagnostic situations. If this series of initial radiographs is normal but the pain persists for 3 weeks, the PA and oblique views should be repeated. If the radiographs are still normal, a bone scan may be considered. Note that a bone scan may not be positive in a scaphoid fracture until 3 days after the injury.

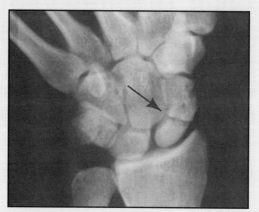

Figure 2
PA radiograph with the wrist in ulnar deviation

Differential diagnosis

deQuervain's tenosynovitis (positive Finkelstein's test)

Fracture of the distal radius (evident on plain radiographs)

Radioscaphoid arthritis (narrow joint space between the radius and scaphoid)

Scapholunate dissociation (Terry Thomas sign on plain radiographs {Figure 3})

Triscaphoid arthritis (an arthritis between the scaphoid, trapezium, and trapezoid)

Adverse outcomes of the disease

Osteoarthritis of the radiocarpal joint, and decreased grip strength and range of motion are possible.

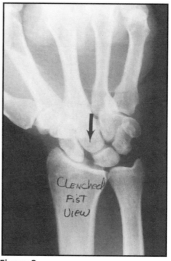

Figure 3
Terry Thomas sign

Treatment

The treatment of acute nondisplaced fractures of the scaphoid is controversial, as there is no agreement on what constitutes the optimum type of immobilization for these fractures. Conflicting data are reported regarding the optimum position of the wrist, whether the elbow should be immobilized, and whether the thumb should be included in the cast.

Based on current data, immobilization for 6 weeks in a long arm thumb spica cast with the wrist in a neutral position is recommended. If radiographs obtained after this time show that the fracture is healing, apply a short arm thumb spica cast. However, if the fracture line appears to be getting wider, indicating absorption at the fracture site, or if the fracture shows any displacement, then further evaluation for possible surgery is indicated.

If the patient has pain over the region of the snuffbox, but the initial radiographs are normal, place the hand and wrist in a thumb spica splint for 3 weeks and then repeat the radiographs. If the radiographs are still normal after this time, but tenderness over the scaphoid persists, order a bone scan. If the scan is positive, treat the hand as for an acute nondisplaced scaphoid fracture.

Because it is often difficult to determine from routine radiographs whether a scaphoid fracture has healed, there is a tendency to discontinue immobilization too soon. Therefore, as a general rule, the closer the fracture line is to the proximal pole, the longer the time for healing. For fractures of the distal pole, the average time for healing is 6 to 8 weeks. Fractures of the middle third take 8 to 12 weeks to heal, while fractures of the proximal pole may take 12 to 24 weeks or longer. If plain radiographs do not clearly reveal that the fracture has healed, order a tomogram to better visualize the degree of healing.

Adverse outcomes of treatment

Chronic pain, loss of motion from prolonged immobilization, and/or loss of grip strength can result.

Referral decisions/Red flags

All patients with displaced fractures of the scaphoid need early further evaluation for possible surgical treatment. Patients who show cystic absorption at the fracture site or displacement of the fracture following immobilization also need evalaution for surgical treatment. Patients with nondisplaced fractures that have not healed after 4 months of immobilization need orthopaedic evaluation.

hand and wrist

Fracture of the Thumb

Synonyms

Bennett's fracture

Rolando's fracture

Definition

Bennett's fracture is an oblique fracture of the base of the thumb metacarpal, which enters the carpometacarpal joint (Figure 1). While the small volar fragment remains attached to the carpus, the major metacarpal fragment subluxates proximally and posteriorly due to the pull of the abductor pollicis longus muscle (Figure 2). The fracture then displaces, creating a subluxation of the carpometacarpal joint. This fracture has two pieces, one large, the other small.

Rolando's fracture, which is much less common than Bennett's fracture, is a term used for a comminuted intra-articular fracture at the base of the thumb metacarpal.

Clinical symptoms

Swelling about the base of the thumb is common following a blow to the end of the thumb, especially if the thumb is flexed. This finding is indicative of a possible Bennett's fracture. Patients typically report pain and limited motion of the thumb.

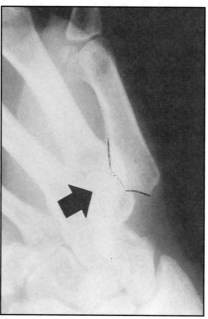

Figure 1
Radiograph showing Bennett's fracture

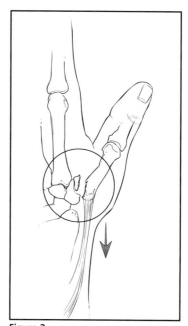

Figure 2
Fracture-dislocation of the base of the thumb metacarpal (Bennett's fracture)

Tests

Exam
Examination reveals that the base of the thumb metacarpal is displaced up and back, while the tip of the thumb is held into the palm. The patient cannot move the thumb without pain.

Diagnostic
Plain AP and lateral radiographs of the thumb will show these fractures.

Differential diagnosis

Arthritis of the carpometacarpal joint

Dislocation of the carpometacarpal joint

Fracture of the scaphoid

Synovitis of the carpometacarpal joint

Adverse outcomes of the disease

Patients have pain at the base of the thumb, along with loss of motion and pinch strength. Posttraumatic arthritis of the carpometacarpal joint is also possible.

Treatment

The goal of treatment is to restore the axial length of the thumb and to replace the metacarpal shaft fragment against the smaller volar lip fragment. Although anatomic reduction of the fracture is the goal, some irregularity (1- to 3-mm offset) can still produce a good functional outcome. Bennett's fracture almost always requires some form of surgical fixation to achieve stability in the joint.

Nondisplaced two-part fractures of the base of the thumb metacarpal can be treated in a thumb spica cast for 4 weeks.

Adverse outcomes of treatment

Following surgery, pin tract irritation and infection can occur, as can tenderness about the surgical plates and screws. Displacement of the fracture (loss of position) and posttraumatic arthritis are also possible.

Referral decisions/Red flags

Patients with displaced fractures of the base of the thumb metacarpal need further evaluation.

Ganglion of the Finger

hand and wrist

Synonyms

Synovial cyst

Mucous cyst

Volar retinacular ganglion

Flexor tendon sheath ganglion

Definition

A ganglion is a small cystic tumor that contains clear, jelly-like viscous fluid and is connected to a synovial sheath or joint capsule. It arises from the joint capsule or a pulley overlying the flexor tendon, but does not move with the tendon. Ganglia are the most common soft-tissue tumors of the hand; they may vary in size and may even temporarily disappear. They typically develop spontaneously and affect individuals between ages 20 and 30 years.

Most finger ganglia arise from the sheath of the flexor tendon and are also known as retinacular cysts or volar retinacular ganglia. Those arising from the interphalangeal (IP) joints are called mucous cysts.

Clinical symptoms

Because ganglia produce little functional disturbance, many patients do not seek medical attention. Patients with flexor tendon sheath cysts often report tenderness when grasping. They also often notice a bump at the base of the finger at the level of the proximal flexion crease.

Mucous cysts typically affect individuals (usually women) between ages 40 and 70 years and most often develop over an arthritic distal IP joint. Patients may have simple striations or furrowing of the fingernail because of pressure of the cyst on the nail matrix. They may also report a cycle of the cyst breaking open, draining a clear, jelly-like fluid, and then healing. Treatment is often sought when the cyst becomes painful, ulcerated, infected, or is cosmetically displeasing.

Tests

Exam

A ganglion of the flexor tendon sheath is characterized by a small, firm, tender mass at the base of the finger in the area of the metacarpophalangeal (MP) flexion crease, most often at the middle or ring fingers (Figure 1). However, these ganglia are sometimes difficult to detect due to their small size; in fact, they may simply feel like a small seed. A ganglion rarely affects motion, but sensory changes occasionally occur due to its proximity to the digital nerve.

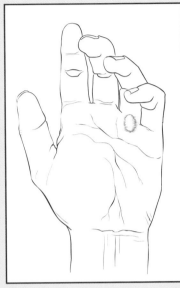

Figure 1
Flexor sheath ganglion

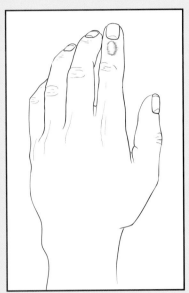

Figure 2
Mucous cyst

Mucous cysts most commonly occur at the distal interphalangeal (DIP) joint, usually lying to one side of the extensor tendon (Figure 2). Initial clinical findings include a mass or a blister. Arthritic nodules may be associated with this joint or other joints.

Diagnostic

Obtain radiographs of the finger to rule out arthritis in the adjacent joints, to identify bony tumors, or to identify calcification in degenerated tissues. A small bone spur is usually seen arising from the dorsum of the phalanges at the DIP joint.

Differential diagnosis

Dupuytren's disease (presence of cords or bands)

Giant cell tumors (different locations)

Lipoma (larger in size)

Nail bed trauma

Paronychia (closer to the fingernail)

Adverse outcomes of the disease

Patients often have pain in the hand and finger and an obvious deformity at the fingernail. Infection can also occur in association with a mucous cyst.

Treatment

Treatment of a ganglion of the tendon sheath consists of needle rupture with or without corticosteroid injections, followed by massage to disperse the contents of the cyst, or injection with 1% or 2% local anesthetic until the cyst pops. Exercise caution when performing a needle rupture due to the proximity of the neurovascular bundle. For this reason, surgery is often the preferred treatment.

Aspiration and/or rupture of a mucous cyst may temporarily reduce the size and symptoms of the lesion (see Ganglion Aspiration).

Adverse outcomes of treatment

Recurrence is quite common. Skin loss is possible, as is injury to the digital nerve.

Referral decisions/Red flags

Persistence of a painful or bothersome cyst after one or two attempts at needle rupture or aspiration indicates that surgical excision should be considered. Increased erythema and pain in a mucous cyst suggests infection that may require surgical treatment.

Acknowledgements

Figures 1 and 2 are adapted with permission from the American Society for Surgery of the Hand: Brochure: Ganglion Cysts. Englewood, CO, 1995.

Ganglion of the Wrist

Synonyms

Synovial cyst

Definition

A ganglion is a cystic structure that arises from a synovial sheath or joint cavity (Figure 1). The cyst contains a thick, slippery fluid identical in composition to joint fluid. Although the exact mechanism of origin is controversial, it has been well established that there is a connection (one-way valve) to the synovial cavity. Thus, synovial fluid can enter the cyst, but it cannot flow freely back into the synovial cavity. This can cause typical variation in size and sometimes significant symptoms from increased pressure within the cavity and on surrounding structures.

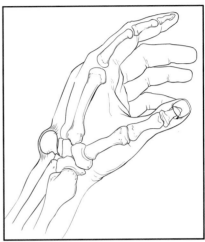

Figure 1

Clinical appearance of wrist ganglion

Clinical symptoms

Patients may have an unexplained lump, which may or may not be painful, on the dorsal or volar radial aspect of the wrist (Figure 2). The pain is often described as aching and is aggravated by extreme flexion or extension. Pain may also radiate proximally or distally and may occasionally be associated with a tingling sensation that radiates into the hand or forearm. Patients often find the lump to be unsightly.

Occasionally, a ganglion will occur in other areas of the wrist, causing compression of the median or ulnar nerve. If this occurs, sensory symptoms in the digits or weakness of the intrinsic muscle may develop. Ganglia in the wrist often increase in size with increased activity; thus, a history of variation in size is a key factor to distinguish a ganglion from other soft-tissue tumors, some of which might be malignant.

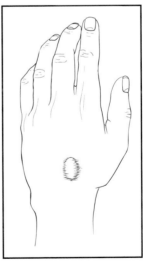

Figure 2A

Dorsal wrist ganglion

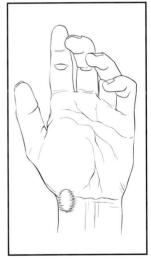

Figure 2B

Volar radial wrist ganglion

hand and wrist

Tests

Exam

A dorsal ganglion is typically a smooth round or multi-lobulated structure on the dorsal aspect of the wrist that becomes more prominent with flexion. It is usually positioned directly over the lunate bone, but may occur more distally even though it is attached at the lunate level.

A volar radial ganglion is usually a less well-defined mass situated between the flexor carpi radialis tendon and the radial styloid. It may extend underneath the radial artery and, in some cases, may adhere to the radial artery. On palpation, the ganglion may appear to pulsate and be confused with an aneursym. Symptoms also become more pronounced with extreme flexion or extension.

Ganglia are usually tender with pressure. If a definite mass is not visible, the possibility of an occult ganglion should be considered. With suspected occult ganglia, swelling may be more subtle. Palpation must be more thorough and careful in order to identify a small tender mass that is different from what is felt on the opposite wrist. A prominent ganglion will often transilluminate if a penlight is placed to the side of the ganglion. Solid tumors will not transilluminate.

Diagnostic

Plain PA and lateral radiographs of the wrist are mandatory to rule out bone pathology, such as an intraosseous ganglion, Kienböck's disease, arthritis, or bone tumor (benign or malignant). In most cases, the common ganglion, which is only a soft-tissue structure, will not be associated with radiographic changes.

Differential diagnosis

Arthritis

Bone tumor

Intraosseous ganglion (evident on radiographs)

Kienböck's disease (collapse of the lunate)

Soft-tissue tumor, benign or malignant

Adverse outcomes of the disease

On rare occasions, ganglia will cause significant compression on the median nerve. Patients may experience a decrease in grip strength; they may also complain about the unsightly bump.

Treatment

If the ganglion appears typical in presentation and physical findings, reassurance to the patient is usually adequate. If the patient has acute, severe symptoms, immobilization of the wrist will relieve symptoms and may cause the ganglion to decrease in size. However, immobilization alone is probably not a permanent solution. Occasionally, aspiration of the cyst will lead to resolution (see Ganglion Aspiration).

If the patient has significant symptoms or is seriously bothered by the appearance of the ganglion, then surgical excision is indicated.

Adverse outcomes of treatment

Recurrence is common in 5% to 10% of patients following surgical excision, and in up to 90% following injection of corticosteroid. Injury to the radial artery can occur as a result of aspiration. Infection following injection of corticosteriod is a risk, but can be largely avoided by use of careful sterile technique.

Referral decision/Red flags

Any mass that does not have the typical findings and appearance associated with a ganglion should be evaluated with further diagnostic tests or excisional biopsy.

Acknowledgements

Figures 2A and 2B are adapted with permission from the American Society for Surgery of the Hand: Brochure: Ganglion Cysts. Englewood, CO, 1995.

hand and wrist

procedure

Ganglion Aspiration

Materials

Sterile gloves

Alcohol

Skin preparation solution (iodinated soap or similar antiseptic solution)

3-mL syringe with a 27-gauge, 3/4-inch needle

18-gauge needle

3 mL of a 2% local anesthetic (plain)

Elastic bandage

Plaster splint

Step 1

Wear protective gloves at all times during this procedure and use sterile technique.

Step 2

Cleanse the skin with alcohol, allow it to dry, then cleanse the area with iodinated soap or similar antiseptic solution.

Step 3

Use a syringe with a 27-gauge needle to infiltrate 2% local anesthetic into the area immediately proximal to and surrounding the ganglion. Remove the 27-gauge needle and change to the 18-gauge needle.

Step 4

Penetrate the ganglion with the 18-gauge needle and attempt to withdraw as much fluid as possible (Figure 1). While aspirating a volar radial ganglion, try to avoid the radial artery, which you can detect by palpation. Injecting corticosteroid has not been shown to influence the recurrence rate and is not recommended.

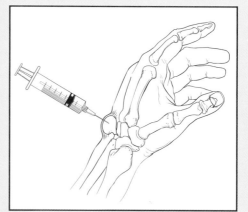

Figure 1
Location for needle insertion

Ganglion Aspiration (continued)

Step 5

Apply a sterile dressing and a volar plaster splint, and wrap with a compression bandage.

Adverse outcomes

Infection and recurrence of the ganglion are possible. Injury to the radial artery can also occur.

Aftercare/patient instructions

Instruct the patient to wear the elastic bandage and wrist splint for 1 week, removing them only for bathing.

Human Bites

hand and wrist

Synonyms

Clenched fist injury

Definition

Human bites to the hand occur from one of two mechanisms. Bites over the knuckles usually occur when a tooth strikes a clenched fist, which typically results in an injury to the extensor tendon and penetration of the joint capsule (Figure 1). Wounds over the fingers occur as a result of a direct human bite (Figure 2). Unlike animal bite wounds, human bite wounds contain higher concentrations of both aerobic and anaerobic bacteria. In addition to the bacterial contamination, human bites may transmit serious infectious diseases such as HIV and hepatitis.

Clinical symptoms

When patients report the mechanism of injury, the diagnosis is obvious. However, patients may be reluctant to admit how the injury occurred, particularly if it was the result of a fight. In addition, patients with clenched fist injuries often do not seek medical attention until 2 to 3 days after the injury. If an extensor tendon or flexor tendon is cut, the patient will not be able to extend or flex the portion of the digit that is distal to the cut (Figure 3). Patients may also report a loss of sensation over the tip of the finger if a nerve has been lacerated.

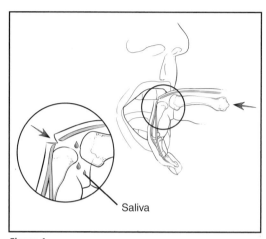

Figure 1
Mechanism of tendon laceration in human bite

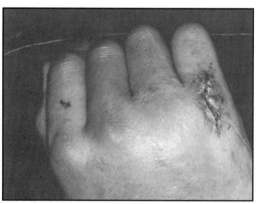

Figure 2
Possible "tooth" injury

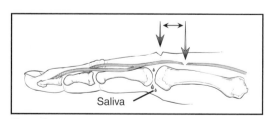

Figure 3
The tendon injury, cut in flexion, retracts as the joint extends

Tests

Exam

A 1-cm laceration over the metacarpophalangeal joint of the ring or little finger is a sign of a possible "tooth" injury. Begin the examination by testing for flexion and extension distal to the laceration. If the cut is on the flexor surface of the finger, test for sensation over the fingertip. The back of the hand is commonly swollen, red, and warm, especially if the injury is more than 2 to 3 hours old. In addition, there may be purulent drainage from the cut. Examine the forearm for red streaks indicative of a lymphangitis, and palpate the inner aspect of the elbow for any enlarged epitrochlear nodes.

Diagnostic

Obtain plain radiographs of the hand, including AP and lateral views, to rule out an underlying fracture or presence of a foreign body. Obtain aerobic and anaerobic cultures if there is any drainage present. A WBC can serve as a baseline to follow the clinical course, but note that it may be within normal limits in the days immediately after the injury.

Differential diagnosis

Laceration from a sharp object

Septic joint due to a retained foreign body

Adverse outcomes of the disease

Tendon rupture and/or laceration can occur as a result of the bite. Infection involving the deep palmar space can develop without treatment. Also, osteomyelitis, joint sepsis, joint stiffness, and possibly septic tenosynovitis can develop.

Treatment

Lacerations caused by human bites should never be closed primarily. Bite wounds that are less than 4 hours old and have no sign of infection can be treated on an outpatient basis if the wound is superficial and does not enter the joint.

Anesthetize the hand with a 1% plain local anesthetic, debride the skin edges of devitalized tissue, and then irrigate the wound with 1,000 mL of saline. Appropriate tetanus prophylaxis should be given as outlined in Table 1, followed by 48 hours of oral antibiotics. Penicillin and a first-generation cephalosporin provide adequate coverage in most instances. Tetracycline can be used for patients who are allergic to penicillin.

Immobilize the hand in a bulky dressing with an incorporated dorsal plaster splint. The wound can be dressed open with an iodoform gauze or a saline-soaked gauze. Advise the patient to return within 24 hours for a recheck to confirm that infection has not developed. After the first 24 hours, daily whirlpool treatment can be started. The wound should be allowed to close by secondary intention.

If the wound extends into the joint or if the bone, nerve, or tendon is involved, surgical consultation is necessary. The same is true if the bite wound is infected.

hand and wrist

Table 1
Guide to tetanus prophylaxis in wound management

History of tetanus toxoid (doses)	Clean, minor wound		Contaminated wound	
	Td	TIG	Td	TIG
Unknown	Yes	No	Yes	Yes*
Fewer than 3 doses	Yes	No	Yes	Yes*
Three or more	Yes, if more than 10 years since last dose	No	Yes, if more than 5 years since last dose	No

Td = combined tetanus and diphtheria toxoid adsorbed (dose = 0.5 mL IM). TIG = tetanus immune globulin (dose = 250 IU IM).
* Td and TIG should be administered at different sites.

Adverse outcomes of treatment

Patients can experience allergic reactions to antibiotics. Infection may develop with primary wound closure.

Referral decisions/Red flags

Wounds that involve the joint, tendon, nerve, or bone indicate the need for surgical consultation. An infected bite wound that is progressing rapidly (6 to 12 hours) despite treatment with antibiotics also needs further evaluation.

Acknowledgements

Figures 1 and 3 are adapted with permission from Carter PR: *Common Hand Injuries and Infections: A Practical Approach to Early Treatment.* Philadelphia, PA, WB Saunders, 1983.

Table 1 is adapted with permission from the Centers for Disease Control and Prevention.

Joint Injuries of the Hand

Synonyms

Sprain

Jammed finger

Corked finger

PIP dislocation

DIP dislocation

MP dislocation

Definition

Sprains and dislocations of the finger are very common. A sprain is characterized by a partial tear of the supporting soft-tissue structures of the joints (collateral ligaments and volar plate) (Figure 1). With a dislocation, these structures are often torn completely. Dislocations of the proximal interphalangeal (PIP) joints are common, but dislocations of the metacarpophalangeal (MP) and the distal interphalangeal (DIP) joints are not. These injuries may be closed or open.

Clinical symptoms

Patients almost always report a history of trauma in which they describe a deformity that developed immediately after the injury. The patient or a well-meaning friend often reduces the dislocation before seeking medical attention.

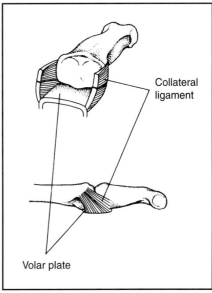

Figure 1
Collateral ligaments and volar plates of the PIP joint

hand and wrist

Tests

Exam

If the joint is swollen, but not grossly deformed, palpate both sides of the joint for tenderness over the collateral ligaments (Figure 2). Next, test stability of the joint by applying medial and lateral stresses to the joint (Figure 3). If the finger angulates under stress, there is a complete tear of the collateral ligament; if the patient has pain but no instability, assume there has been a sprain. Dislocations are obvious on inspection. With PIP dislocations, the middle phalanx most commonly dislocates dorsally in relation to the proximal phalanx.

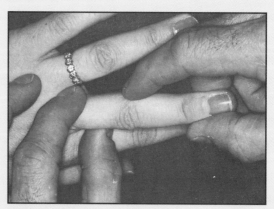

Figure 2
Palpation of the collateral ligaments

Diagnostic

Obtain plain AP and lateral radiographs to rule out a fracture in association with a dislocation (Figure 4).

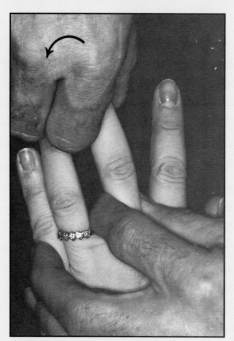

Figure 3
Applying lateral stress to the PIP joint

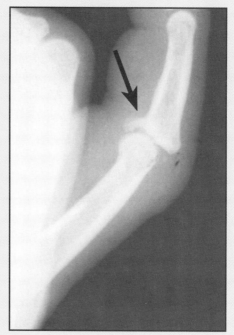

Figure 4
Fracture-dislocation of the PIP joint

Differential diagnosis

Extensor tendon rupture (boutonnière deformity, inability to extend the PIP joint)

Fracture-dislocation of the joint (fracture evident on radiographs)

Adverse outcomes of the disease

Patients may continue to have limited motion, stiffness, chronic pain, and swelling. Chronic hyperextension of the PIP joint or flexion contracture may occur.

Treatment

Closed reduction of a joint dislocation should be done under a digital block anesthetic [see Digital Anesthetic Block (Hand)]. To reduce the dislocation, grasp the distal portion of the finger and apply longitudinal traction while stabilizing the finger or hand proximal to the dislocation. Apply gentle pressure over the dorsum of the deformity to guide the reduction. After reduction, move the finger through a range of motion. If the joint seems stable, the finger can be "buddy splinted" (Figure 5) to the adjacent finger for 2 weeks, followed by active motion.

If the dislocation cannot be reduced with adequate anesthesia, there may be soft tissue interposed and open reduction may be necessary. If the joint has full range of motion after the reduction but tends to dislocate during the last 20° of extension, apply a dorsal extension block splint. This type of splint blocks the last 20° to 30° of extension (Figure 6). Use the splint for 3 weeks, then buddy splint the finger to an adjacent finger for an additional 3 weeks.

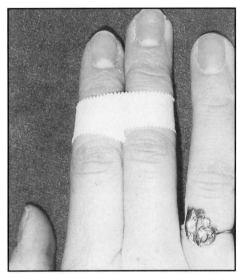

Figure 5
"Buddy splinting"

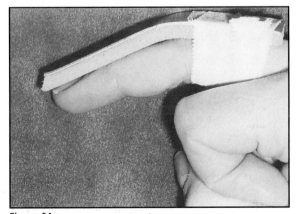

Figure 6A
Dorsal extension block splint for PIP dislocations

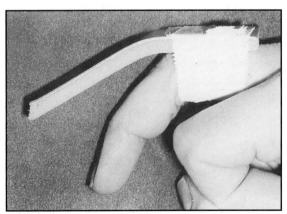

Figure 6B
Dorsal extension block splint for DIP dislocations

DIP dislocations are typically dorsal or dorsolateral and often are open. With open injuries, suspect an associated tear of the extensor tendon. After adequate digital block anesthesia, apply longitudinal traction to reduce the dislocation; open dislocations tend to be stable after reduction. Next, apply a dorsal aluminum splint over the middle and distal phalanges for 1 to 2 weeks. If the fingertip droops after the reduction and the patient cannot actively extend the distal phalanx, treat the injury as a mallet finger.

MP dislocations of the thumb are relatively common, while MP dislocations of the finger are rare. The latter are usually dorsal and usually cannot be reduced by closed means because of an interposition of the palmar plate or an entrapment of the metacarpal head between the lumbrical tendon and a flexor tendon (Figure 7).

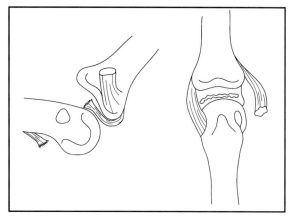

Figure 7A
Interposition of the palmar plate

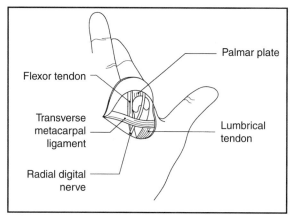

Figure 7B
Entrapment of the metacarpal head between the lumbrical and extrinsic flexor tendon

Adverse outcomes of treatment

Instability, joint stiffness, persistent hyperextension deformity, and/or residual flexion deformity can develop. Arthritis may also develop with an inadequate reduction.

Referral decisions/Red flags

Patients whose dislocations cannot be reduced are ideal candidates for open reduction. In addition, patients with fracture-dislocations and open dislocations need further evaluation. Open dislocations are best treated surgically where adequate debridement and repair can be accomplished.

Acknowledgements

Figure 1 is reproduced with permission from American Society for Surgery of the Hand: *The Hand: Examination and Diagnosis,* ed 3. New York, NY, Churchill Livingstone, p 54.

Figure 7 is reproduced with permission from American Society for Surgery of the Hand: *Hand Surgery Update.* Rosemont, IL, American Academy of Orthopaedic Surgeons, 1994, p 22.

Kienböck's Disease

Synonyms

Wrist sprain

Lunatomalacia

Osteonecrosis of the carpal lunate

Definition

Kienböck's disease, or osteonecrosis of the carpal lunate, is an idiopathic condition in which the blood supply to the lunate bone is lost, resulting in necrosis (Figure 1). This process is seen initially on plain radiographs as increased density or whiteness of the lunate bone when compared with the surrounding carpal bones. In later stages, the dead bone will fragment and collapse, resulting in generalized degenerative arthritis of the wrist (Figure 2). Patients may have a history of trauma prior to this condition, but a cause-and-effect relationship has not been established. It more commonly affects men and usually occurs between ages 20 and 40 years.

Clinical symptoms

Pain, stiffness, and diffuse swelling over the dorsal aspect of the wrist are common. The patient may complain of weakness or inability to grasp heavy objects.

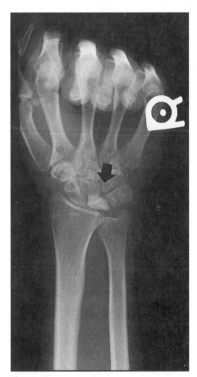

Figure 1
Radiograph showing Kienböck's disease

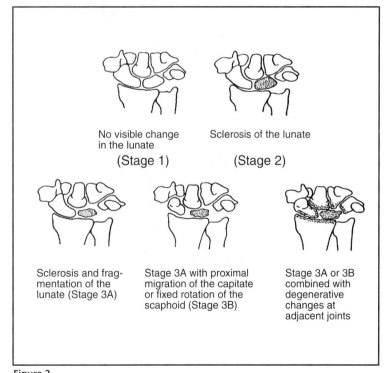

No visible change in the lunate
(Stage 1)

Sclerosis of the lunate
(Stage 2)

Sclerosis and fragmentation of the lunate (Stage 3A)

Stage 3A with proximal migration of the capitate or fixed rotation of the scaphoid (Stage 3B)

Stage 3A or 3B combined with degenerative changes at adjacent joints

Figure 2
Classification of Kienböck's disease

Tests

Exam

Examination typically reveals tenderness directly over the lunate bone (mid-dorsal wrist area, just below the radius), occasional diffuse swelling dorsally, and stiffness (limited movement) of the wrist. In addition, grip and pinch strength are usually reduced.

Diagnostic

PA and lateral radiographs of the wrist are indicated for all patients with dorsal wrist pain. Failure to obtain radiographs is the most common reason for delay in diagnosis. If initial radiographs are normal, and pain persists after 3 weeks of splinting, order a three-phase bone scan.

Differential diagnosis

Arthritis (radiograph will differentiate)

Fracture of the distal radius

Fracture of the scaphoid (navicular)

Ganglion (discrete mass)

Scapholunate dissociation (Terry Thomas sign evident on radiograph)

Tenosynovitis of the extensor tendon

Adverse outcomes of the disease

Untreated Kienböck's disease will almost always result in severe collapse and destruction of the entire wrist joint, with stiffness and decreased strength and motion. In addition, osteoarthritis of the wrist may eventually develop.

Treatment

Patients with abnormal plain radiographs need further evaluation for possible surgery.

If radiographs are normal, splint the wrist in a neutral position for 3 weeks. NSAIDs may make the patient more comfortable. If the pain persists after 3 weeks, and results of a bone scan are positive, further evaluation is indicated.

Adverse outcomes of treatment

Loss of motion, chronic pain, and decreased grip strength are all possible.

Referral decisions/Red flags

Patients whose plain radiographs show any abnormality of the lunate need further evaluation. Persistent dorsal wrist pain, despite 3 weeks of immobilization, is also an indication for further evaluation.

Acknowledgements

Figure 1 and 2 are reproduced with permission from American Society for Surgery of the Hand: *Hand Surgery Update*. Rosemont, IL, American Academy of Orthopaedic Surgeons, 1994, p 86.

Mallet Finger

Synonyms

Baseball finger

Extensor tendon injury

Definition

The most common extensor tendon injury is mallet finger deformity, caused by rupture of the extensor tendon at its insertion at the base of the distal phalanx, above the nail bed. Sometimes, instead of the tendon tearing, the injury avulses a piece of distal phalanx at the tendinous attachment (Figure 1).

Clinical symptoms

Patients cannot straighten the fingertip.

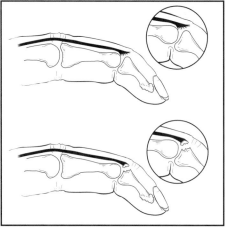

Figure 1
Mallet finger caused by rupture of the extensor tendon at its insertion (top); Mallet finger caused by avulsion of a piece of distal phalanx (bottom)

Tests

Exam
Examination reveals that the distal interphalangeal (DIP) joint is in flexion, and the patient is unable to extend the joint. The dorsal area is often tender, painful, slightly swollen, and red initially, but about 2 weeks after the injury occurs, the fingertip is usually not painful.

Diagnostic
AP and lateral radiographs may reveal a small bony avulsion from the dorsal side of the distal phalanx. The distal phalanx may have subluxated in a volar direction from the unopposed pull of the flexor tendon.

Differential diagnosis

Fracture of the distal phalanx

Intra-articular fracture

Adverse outcomes of the disease

Permanent flexion of the DIP joint is possible.

Treatment

Application of a splint on either the volar or the dorsal surface of the finger is indicated to keep the DIP joint in extension (Figure 2). With acute injuries, the splint should be worn for 6 weeks. If the injury is old, a splint should be worn for a minimum of 8 weeks.

Advise patients to maintain the DIP joint in extension when the splint is removed for cleaning. If the fingertip droops at any time after the splint is applied, the period of splinting starts over. At 4 to 5 days after the splint is applied, check the dorsal skin for maceration or pressure spots. If the joint does not come into full extension by the second visit, consider surgical pinning.

Weekly follow-up visits are helpful to monitor progress and usually lead to a better outcome than if the splint is applied and the patient seen 6 to 8 weeks later. At the end of the splinting period, if there is no extensor lag evident, guarded active flexion is started with splinting continued at night for 2 to 4 weeks.

Certain occupations make splint wear difficult, and these patients should be evaluated for possible percutaneous pinning.

Surgical treatment is appropriate if there is subluxation of the distal phalanx volarly and/or if an avulsed bony fragment involves more than one third of the joint.

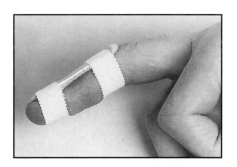

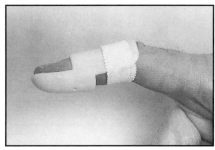

Figure 2
Types of splints used in treatment of mallet finger: dorsal aluminum splint (top), commercial splint (bottom)

Adverse outcomes of treatment

Persistent deformity associated with flexion of the fingertip is possible despite treatment.

Referral decisions/Red flags

Patients with volar subluxation of the distal phalanx and/or an avulsed bony fragment that involves more than one third of the joint surface need further evaluation.

Acknowledgements

Figures 1A and 1B are adapted with permission from Rockwood CA, Green DP, Bucholz RW, et al (eds): *Rockwood and Green's Fractures in Adults,* ed 4. Philadelphia, PA, Lippincott-Raven, 1996, vol 2, p 617.

Figures 2A and 2B are reproduced with permission from Culver JE: Office management of athletic injuries of the hand and wrist, in Barr JS (ed): *Instructional Course Lectures XXXVIII.* Park Ridge, IL, American Academy of Orthopaedic Surgeons, 1989, pp 473–482.

hand and wrist

Tenosynovitis of the Wrist

Synonyms

deQuervain's
tenosynovitis

Stenosing tenosynovitis

Definition

deQuervain's tenosynovitis is characterized by irritation or swelling of the tendons on the thumb side of the wrist (Figure 1). The inflammation thickens the tendon sheath (tenosynovium) and constricts the tendon as it glides in the sheath. This may cause a "triggering" phenomenon, with the tendon seeming to lock or "stick" as the patient moves the thumb. Typically, the first extensor compartment is involved.

Clinical symptoms

Patients report pain and swelling over the radial styloid that is aggravated by attempts to move the thumb or make a fist. They may also notice "creaking" as the tendon moves.

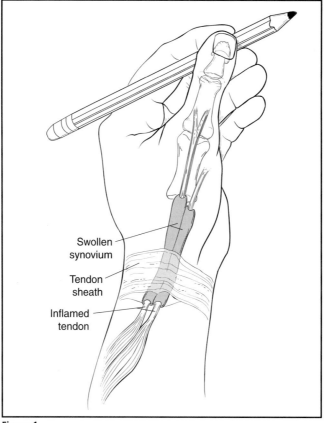

Swollen synovium

Tendon sheath

Inflamed tendon

Figure 1
deQuervain's tenosynovitis of the first extensor compartment

hand and wrist

Tests

Exam

Examination reveals swelling and tenderness over the synovial compartment in the region of the distal radius. Crepitation may be palpable as the patient moves and actively flexes and extends the digit. Full flexion of the thumb into the palm, followed by ulnar deviation of the wrist (Finkelstein's test) will produce pain and is diagnostic for deQuervain's tenosynovitis (Figure 2).

Diagnostic

Even though this is largely a clinical diagnosis, PA and lateral radiographs of the wrist should be obtained to rule out any bony pathology. Calcification associated with tendinitis can be seen on radiographs. Obtaining serum levels of uric acid may help to determine if gout is a possible cause of the tenosynovitis.

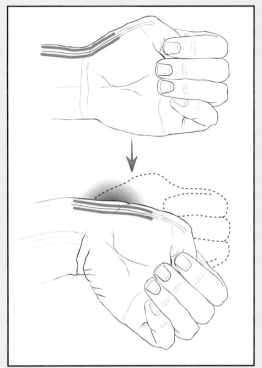

Figure 2
Finkelstein's test

Differential diagnosis

Basal joint arthritis of the thumb (positive Watson's stress test)

Dorsal ganglion

Fracture of the scaphoid (tenderness over the anatomical snuffbox)

Kienböck's disease (radiographs show lunate changes)

Radioscaphoid degenerative arthritis (tenderness just distal to the radial styloid)

Triscaphoid arthritis (evident on plain radiographs)

Adverse outcomes of the disease

Chronic pain, loss of strength, and loss of thumb motion may occur; tendon rupture is possible, but it is rare.

Treatment

Initial treatment should consist of immobilization of the thumb with a thumb spica splint that immobilizes both the wrist and thumb. A 2-week course of NSAIDs is also helpful for pain relief. If immobilization fails, inject the tendon sheath with a corticosteroid preparation (see deQuervain's Injection). The patient should have no more than three injections, then surgical treatment should be considered.

Adverse outcomes of treatment

NSAIDs may cause gastric, renal, or hepatic complications. The patient may experience some discomfort from wearing the splint. If the splint produces pressure over the radial styloid, the patient may cease wearing it. Injury to the radial sensory nerve is also possible. In addition, corticosteroids can sometimes cause subcutaneous atrophy and loss of pigmentation that can be unsightly. Infection following an injection is also a risk, but can be largely avoided by use of careful sterile technique.

Referral decisions/Red flags

Failure to respond to splinting and injection indicates the need for further evaluation.

Acknowledgements

Figure 2 is adapted with permission from the American Society for Surgery of the Hand: Brochure: deQuervain's Stenosing Tenosynovitis. Englewood, CO, 1995.

procedure

deQuervain's Injection

The extensor and abductor tendons to the thumb cross at a point approximately 2″ proximal to the radial styloid.

Materials

Sterile gloves

Skin preparation solution (iodinated soap or similar antiseptic solution)

3-mL syringe with a 25-gauge, 7/8-inch needle

3-mL syringe with a 27-gauge, 7/8-inch needle

2 mL of 40 mg/mL corticosteroid preparation

2 mL of a 1% local anesthetic

Adhesive bandage

Step 1

Wear protective gloves at all times during this procedure and use sterile technique.

Step 2

Cleanse the skin with iodinated soap or similar antiseptic solution.

Step 3

Insert the 27-gauge needle at a 45° angle to the skin in line with the two tendons (Figure 1).

Step 4

Create a skin wheal with 0.5 mL of 1% local anesthetic and advance the needle until it strikes one of the underlying tendons. Inject the remaining anesthetic while slowly withdrawing the needle, until the anesthetic flows freely.

Step 5

Wait 2 to 3 minutes, then insert the 25-gauge needle. Ask the patient to move the thumb. If the needle moves with the thumb, pull the needle back out of the tendon 1 mm. Inject 2 mL of corticosteroid preparation into the tendon sheath.

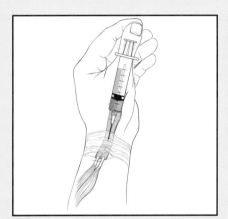

Figure 1
Location for needle insertion

procedure

deQuervain's Injection (continued)

Step 6

Dress the puncture wound with a sterile adhesive bandage.

Adverse outcomes

Although rare, infection is possible. Subcutaneous fat atrophy may follow subcutaneous infiltration of the cortisone, leading to a waxy appearing depression in the skin.

Aftercare/patient instructions

Apply a thumb spica splint for night use. This is often helpful for the first week following injection. Increased pain is not uncommon following the injection, especially in the first 24 to 48 hours. At least 33% of patients will experience increased discomfort.

Trigger Finger

Synonyms

Tendinitis

Locked finger

Stenosing tenosynovitis

Digital flexor tenosynovitis

Definition

Trigger finger is an inflammatory process of the flexor tendon sheath that most commonly affects the thumb, ring, and long fingers and is associated with repetitive trauma. Patients older than age 40 years and those who have a history of diabetes and rheumatoid arthritis are most commonly affected.

A normal tendon glides back and forth under a restraining pulley. When a nodule develops in the tendon, it passes under the pulley as the finger flexes but becomes "stuck" on the palmar side of the pulley as the finger extends. The finger is then "locked" in flexion (Figure 1).

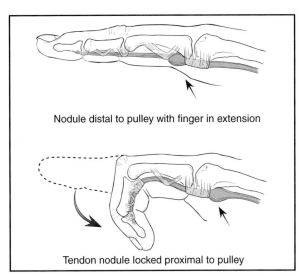

Nodule distal to pulley with finger in extension

Tendon nodule locked proximal to pulley

Figure 1

Nodule or thickening in flexor tendon, which strikes the proximal pulley, making finger extension difficult

Clinical symptoms

Patients typically report pain and "catching" when they flex their finger. They may awaken with the finger locked in the palm, which gradually unlocks as the day progresses. The proximal interphalangeal (PIP) joint of the finger or the interphalangeal (IP) joint of the thumb, rather than the MP joint, is typically identified as the site of the pain. However, some patients have a painful nodule in the distal palm, usually at the level of the distal flexion crease, yet no history of triggering. Other patients' only symptoms are swelling and/or stiffness in the fingers, particularly in the morning. Patients with diabetes mellitus may have several fingers involved.

hand and wrist

hand and wrist

Tests

Exam
Examination reveals a tender nodule palpable in the palm at the level of the distal palmar crease, usually overlying the MP joint. The nodule may move and the finger may lock when the patient flexes and extends the affected finger. This maneuver is almost always painful for the patient. Fusiform swelling of the finger may also be noted.

Diagnostic
This is a clinical diagnosis; radiographs are not needed.

Differential diagnosis

Anomalous muscle belly in the palm

Diabetes mellitus (single and multiple trigger fingers)

Dupuytren's disease

Ganglion of the tendon sheath

Rheumatoid arthritis

Adverse outcomes of the disease

Flexion contracture of the PIP joint may develop, as may loss of extension or flexion.

Treatment

Initial treatment can involve splinting the MP joint in extension for 10 to 14 days and a short course of NSAIDs or injection of corticosteroid into the tendon sheath (see Trigger Finger Injection). If symptoms persist, a second injection in 3 to 4 weeks is indicated. Since patients with rheumatoid disease are already at increased risk for tendon rupture, only one injection is indicated for these patients before surgical release should be considered. If two injections fail to solve the problem, consider surgical release.

Adverse outcomes of treatment

NSAIDs may cause gastric, renal, or hepatic complications. Repeated corticosteroid injections may lead to rupture of the flexor tendon. Injections can also injure the digital sensory nerve. Infection is also a risk, but can be largely avoided by use of careful sterile technique.

Referral decisions/Red flags

Failure of conservative treatment, development of contractures in the PIP joint, and/or a locked finger (in flexion or extension) indicate the need for further evaluation. Patients with rheumatoid arthritis whose problem does not resolve after a single injection also need additional evaluation.

hand and wrist

Acknowledgements

Figure 1 is adapted with permission from the American Society for Surgery of the Hand: Brochure: Trigger Finger. Englewood, CO, 1995.

procedure

Trigger Finger Injection

The flexor tendons pass beneath a pulley situated just distal to the distal palmar crease. Palpating this area as the patient flexes and extends the finger reveals a lump that catches as it passes beneath the pulley.

Materials

Sterile gloves

Skin preparation solution (iodinated soap or similar antiseptic solution)

3-mL syringe with a 27-gauge, 3/4-inch needle

2 to 3 mL of a 1% local anesthetic (without epinephrine)

1-mL syringe with a 25-gauge needle

1 mL of a 40 mg/mL corticosteroid preparation

Adhesive dressing

Step 1

Wear protective gloves at all times during this procedure and use sterile technique.

Step 2

Cleanse the palm with iodinated soap or similar antiseptic preparation.

Step 3

Identify the lump on the tendon and infiltrate the skin at the distal palmar crease, which directly overlies the tendon.

Step 4

Use the 27-gauge needle to inject 2 mL of a 1% anesthetic solution (Figure 1).

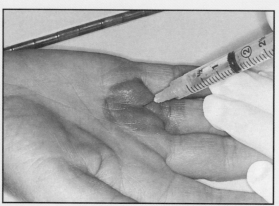

Figure 1
Location for needle insertion

Trigger finger injection (continued)

Step 5

Change to the 25-gauge needle/syringe combination filled with 1 mL of corticosteroid preparation. Insert the needle, bevel toward the palm, through the same needle tract (Figure 2). Continue to insert the needle as the patient moves the affected finger through a small arc of flexion and extension. When the needle touches the moving tendon, there will be a scratchy sensation. If the needle moves, it has penetrated the tendon and should be partially withdrawn until the scratchy sensation occurs. At this point, the needle tip is inside the flexor tendon sheath and external to the tendon.

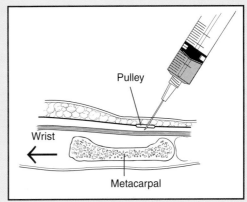

Figure 2
Insert the needle through the pulley

Step 6

Inject the corticosteroid preparation.

Step 7

Dress the puncture wound with a sterile adhesive bandage.

Adverse outcomes

Injection of corticosteroid into the subcutaneous tissues may lead to local fat atrophy and a tender, unsightly depression beneath the skin.

Aftercare/patient instructions

Advise the patient that there may be significant discomfort for 1 to 2 days following any injection of a corticosteroid. Also, the finger may be numb for 1 to 2 hours until the local anesthetic wears off. Instruct the patient to return to your office if swelling, redness, or inordinate pain occur. The patient should be able to use the finger in a normal fashion after the injection.

hand and wrist

Tumors of the Hand and Wrist

hand and wrist

Synonyms

Ganglion cyst

Flexor sheath ganglion

Mucoid cyst

Epidermal inclusion cyst

Fibroma

Enchondroma

Giant cell tumor of the tendon sheath

Fibroxanthoma

Pigmented villonodular synovitis

Glomus tumor

Squamous cell carcinoma

Subungual melanoma

Definition

More than 90% of tumors in the hand and wrist are benign, as primary malignant tumors and skeletal metastases below the elbow are rare. Ganglia are the most common soft-tissue tumors of the hand and wrist, followed by giant cell tumors and epidermal inclusion cysts. Enchrondromas are the most common primary neoplasms of the bones of the hand and are benign. Squamous cell carcinomas are the most common malignant neoplasm of the hand.

Clinical symptoms

Many tumors of the hand are painless; the mass is often present for some time but does not cause problems until the patient sustains some type of accidental trauma. Masses located near joints may become painful because of joint motion.

Tests

Exam

Careful examination is necessary to pinpoint the position and characteristics of the mass, as these factors help to narrow the diagnosis considerably.

A ganglion cyst is characterized as a mass located over the dorsal or volar radial aspect of the wrist, the flexion crease of the finger at the level of the web space, or over the top of the DIP joint of the finger. Epidural inclusions cysts typically occur around the end of the finger or over the end of a finger amputation stump. Inclusion cysts are commonly located on the flexor surface of the fingers or palm and are commonly fixed with any attempt to move them. Pressing a small flashlight against the skin over an inclusion cyst will not transilluminate the mass; this same maneuver will transilluminate a ganglion cyst. Cysts located near the fingertip will often cause the nail plate and bed to arch upward.

A giant cell tumor is characterized by a multi-nodular, firm mass located around a finger. If a blue or red area is visible under the finger nail, consider a glomus tumor, a subungual hematoma, or a foreign body. Increased pain with pressure over the nail or immersion of the fingertip in cold water supports the diagnosis of a glomus tumor.

Lipoma is a possibility if the mass is located in the thenar area of the hand, especially if the mass is well defined, superficial, soft, and nontender on palpation. However, if the mass is located anywhere in the palm or flexor surface of the wrist, conduct a careful neurologic examination, as a lipoma can cause nerve entrapment particularly at the ulnar nerve and motor branch of the median nerve.

Symptomatic enchondroma is characterized by tenderness to palpation over the involved phalanx (usually the proximal) most likely due to the presence of a pathologic fracture. In addition, the finger will appear swollen.

A hard mass at the base of the second and third metacarpals is called a carpal boss, which is a hard prominence composed of dorsal bone projections off these two metacarpals, the trapezoid, and trapezium. A ganglion is sometimes associated with a carpal boss.

Diagnostic
Plain PA and lateral radiographs of the involved finger or hand should be obtained to pinpoint the diagnosis. Any bony involvement or calcification within the tumor is easily seen from these studies. If the diagnosis is not clear from the history, examination, and plain radiographs, consider an MRI scan even though it is rarely needed to make a correct diagnosis.

Differential diagnosis

See Figure 1 and Table 1 for a complete listing.

Adverse outcomes of the disease

With enchondroma, fracture can occur. Nerve compression can develop as a result of lipoma. Nail changes, skin atrophy, and infection can develop as a result of a mucoid cyst. Drainage is a problem associated with epidermal mucoid cysts. Ganglions can result in limited joint motion. Patients with giant cell tumors can have limited tendon function due to peritendinous adhesions.

Treatment

Treatment is based on the diagnosis. Histologic examination and surgical excision are required for most expanding symptomatic masses to confirm the diagnosis.

Adverse outcomes of treatment

The recurrence rate of giant cell tumors is high following surgical excision. Ganglia may recur at the same site in 5% to 10% of patients. Joint stiffness may develop following treatment of pathologic fractures due to enchondromas.

Referral decisions/Red flags

Patients with a painful or expanding mass, one that interferes with function, or one believed be malignant, need further evaluation.

Table 1
Common benign tumors of the hand and wrist

Type of tumor*	Location	Patient age and sex	Signs and symptoms	Radiographic findings
Giant cell tumor	Digits on palmar surface	> 30 years; ratio of men to women 2:3	Slowly enlarging painless mass	20% show cortical erosion
Epidermal inclusion cyst	Fingertip or anywhere from penetrating injury	Teens to middle age; more common in men	Painless, slow growing; does not transilluminate	Round soft-tissue mass, also in distal phalanx
Glomus tumor	50% occur under fingernail	30 to 50 years; ratio of women to men 2:1	Triad of symptoms: marked pain, cold intolerance, very tender, blue discoloration of nail	Some show erosion on lateral view
Lipomas	Thenar area in palm and first web space	30–60 years; slight predominance in women	Painless, slow growing; may cause nerve entrapment	No bony involvement, soft-tissue mass
Enchondroma	In proximal phalanges	10–60 years; affects men and women equally	May become painful after trauma due to fracture	Radiolucent expansive lesion, cortex thin, may see a fracture and areas of calcification
Carpal boss	Base of second and third metacarpals	30–50 years; more common in women	Prominence due to a bony ossicle, a ganglion, and/or bone spur	Oblique lateral view will show bone spur and/or ossicle

Ganglia
(See Ganglion of the Finger, p. 224, and Ganglion of the Wrist, p. 227.)

* See Figure 1.

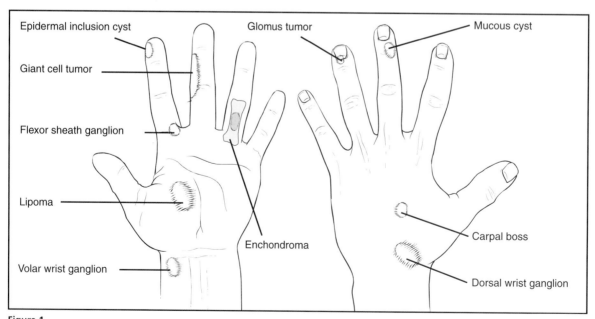

Figure 1
Typical locations and types of benign hand tumors

Ulnar Collateral Ligament Tear

Synonyms

Gamekeeper's thumb

Skier's thumb

Definition

The ulnar collateral ligament of the thumb is the prime stabilizer of the metacarpophalangeal (MP) joint of the thumb. When this ligament is torn, the thumb deviates outward or radially when pinching with the thumb and index finger (Figure 1).

The eponym "gamekeeper's thumb" comes from the practice of early European gamekeepers who used to kill game by grabbing it by the neck between the thumb and the index finger. Over time, the ulnar collateral ligaments would stretch, leaving the gamekeepers with chronically unstable thumbs; this condition is perhaps one of the first work-related diseases. Today the most common mechanism of injury is from a ski pole injury.

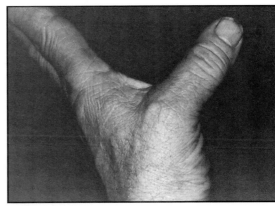

Figure 1
Unstable thumb due to tear of the ulnar collateral ligament

Clinical symptoms

Patients almost always remember the instant of injury. Frequently, the thumb is forced radially at the MP joint from the strap on a ski pole or from the pole itself during a skiing accident. Pain and swelling occur on the inside (ulnar side) of the thumb at the MP joint. With attempted pinch, the thumb deviates to the radial side, and pinch strength is markedly decreased.

Tests

Exam
Examination reveals swelling and discoloration over the ulnar side of the MP joint of the thumb, and gentle palpation elicits tenderness. The instability is easily appreciated when gentle pressure is applied to the thumb by pushing it away from the palm. The opposite, uninjured thumb should be examined for comparison.

hand and wrist

The Essentials of Musculoskeletal Care 257

hand and wrist

Diagnostic

Obtain plain radiographs of the thumb to rule out a fracture. If present, an avulsion fracture off the base of the proximal phalanx can be seen, but this is not common. The fragment is most commonly attached to the end of the ulnar collateral ligament, and if the bone fragment is minimally displaced (less than 3 mm), the tear will heal with splinting alone.

Next, obtain stress radiographs of the MP joint of the thumb to determine if the tear is complete (Figure 2). If the patient is too uncomfortable to undergo the stress test, instill 2 mL of 1% local anesthetic into the joint before doing the stress test. The radiograph can then be obtained by holding the thumb metacarpal with one hand and by

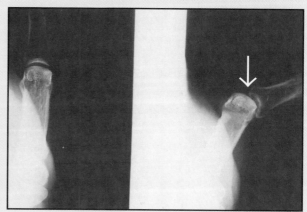

Figure 2
Radiograph of unstressed thumb (left), radiograph of stressed thumb in radial deviation (right) (Over 30° of angulation suggests a complete ligament tear)

applying a radially directed force to the thumb tip with the opposite hand while the radiograph is taken. A stress view of the uninjured thumb can be obtained for comparison if there is any question about the degree of opening.

Differential diagnosis

Dislocation of the thumb (obvious deformity)

Fracture of the thumb (evident on radiograph)

Adverse outcomes of the disease

Weakness, pain, and instability in pinch can occur. Arthritis of the MP joint may also occur.

Treatment

With complete tears, the ulnar collateral ligament tends to be trapped behind the sagittal band, which prevents it from reapproximating to its site of origin. These lesions will not heal, and surgical repair is the preferred treatment.

With incomplete tears, place the thumb in a thumb spica splint with the MP joint flexed 20°. After 3 weeks of immobilization, initiate flexion and extension exercises out of the splint, but continue the protection of the splint for another 2 to 3 weeks until swelling and tenderness have subsided.

Adverse outcomes of the treatment

Nonsurgical treatment of a complete lesion can result in weakness of pinch.

Referral decisions/Red flags

Patients with complete tears of the ulnar collateral ligament should be evaluated for possible surgical treatment.

Ulnar Nerve Entrapment at the Wrist

hand and wrist

Synonyms

None

Definition

Entrapment of the ulnar nerve at the wrist is rare, usually the result of a space-occupying lesion such as a lipoma, ganglion, or ulnar artery aneurysm (Figure 1). Repetitive trauma, such as operating a jackhammer, may cause ulnar neuropathy at the wrist. Ulnar nerve entrapment at the wrist is less common than ulnar nerve entrapment at the elbow.

Clinical symptoms

Patients typically have no pain, but they often report weakness or numbness. These symptoms may be due to direct pressure on the hypothenar eminence as a result of the patients' occupation, such as operating a jackhammer or using the bone of the hand as a "hammer."

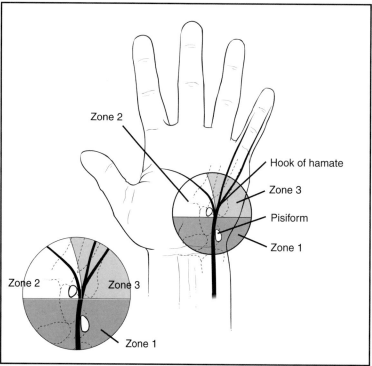

Figure 1

Distal ulnar tunnel showing the three zones of entrapment. Lesions in zone 1 give motor and sensory symptoms; lesions in zone 2 cause motor deficits; and lesions in zone 3 create sensory deficits

Tests

Exam

Ulnar neuropathy at the wrist is detectable primarily with Tinel's sign, sensory loss at the tip of the small finger, atrophy of the hypothenar and intrinsic muscles, or motor weakness of the ulnar innervated intrinsics (finger spreaders) (Figure 2). In some patients, only the motor branch of the ulnar nerve may be affected, sparing the sensory branches; however, if there is sensory involvement, tapping over the ulnar nerve in the hypothenar region will produce tingling in the ring and little fingers (Tinel's sign). Sensation over the dorsal and ulnar aspects of the hand is normal. If the ulnar nerve is involved at the elbow, almost all patients will have both sensory and motor involvement, with numbness over the dorsal and ulnar side of the hand. Generally, this condition is not painful.

Diagnostic

Nerve conduction studies may be abnormal and may differentiate ulnar entrapment at the wrist from the more common entrapment at the elbow.

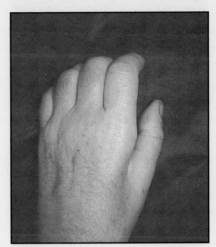

Figure 2
Instrinsic muscle wasting

Differential diagnosis

Carpal tunnel syndrome

Cervical (C7–C8) radiculopathy

Peripheral neuropathy (from diabetes, alcoholism, or hypothyroidism)

Pisotriquetral arthritis

Thoracic outlet syndrome

Ulnar artery thrombosis in the hand

Ulnar neuropathy at the elbow

Adverse outcomes of the disease

Loss of intrinsic muscle function is a serious impairment, producing decreased grip strength and pinch. Sensory loss, when present, involves the ring and little fingers.

Treatment

Since the usual cause of ulnar entrapment at the wrist is extrinsic compression (lipoma, ganglia, tumors, etc), treatment is usually surgical. If the obvious cause is external pressure, such as resting the hypothenar area on a keyboard or desk, then use of padding or a change in position may help.

Adverse outcomes of treatment

Postoperative infection and/or persistent symptoms are possible.

Referral decisions/Red flags

Patients with ulnar weakness and neuropathy need further evaluation.

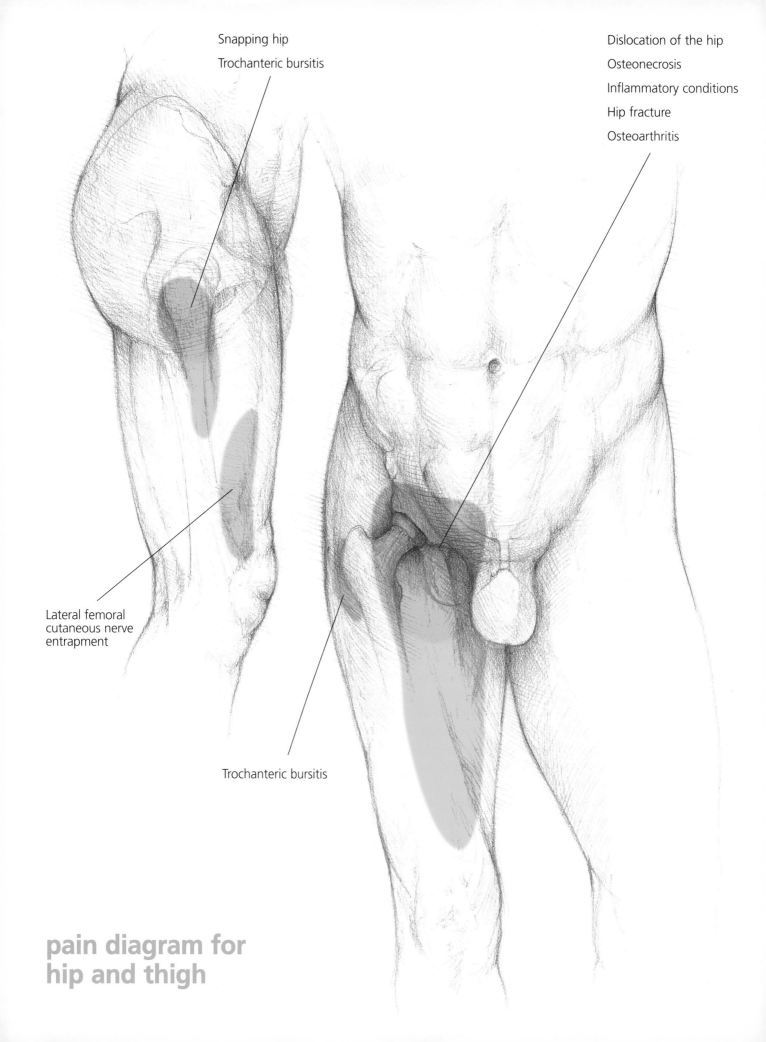

Snapping hip

Trochanteric bursitis

Dislocation of the hip

Osteonecrosis

Inflammatory conditions

Hip fracture

Osteoarthritis

Lateral femoral
cutaneous nerve
entrapment

Trochanteric bursitis

**pain diagram for
hip and thigh**

section five

hip and thigh

Section Editor

Jay R. Lieberman, MD
Assistant Professor of Orthopaedic Surgery
UCLA Medical Center
Los Angeles, California

Daniel J. Berry, MD
Consultant, Orthopaedic Surgery
Assistant Professor, Orthopaedics
Mayo Clinic
Rochester, Minnesota

James V. Bono, MD
Assistant Clinical Professor of Orthopaedic Surgery
Tufts University School of Medicine
Staff Orthopaedic Surgeon
New England Baptist Hospital
Boston, Massachusetts

J. Bohannon Mason, MD
Charlotte Hip and Knee Center
Charlotte, North Carolina

hip and thigh—
an overview

This section of *Essentials of Musculoskeletal Care* focuses specifically on common conditions that affect the hip and thigh. When evaluating patients with hip pain, you can generally assume the pain comes from one of five likely sources: 1) the hip joint; 2) the soft tissues around the hip and pelvis; 3) the pelvic bones; 4) the sacroiliac joint; and 5) referred pain from the lumbar spine. Diagnosing pathology involving the hip joint and pelvis is often possible with a careful history and physical examination. In some cases, plain radiographs, a bone scan, or even an MRI scan will be required.

Radiographic examination of the hip should include AP pelvis and AP and lateral radiographs of the hip with the patient in a supine position. All radiographs should be carefully evaluated for changes in the bony architecture, including nondisplaced fractures, and lytic and blastic lesions. The joint spaces (hip and sacroiliac) should be carefully assessed for narrowing.

Type of pain

The hip joint refers specifically to the ball-and-socket joint that consists of the femoral head articulating with the acetabulum (Figure 1). However, patients who report hip pain will most often point to the lateral aspect of the proximal thigh, the buttock, or the groin. In general, when thinking about hip problems, remember that the hip joint is just one part of the pelvic girdle. The pelvic girdle is composed of three different bones (the ilium, pubic ramus, and sacrum) and contains two different joints (the hip joint and the sacroiliac joint).

The hip joint is mobile; therefore, pathology affecting this joint will often manifest itself as either pain with ambulation, weight bearing, or limited motion. In contrast, the sacroiliac joint is practically immovable. Although it may be involved in various pathologic conditions, the sacroiliac joint is less likely to restrict function or motion.

Specific problems with the hip joint include osteoarthritis, osteonecrosis, inflammatory conditions, and fractures and dislocations. Other problems with the hip joint include developmental dysplasia, infection, and rheumatoid arthritis.

Problems involving the bony pelvis include stress fractures, tumors, and a number of inflammatory conditions such as seronegative arthritides that occur with irritation of the sacroiliac joint.

Problems involving both the hip joint and bony pelvis will often manifest as pain in the groin, buttock, or lateral thigh.

hip and thigh

Conditions affecting the soft tissues around the hip are also included in this section, specifically trochanteric bursitis, lateral femoral cutaneous nerve impingement, and snapping hip syndrome. These patients will usually have pain on the lateral aspect of the hip.

Finally, pathology in the lumbar spine can present itself as referred pain to the region of the hip. This pain will most typically be referred to the buttock and, at times, will radiate into the thigh or even the lower leg. Patients with disk herniation, facet joint arthritis, and spinal stenosis may have pain specifically around the buttock. However, these patients usually do not have groin pain, significant discomfort with either internal or external rotation of the hip, or limited range of motion of the hip joint.

Gait

A brief examination of the patient's gait can be very helpful in making a diagnosis. Simply have the patient walk up and down the hall several times at a brisk pace. An abductor or gluteus medius lurch is manifested by a lateral shift of the body to the weightbearing side with ambulation. This type of gait often occurs in patients who have intra-articular hip pathology (osteoarthritis, inflammatory arthritis, or osteonecrosis of the hip).

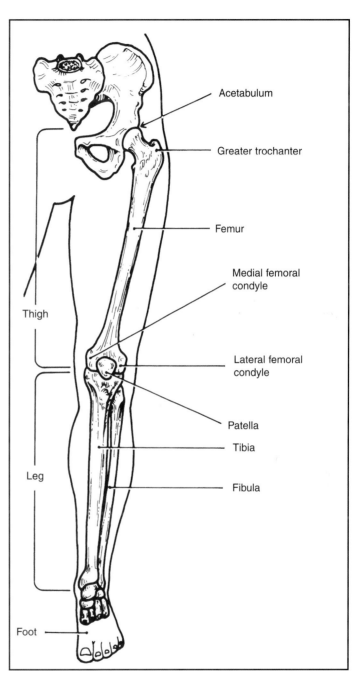

Figure 1
Bones of the hip and leg

hip and thigh—
physical exam

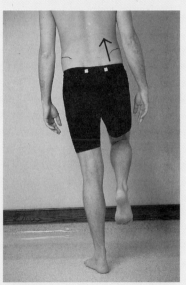

IA

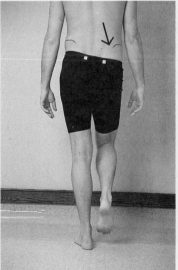

IB

Figure I
Trendelenburg test

With the patient standing, ask the patient to raise one knee. Watch the pelvis. If it elevates on the raised knee side, the test result is normal and indicates adequate hip abductor strength on the straight knee side (A). If the pelvis drops on the raised knee side, the test result is abnormal and indicates weakness in hip abductors on the straight knee side (B). These patients will have an associated limp, and usually sway their body over the weak hip.

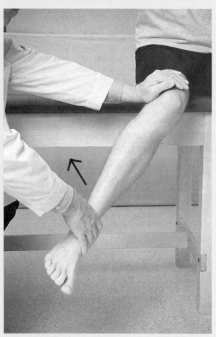

2

Figure 2
Internal rotation

With the patient seated, stabilize the knee and move the foot away from the midline to internally rotate the femur. With restricted unilateral motion, one hip will stop moving and the other will continue. With internal femoral torsion, internal rotation will exceed external rotation by 30° or more.

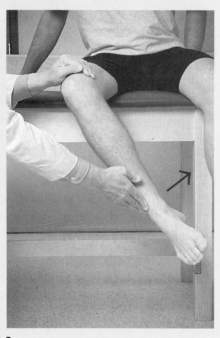

3

Figure 3
External rotation

With the patient seated, stabilize the knee and move the foot toward the midline to externally rotate the femur. With restricted unilateral motion, one hip will stop moving and the other will continue. With external femoral torsion, external rotation will exceed internal rotation by 30° or more.

hip and thigh

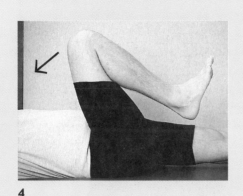

4

Figure 4
Flexion

With the patient supine, ask the patient to grasp the bent knee and pull it toward the chest. Have the patient stop when the back flattens or when the opposite hip starts to flex as the pelvis moves. Measure this as hip flexion.

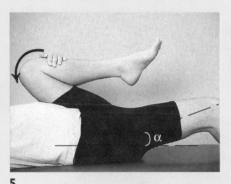

5

Figure 5
Flexion contracture

Ask the patient to grasp the opposite knee and pull it toward the chest. Once the back is flattened against the examination table, record the amount of flexion.

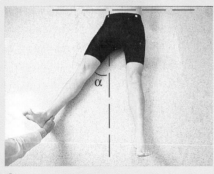

6

Figure 6
Abduction range of motion

With the patient supine, hold the ankle and gently abduct the leg. Do not allow the pelvis to tilt. You may want to stabilize the pelvis by placing your hand on the opposite anterior superior iliac spine.

hip and thigh

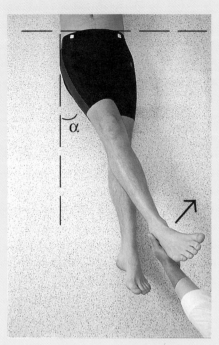

7

Figure 7
Adduction range of motion

With the patient supine, identify the anterior superior or iliac spine by palpation or marking. Adduct the leg and stop when the pelvis begins to tilt. Measure at this point.

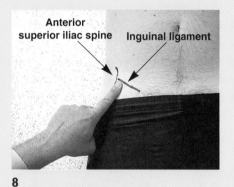

Anterior superior iliac spine   Inguinal ligament

8

Figure 8
Meralgia

With the patient supine, press immediately medial to the anterior superior iliac spine at the level of the inguinal ligament to elicit tenderness. Maintain pressure there to reproduce distal thigh numbness or paresthesias. Sensory hypesthesia over the distal lateral thigh coupled with burning are more reliable indicators of entrapment of the lateral femoral cutaneous nerve.

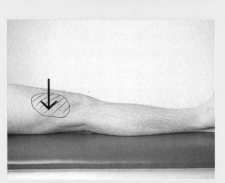

9

Figure 9
Meralgia paresthetica

Check for sensory hypesthesia to light touch or pinprick over the distal lateral thigh.

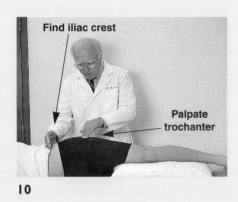

Find iliac crest

Palpate trochanter

10

Figure 10
Trochanteric bursitis

With the patient lying on the unaffected side, press over the trochanter to elicit tenderness in the trochanteric bursa. Tenderness at the proximal tip of the trochanter may indicate a tendinitis of the gluteus medius tendon; the patient may also have a limp. Tenderness at the posterior margin of the trochanter may indicate tendinitis of the external rotator muscles. In obese patients, identify the iliac crest with one hand to assist in locating the trochanter, which lies about 8″ below the pelvic brim.

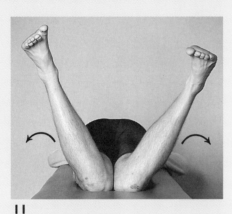

11

Figure 11
Internal rotation

With the patient prone, flex the knees (keeping them together) and move the feet apart. With restricted unilateral motion, one hip will stop moving and the other will continue. With internal femoral torsion, internal rotation will exceed external rotation by 30° or more.

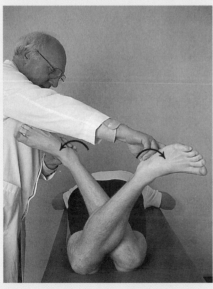

12

Figure 12
External rotation

With the patient prone, flex the knees (keeping them together) and move the feet together, crossing the legs. With restricted unilateral motion, one hip will stop moving and the other will continue. With external femoral torsion, external rotation will exceed internal rotation by 30° or more.

hip and thigh

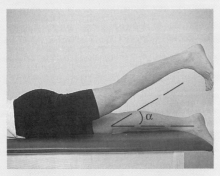

13

Figure 13
Extension

With the patient prone, ask the patient to raise the entire leg from the table. This test can also be done with the patient standing. Measure the angle formed by the leg and the table.

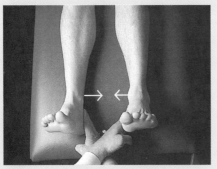

14

Figure 14
Adductor strength

With the patient supine, ask the patient to keep the feet together while you attempt to spread the legs apart. Asymmetrical strength will usually cause the pelvis to tilt, or one leg to move more quickly apart than the other.

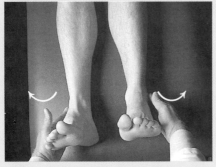

15

Figure 15
Abductor strength

This test is best done with the Trendelenburg test, but can be done with the patient supine. With the patient's feet together, resist the patient's effort to spread the legs apart.

hip and thigh

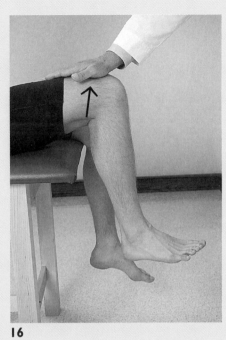

16

Figure 16
Flexor strength

With the patient seated, ask the patient to flex the hip as you resist the effort.

Dislocation of the Hip

Synonyms

Posterior dislocation of the hip

Anterior dislocation of the hip

Definition

The hip is a ball-and-socket joint, and a dislocation occurs when the femoral head (ball) is displaced from the acetabulum (socket). The usual direction of dislocation is posterior, leaving the hip adducted, flexed, and internally rotated. Anterior dislocations place the hip in abduction, flexion, and external rotation.

Clinical symptoms

Patients will have pain, be unable to move the lower extremity, and may also experience numbness throughout the limbs. The causative injury is usually high-energy trauma, such as a motor vehicle accident, industrial trauma, or fall from a height. Patients often have multiple injuries and may be unconscious from associated head trauma.

Tests

Exam

An associated ipsilateral fracture of the femur will alter the classic findings in hip dislocations.

Posterior hip dislocations are most common. The classic presentation is severe pain with the hip in a fixed position of flexion, internal rotation, and adduction (Figure 1). Evaluate sciatic nerve function (8% to 20% incidence of injury) in conscious patients by asking them to actively move their toes and ankle joints up and down and by assessing sensation on the plantar and dorsal aspects of the foot.

With anterior dislocations, the hip assumes a position of mild flexion, abduction, and external rotation. Evaluate femoral nerve function with quadriceps contraction, thigh sensation, and knee extension, if possible.

Abrasions or swelling about the knee may indicate significant knee ligament injury since most hip dislocations usually occur from a direct blow to the flexed hip and knee.

Figure 1
The clinical appearance of a posterior dislocation of the right hip

Diagnostic

Obtain AP pelvis, groin, or frog lateral views of the hip, and AP and lateral views of the femur to include the knee. The femoral heads should appear symmetric in size, and the joint spaces should be symmetric when comparing the left and right hips. When a posterior hip dislocation has occurred, the affected femoral head will appear smaller on the AP radiograph than the normal hip (Figure 2); the femoral head will appear larger than the opposite hip with an anterior hip dislocation. Rule out the presence of a femoral neck fracture prior to any manipulative reduction. With some dislocations, a piece of the femoral head may have been sliced off at the time of dislocation and will appear as a free fragment in the acetabulum. The posterior rim of the acetabulum may have been fractured as the head exited the socket. A CT scan is necessary after reduction to rule out the presence of bony fragments in the joint.

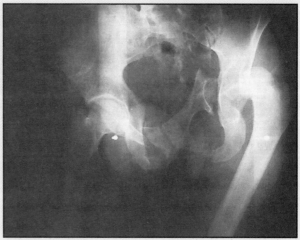

Figure 2
Posterior dislocation of the left hip with associated fractures of the acetabulum and the pubic rami

Differential diagnosis

Central fracture-dislocation of the hip

Fracture of the hip or shaft of the femur

Fracture of the posterior acetabular rim

Adverse outcomes of the disease

Osteonecrosis may occur (in 10% of patients) since the dislocation tears the hip capsule and alters the blood supply to the femoral head. Delay in reduction increases the risk of osteonecrosis. Posttraumatic arthritis, chondrolysis (dissolution of articular cartilage), limp, thrombophlebitis (pulmonary embolism or deep vein thrombosis), sciatic or femoral nerve injury, and chronic pain can also occur.

Treatment

A hip dislocation is an orthopaedic emergency. After confirming the absence of associated fractures or intra-articular loose bodies, the dislocation is reduced under IV sedation by longitudinal traction in the position of deformity. It is essential that reduction be performed in an atraumatic fashion to avoid fracture of the femoral head or the acetabulum. If this fails, the patient will require a general anesthetic and possibly an open reduction. A CT scan may be necessary to identify intra-articular bony fragments or an occult acetabular fracture. Repeat radiographs are needed following reduction to identify unrecognized intra-articular bony fragments and to confirm reduction. Loose bony fragments require surgical excision, or reduction and internal fixation. Persistent instability is either from tissue (bony or soft) interposed in the joint, or bony deficiency of the acetabulum or femoral head from an associated fracture. Evaluate sciatic nerve function (posterior dislocation) or femoral nerve function (anterior dislocation) following reduction.

The postreduction treatment of a dislocated hip is early crutch-assisted ambulation with weightbearing as tolerated until patients are free of pain, usually 2 weeks after the injury. Patients should start hip abduction and extension exercises and use a walking aid in the hand opposite the involved hip until they walk without a limp.

Adverse outcomes of treatment

Redislocation, infection, posttraumatic arthritis, and/or a missed loose body or fracture are all possible. A dislocation that is missed for approximately 3 to 4 months may still be salvaged by an open reduction.

Referral decision/Red flags

Patients with any associated fracture, dislocation, open dislocation, or associated nerve palsy should be evaluated further. Immediate consultation is recommended if the physician is unfamiliar with techniques of reduction or assessment.

Acknowledgements

Figure 1 is reproduced with permission from Heckman JD (ed): *Emergency Care and Transportation of the Sick and Injured,* ed 4. Park Ridge, IL, American Academy of Orthopaedic Surgeons, 1987, p 208.

Figure 2 is reproduced with permission from Kasser JR (ed): *Orthopaedic Knowledge Update 5.* Rosemont, IL, American Academy of Orthopaedic Surgeons, 1996, pp 365–378.

hip and thigh

Fracture of the Hip

Definition

Hip fractures are a common problem among the elderly (primarily women) and generally involve either the femoral neck region (above the intertrochanteric line) or the intertrochanteric region. Both types occur with approximately the same frequency and affect the same patient population. The frequency of hip fractures in general doubles with each decade beyond age 50 years. Treatment is determined primarily by location (femoral neck or intertrochanteric area) and by type (displaced or nondisplaced).

Clinical symptoms

Most patients report a fall followed by the inability to walk. A few can walk with assistance (crutches, cane, or walker), but have groin or buttock pain on weightbearing that seems to get worse as they walk. Occasionally, patients report pain referred to the knee. Elderly patients with hip pain after a fall should be treated as if they have a hip fracture until proven otherwise. Decreased proprioceptive function and loss of protective responses increase the likelihood that patients will strike the ground with the lateral thigh and hip region first when they fall. Dizziness, stroke, syncope, peripheral neuropathies, medication changes, or other conditions that may compromise balance predispose patients to hip fractures and must be investigated as part of the fracture evaluation.

Age is the single strongest risk factor for a hip fracture. Caucasian women are two to three times more likely to be affected than African-American or Hispanic women. Other risk factors include living in an urban area, sedentary lifestyle, alcoholism, use of psychotropic medication, and dementia.

Tests

Exam

The classic presenting position for patients with a femoral neck fracture or an intertrochanteric fracture is lying down with the limb externally rotated, abducted, and shortened. Patients with stress fractures or nondisplaced fractures of the femoral neck may have no obvious deformity. Attempts to roll the limb (like a rolling pin) are painful. With few exceptions, patients either report pain with walking or are unable to walk.

Diagnostic

AP pelvis and groin lateral radiographs are usually adequate to diagnose hip fractures (Figure 1). Patients may have hip pain and a history consistent with hip fracture, but negative radiographs. When it does occur, an MRI scan is useful to demonstrate acute occult hip fractures (Figure 2).

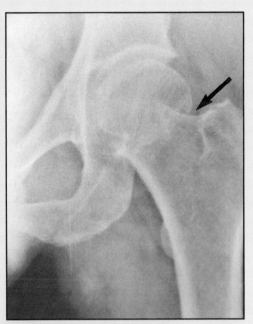

Figure 1
An impacted fracture of the femoral neck

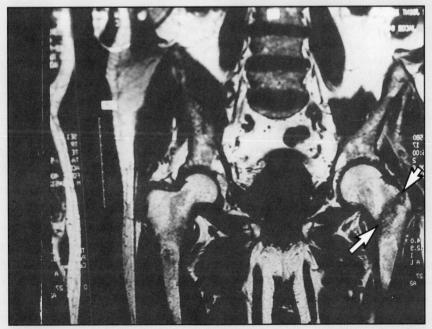

Figure 2
MRI scan of hips demonstrating increased signal (arrow) along intertrochanteric line (left hip) consistent with a fracture

Differential diagnosis

Pathologic fracture (underlying or associated tumor, benign or malignant)

Pelvic fracture (normal hip joint motion; pain on external rotation)

Adverse outcomes of the disease

Inability to walk, thrombophlebitis, decubitus ulcer, pneumonia, painful nonunion of the fracture, and even death are possible.

Treatment

A small number of patients can be managed without surgical intervention. Patients who are not ambulatory prior to injury, and who have minimal pain, can be considered candidates for conservative treatment. It should be stressed that these patients must be mobilized as quickly as possible to avoid complications from a bedridden condition. These patients usually have severe dementia.

Surgical management is required for most patients. Fixation of femoral neck fractures depends on the type of fracture and the degree of displacement of the fragments. Generally, younger, more active patients undergo reduction and internal fixation, despite the degree of displacement. Patients under age 50 years usually sustain fractures from major trauma; thus, their treatment is considered a medical emergency, especially with fractures of the femoral neck. In the older, more sedentary population, nondisplaced or minimally displaced fractures of the femoral neck are usually stabilized with internal fixation devices, whereas displaced fractures are treated by replacement arthroplasty. The results of total hip arthroplasty in the setting of an acute fracture are less rewarding. Intertrochanteric hip fractures are fixed using a compression screw and side plate.

Most patients undergo surgery within the first 24 to 48 hours after their injury. However, timing of the surgery depends on the fracture type and the health of the patient. Patients with multiple medical conditions require thorough preoperative medical evaluation. Postoperative physical therapy and rehabilitation help restore function.

Adverse outcomes of treatment

Osteonecrosis of the femoral head, malunion, and failure of fixation of the surgical device (especially in patients with severe osteoporosis) are possible. Other complications include thrombophlebitis, infection, and decubitus ulcer; death is also a possibility.

Referral decisions/Red flags

A femoral neck fracture in a patient younger than age 60 years constitutes a surgical emergency. Because all undisplaced hip fractures have the potential to displace, they require surgical evaluation. Hip fractures in alert, ambulatory patients are best treated surgically if the patients' medical condition allows.

Acknowledgements

Figures 1 and 2 are reproduced with permission from Callaghan JJ, Dennis DA, Paprosky WG, Rosenberg AG (eds): *Orthopaedic Knowledge Update: Hip and Knee Reconstruction.* Rosemont, IL, American Academy of Orthopaedic Surgeons, 1995, pp 97–108.

Inflammatory Conditions

Synonyms

Synovitis of the hip

Definition

Most inflammatory conditions that involve the hip are local manifestations of systemic disorders; however, because the hip is a major weightbearing joint, these conditions may first present with symptoms referable to the hip. With few exceptions, the pathophysiology of inflammatory arthropathies results from an immunologic host response to antigenic challenge. The exact cause of many inflammatory conditions remains unclear, but epidemiologic and genetic evidence supports a genetic component to many inflammatory arthritides.

Clinical symptoms

Inflammatory arthritis of the hip may be characterized by a dull, aching pain in the groin, lateral thigh, or buttocks region. The pain is often episodic, with patients experiencing morning stiffness, improvement with moderate activity, and frequent stiffness during motion of the hip joint.

Tests

Exam
Restricted or asymmetric internal rotation is a reliable physical finding in hip joint disease and is easily elicited with the patient seated on the edge of the examining table. Synovial inflammation can be detected by placing the patient in a prone position with the knee flexed and applying gentle rotation like a rolling pin to the extremity, moving only the hip. Antalgic gait (short stance phase on the affected side) or a limp is common.

Diagnostic
AP pelvis and groin lateral or frog lateral radiographs obtained in the early stages of inflammatory conditions may show decreased bone mineralization in the affected hip or a joint effusion. In later stages, and in inflammatory conditions such as rheumatoid arthritis, symmetric joint space loss and periarticular bone erosions may be seen (Figure 1).

Laboratory studies should include rheumatoid factor, CBC, acute phase reactants (erythrocyte sedimentation rate, C-reactive protein), and antinuclear antibody test. When an effusion is present, aspiration performed with radiographic assistance can be considered. The aspirate should be sent for culture, sensitivities, cell count with differential analysis, and inspection for crystalline deposits.

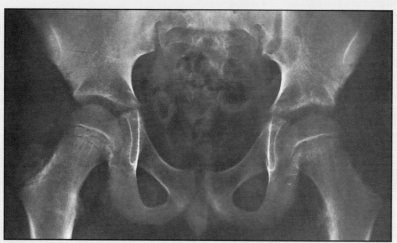

Figure 1
AP radiograph of the pelvis (note symmetric joint space commonly seen in RA of the hip joint and reduced density of the femoral head on the right side of the radiograph)

Differential diagnosis

Ankylosing spondylitis

Calcium pyrophosphate deposition disease (CPDD)

Gout

Hemophilic arthropathy

Infection

Inflammatory bowel disease

Reiter's syndrome

Rheumatoid arthritis

Stress fracture

Systemic lupus erythematosus

Trochanteric bursitis

Adverse outcomes of the disease

Adverse outcomes include complete destruction of the joint, a limp, and if the disease is systemic, severe generalized disability and immobility.

Treatment

Treatment depends on the individual diagnosis. Local injections of corticosteroids into the hip are of limited use. Most patients with hip pain and a limp will benefit by use of a cane in the hand opposite the symptomatic hip. Bilateral pain may require use of a walker or bilateral crutches. A physical therapy program emphasizing range of motion and strengthening the muscles around the hip may be helpful.

For noninfectious inflammatory arthritis, first-line agents and second-line medications may be useful. First-line agents, such as arachidonic acid, cascade inhibition with NSAIDs, or aspirin may be quite effective. Second-line medications usually include immunosuppressive or antimalarial agents.

In the early stages of hip disease prior to erosion of articular cartilage, synovectomy may be effective. Although issues related to implant fixation and a slightly increased risk of infection have been recognized, total hip arthroplasty remains a highly successful method of relieving pain and restoring function in patients with advanced disease.

Infection in the hip joint mandates immediate surgical debridement. The rapid loss of mucopolysaccharides and hyaluronic acid in suppurative arthritis leading to subsequent joint destruction is well documented.

Adverse outcomes of treatment

NSAIDs may cause gastric, renal, or hepatic complications. Adverse outcomes include osteonecrosis of the femoral head following oral corticosteroid treatment, postoperative infection, loosening of prosthetic implants, and thrombophlebitis.

Referral decisions/Red flags

Failure of conservative treatment, development of osteonecrosis, pain at rest, night pain, severe limp, or loss of motion all indicate the need for further evaluation.

Acknowledgements

Figure 1 is reproduced with permission from *Orthopaedic In Training Examination 1996*. Rosemont, IL, American Academy of Orthopaedic Surgeons, 1996.

hip and thigh

Lateral Femoral Cutaneous Nerve Entrapment

hip and thigh

Synonyms

Meralgia paresthetica

Femoral cutaneous nerve syndrome

Definition

Lateral femoral cutaneous nerve entrapment is characterized by pain, burning (dysesthesia), or hypoesthesia over the lateral thigh. There is no motor nerve dysfunction as this nerve is a sensory nerve. Nerve entrapment may occur due to a number of factors, including obesity, wearing tight clothing or a tool belt around the waist, local surgery, significant trauma (especially involving hip extension), or mild repetitive trauma over the course of the lateral femoral cutaneous nerve, which seems most susceptible as it exits the pelvis near the anterior superior iliac spine.

Clinical symptoms

Symptoms associated with this condition are entirely sensory and include pain and dysesthesia in the anterolateral or lateral thigh that sometimes extends to the lateral knee. Patients may report aching in the groin area, and if the condition is acute, pain radiating to the sacroiliac joint area.

This condition commonly affects young, muscular women (especially cheerleaders, who extend their hips doing splits) or women with scoliosis. Joggers often report symptoms, usually after running a short distance, and describe the pain as an "electric jab" each time the affected hip extends.

Pathologic intrapelvic or abdominal processes (cecal tumors) can cause compression in this syndrome, but this is rare.

Tests

Exam

Hypoesthesia or dysesthesia in the distribution of the lateral femoral cutaneous nerve is typical, with the most reproducible spot of hypoesthesia above and lateral to the knee (Figure 1). Burning is most consistent in this area. Pressure over the nerve as it exits the pelvis just medial to or directly over the anterior superior iliac spine may produce tenderness or reproduce paresthesias along the distribution of the nerve. The nerve lies medial to the anterior superior iliac spine more than 75% of the time. Motor nerve dysfunction, reflex changes, and marked tenderness over the greater trochanter are typically absent. If this condition is suspected, abdominal and pelvic exams are needed to exclude the intra-abdominal problems.

Diagnostic

An AP pelvis radiograph will rule out any abnormality, and AP and lateral radiographs of the hip may be appropriate if the patient has restricted internal rotation of the hip and groin pain. A CT scan or MRI scan may be appropriate to investigate an intrapelvic mass.

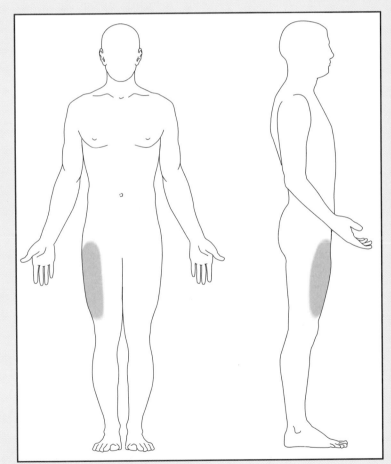

Figure 1
Hypoesthesia or dysesthesia associated with lateral femoral cutaneous nerve entrapment

Differential diagnosis

Diabetes mellitis or other causes of peripheral neuropathy

Hip arthritis (limited internal rotation, a limp)

Intra-abdominal tumor (pelvic or abdominal mass, hematochezia, weight loss)

Lumbar disk herniation (L2–L4 levels, possible quadriceps weakness, decreased knee reflex)

Trochanteric bursitis (tenderness over trochanter, stiffness when rising)

Adverse outcomes of the disease

Pain and dysesthesia will continue if the patient is not treated.

Treatment

While numbness is often well tolerated, burning may become intolerable. Removing the source of compression, such as a tight waist band or mild repetitive trauma to the nerve, may relieve symptoms. In obese patients, significant weight loss often relieves symptoms. Infiltration of the area around the nerve as it exits the pelvis through or beneath the inguinal ligament near the anterior superior iliac spine with a corticosteroid preparation may reduce symptoms. Surgery is most commonly needed in patients who are having burning dysesthesias.

Adverse outcomes of treatment

In some instances, symptoms persist despite treatment.

Referral decisions/Red flags

Suspected pelvic or abdominal mass signals the need for immediate further evaluation. The presence of intolerable symptoms that have failed to respond to conservative treatment also indicates the need for further evaluation.

Osteoarthritis of the Hip

Synonyms

Degenerative arthritis of the hip

Coxarthrosis

Coxalgia

Osteoarthrosis of the hip

Definition

Osteoarthritis of the hip is characterized by loss of articular cartilage of the hip joint, whether from trauma, infection, heredity, or for idiopathic reasons.

Clinical symptoms

The classic presentation is a gradual onset of pain in the groin. Some patients have pain in the buttock or the lateral aspect of the thigh. Initially, pain occurs only with activity, but gradually the frequency and intensity of the pain increases to the point that rest will not relieve it. Night pain is usually associated with severe arthritis. As the arthritis progresses, patients will have limited motion and a limp. Occasionally, patients will have a severe limp and stiffness, but little pain.

Patients with osteoarthritis of the hip may have other coexisting conditions, as listed in the differential diagnosis.

Tests

Exam

The earliest physical sign of osteoarthritis of the hip is loss of internal rotation on physical examination. Gradually, as the arthritis worsens, patients will lose flexion and extension. The joint contracture makes their gait more awkward since they must sway their back to get the hip straight below them. In addition, an antalgic gait (short stance time on the painful leg) and an abductor lurch (swaying the trunk far over the affected hip) will develop as the body tries to compensate for the pain and secondary weakness in the hip abductor muscles (Figure 1).

Young patients with limited internal rotation must have radiographs to determine the cause. Younger patients who have hip pain will often have hip dysplasia, residual effects of Legg-Calvé-Perthes disease, or slipped capital femoral epiphysis (SCFE) as the etiology of their untimely arthritis. These high-risk patients must be identified because they may benefit from an early osteotomy, which can halt progression of the disease.

hip and thigh

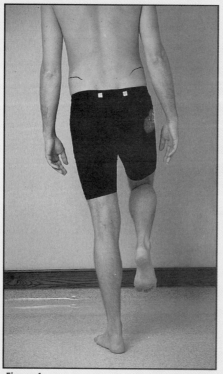
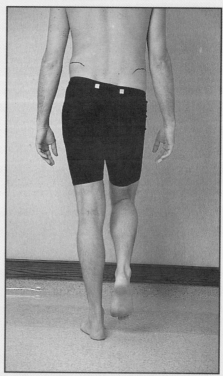

Figure 1
Normal Trendelenburg test (left). Positive sign indicates weak hip abductors (right)

Diagnostic

Obtain AP pelvis and AP and lateral radiographs of the hip. The classic radiographic features of osteoarthritis of the hips are joint space narrowing, osteophytes, cyst formation, and sub-chondral sclerosis (Figure 2).

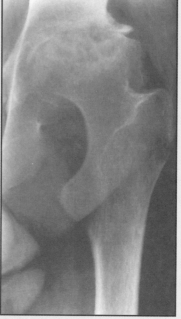

Figure 2
Radiograph of left hip with severe joint space narrowing, osteophytes, and cyst formation

Differential diagnosis

Degenerative lumbar disk disease (normal hip motion)

Femoral cutaneous nerve entrapment (sensory changes, burning, normal motion)

Herniated lumbar disk (diminished knee reflex, sensory changes)

Hip dysplasia (developmental abnormalities of the hip)

Osteonecrosis of the femoral head (evident on radiographs)

Trochanteric bursitis (local tenderness, normal motion)

Tumor of the pelvis or spine

Adverse outcomes of the disease

Osteoarthritis of the hip is a progressive condition for which there is no cure. The usual clinical course is worsening gait, increasing pain and stiffness, and eventually, pain at rest and at night.

Treatment

Treatment depends on the stage of the disease and the age of the patient. Begin with NSAIDs and physical therapy to improve strength and range of motion. As the disease progresses, patients may benefit from use of a cane held in the hand opposite the affected hip. When patients have pain even at rest, a total hip replacement arthroplasty is appropriate treatment. Patients with osteoarthritis in the third to fourth decades of life may be treated with either an osteotomy, hip fusion, or total hip arthroplasty. Total hip arthroplasty should be delayed as long as possible in younger patients due to the concerns about durability of total hip arthroplasties in high-demand patients.

Adverse outcomes of treatment

NSAIDs may cause gastric, renal, or hepatic problems. Postoperative thrombophlebitis, infection, and failure of the prosthetic components are complications associated with total hip arthroplasty. Younger patients will require multiple revisions during their lifetime. Osteoarthritis at the spine or knee joint may develop over time following hip fusion. Patients with osteotomies can have persistent hip pain and a limp.

Referral decisions/Red flags

Patients in the second, third, and fourth decades of life who have hip pain require orthopaedic evaluation because an osteotomy may slow progression of the disease. All patients with hip pain at rest require further evaluation.

Acknowledgements

Figure 2 is reproduced with permission from Callaghan JJ, Dennis DA, Paprosky WG, et al (eds): *Orthopaedic Knowledge Update: Hip and Knee Reconstruction.* Rosemont, IL, American Academy of Orthopaedic Surgeons, 1995, pp 79–86.

hip and thigh

Osteonecrosis of the Hip

Synonyms

Avascular necrosis of the hip

Aseptic necrosis of the hip

Definition

Osteonecrosis of the hip involves the death of varying amounts of trabecular bone in the femoral head. The cause is unknown but the condition occurs with greater frequency in the third through fifth decades, following trauma (hip dislocation or femoral neck fracture), and in patients with a history of alcohol abuse, corticosteroid use, rheumatoid arthritis, or systemic lupus erythematosus. Osteonecrosis affects 10,000 to 20,000 new patients per year in the United States and is often bilateral.

Clinical symptoms

Patients usually report a dull ache or a throbbing pain in the groin, lateral to the hip, or in the buttock. The pain is usually gradual in onset and duration, but may begin suddenly with collapse of the necrotic femoral head. Thus, the patient's complete history must be obtained. A complete listing of common clinical entities that have been associated with osteonecrosis are outlined in Table 1. Patients can develop osteonecrosis even after receiving only one or two doses of intravenous steroids. Some patients also have a tendency to minimize their use of alcohol.

Table 1
Risk factors for osteonecrosis

Corticosteroid use	Gaucher's disease
Alcohol	Radiation
Trauma	Chronic pancreatitis
Rheumatoid arthritis	Crohn's disease
Sickle cell disease	Caisson's disease
Myeloproliferative disorders	Systemic lupus erythematosus

Tests

Exam

Patients with osteonecrosis without collapse may have discomfort with either internal or external rotation or abduction at the hip. Patients whose femoral head has collapsed will complain of pain with internal and external rotation and usually have diminished internal rotation, flexion, and abduction of the joint. They often have an antalgic limp (short stance phase).

Diagnostic

Obtain AP pelvis and frog lateral radiographs of the hip. One of the earliest signs of osteonecrosis is sclerosis of the femoral head (Figure 1). If the radiographs demonstrate evidence of significant collapse of the femoral head or degenerative arthritis of the hip joint, then no further evaluation of that hip is usually necessary. If the patient has risk factors for osteonecrosis but no changes in the femoral head, if there are only sclerotic changes seen in the femoral head on plain radiographs, or if there is evidence of osteonecrosis of the opposite hip, then MRI scans of both hips are indicated. A single band-like area of low intensity on the T1 image or double-line signs seen on the T2-weighted image can be accepted as a diagnosis for osteonecrosis (Figure 2).

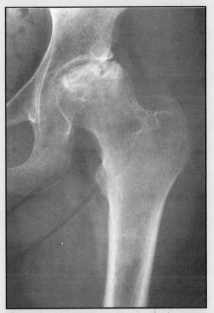

Figure 1
Sclerosis of the femoral head

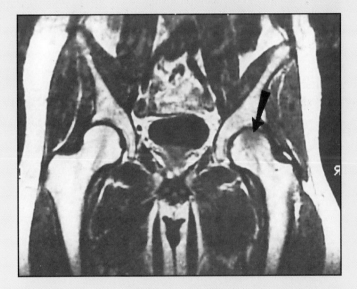

Figure 2
MRI scan of hips consistent with osteonecrosis of left hip (arrow). AP view reveals significant involvement of the weightbearing surface.

hip and thigh

Differential diagnosis

Fracture of the femoral neck (radiographic findings, MRI findings)

Lumbar disk disease (back pain, reflex changes, radiation below knee)

Muscle strain or groin pull (normal radiographs, intermittent limp)

Osteoarthritis of the hip (radiographic osteoarthritis)

Septic arthritis of the hip (fever, constitutional symptoms)

Transient osteoporosis of the hip (disabling pain without previous trauma, osteopenia, in women in third trimester of pregnancy or middle-aged men)

Adverse outcomes of the disease

Collapse of the femoral head, secondary osteoarthritis, pain, limp, and disability may occur.

Treatment

Conservative treatment is rarely appropriate for this disease. Protective weightbearing should be considered only as a temporary treatment until a more definitive work-up and treatment plan can be established.

Surgical treatment for an uncollapsed hip is controversial. There is a race between revascularization and replacement of bone in the femoral head and the forces causing collapse of the necrotic trabeculae. Surgical procedures in these circumstances are directed at accelerating the revascularization and bone formation processes, and surgeons claim varying success with different techniques.

Because revascularization takes months and sometimes years to occur, collapse often ensues. The prognosis depends on the extent of the osteonecrosis and its location (degree of involvement of weightbearing surface). Once collapse occurs, total hip arthroplasty is the procedure of choice for pain relief and restoration of function. However, the timing of this procedure depends on the patient's age, diagnosis, and symptoms.

Adverse outcomes of treatment

Failure of the treatment to revascularize the femoral head before collapse can occur. Fracture of the proximal femur may follow any procedure that invades the femoral cortex. Postoperative infection and/or thrombophlebitis can also develop.

Referral decisions/Red flags

Suspicion or confirmation of the diagnosis indicates the need for immediate further evaluation. A patient with one of the risk factors and hip pain requires evaluation of the hip.

Acknowledgements

Figure 1 is reproduced with permission from Cabanela ME: Hip arthroplasty in osteonecrosis of the femoral head, in Jones JP, Urbaniak M (eds): *Osteonecrosis.* Rosemont, IL, American Academy of Orthopaedic Surgeons, 1997.

Figure 2 is reproduced with permission from Poss R (ed): *Orthopaedic Knowledge Update 3.* Park Ridge, IL, American Academy of Orthopaedic Surgeons, 1990, p 540.

Snapping Hip

Synonyms

Snapping iliopsoas tendon

Snapping iliotibial band

Definition

Snapping hip is characterized by a snapping sensation that occurs as tendons around the hip subluxate over bony prominences. The most common occurrence is the iliotibial band snapping over the greater trochanter. Snapping can also occur when the iliopsoas tendon subluxates over the pectineal eminence of the pelvis or from problems within the hip joint itself.

Clinical symptoms

Iliotibial band subluxation usually occurs with walking or rotation of the hip. Patients will point to the trochanteric area (Figure 1). Some patients notice the snapping when they lie with the affected side up and rotate the leg. If a trochanteric bursitis develops related to this condition, patients will have increased pain when first rising, pain at night, and difficulty lying on the affected side.

Snapping due to subluxation of the iliopsoas tendon is usually felt in the groin as the hip extends from a flexed position, as when rising from a chair. Many patients feel the snapping but have no disability. In a few, the snapping is either annoying or painful.

Snapping from intra-articular causes is more disabling and more likely to cause patients to grab for support.

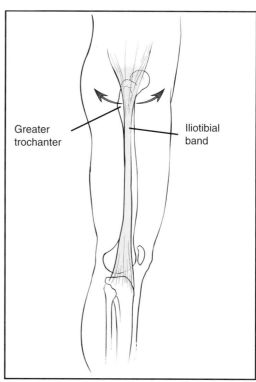

Greater trochanter

Iliotibial band

Figure 1
Iliotibial band slips anteriorly and posteriorly over the prominent greater trochanter

Tests

Exam

Iliotibial band subluxation can be recreated by having the patient stand, adduct, and rotate the hip. The snap is palpated over the lateral hip as the iliotibial band subluxates over the trochanter.

Snapping of the iliopsoas tendon may be palpated as the hip extends from a flexed position and the tendon subluxates over the pectineal eminence of the pelvis.

Restricted internal rotation of the involved hip, a limp, or shortening of the limb suggest problems within the hip joint.

Diagnostic

Obtain AP pelvis and lateral hip radiographs to exclude bony pathology or intra-articular hip disease. Radiographs are typically normal for patients with a snapping hip. Either a CT scan to rule out intra-articular loose bodies or an MRI scan to rule out a labral tear may be necessary.

Differential diagnosis

Osteoarthritis of the hip (limited internal rotation)

Osteochondral loose body (a fragment of bone and cartilage within the joint, pain with hip motion)

Osteonecrosis of the femoral head (compromised blood supply of the femoral head, groin pain)

Tear of the acetabular labrum (fibrocartilage rim at periphery of the acetabulum, pain or instability with hip motion)

Adverse outcomes of the disease

Pain and annoyance are the two most common complaints.

Treatment

Snapping hip is often painless, and once the diagnosis is made with certainty, patients often require only an explanation of the source of the symptoms for reassurance. Patients who are significantly bothered by the symptoms should be advised to avoid provocative maneuvers and activities so that the symptoms may subside. Physical therapy, consisting of stretching exercises and ultrasound, and a short course of NSAIDs may also reduce the discomfort associated with tendon snapping and secondary bursitis. Corticosteroid injection into the greater trochanteric bursa (for snapping iliotibial band) or into the psoas sheath (for snapping iliopsoas tendon) can reduce pain. Surgery is reserved for rare disabling cases that fail to resolve with conservative management.

hip and thigh

Adverse outcomes of treatment

NSAIDs may cause gastric, renal, or hepatic complications. Postoperative infection or persistent pain is also possible.

Referral decisions/Red flags

Unclear diagnosis, intra-articular pathology, and/or failure of conservative measures indicate the need for further evaluation.

Trochanteric Bursitis

Synonyms

Greater trochanteric bursitis

Greater trochanteric pain syndrome

Definition

Trochanteric bursitis is characterized by pain and tenderness over the greater trochanteric bursa that may radiate down the lateral leg as far as the ankle and simulate sciatica (Figure 1). Inflammation of the gluteal tendons may also cause the same pain pattern.

Clinical symptoms

Patients usually have pain in the area of the lateral hip that may radiate distally to the knee or ankle (but not onto the foot), or proximally into the buttock. The pain is worse when first rising from a seated or recumbant position, feels somewhat better after a few steps, and recurs after walking for half an hour or more. Patients report night pain and are unable to lie on the affected side.

Trochanteric bursitis may occur in association with lumbar spine disease, intra-articular hip pathology, significant leg length inequalities, previous surgery around the lateral hip (particularly if internal fixation devices are in place in the trochanter), and rheumatoid arthritis.

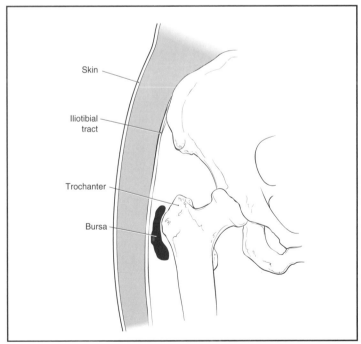

Skin

Iliotibial tract

Trochanter

Bursa

Figure 1
Relationship of trochanteric bursa between the iliotibial band and the greater trochanter

hip and thigh

Tests

Exam

Point tenderness over the lateral greater trochanter is the essential finding (Figure 2). Tenderness above the trochanter suggests tendinitis of the gluteus medius tendon and may be associated with a positive Trendelenburg's sign and a limp.

Patients may be uncomfortable with external rotation, but significant pain with internal rotation of the hip is unusual.

Diagnostic

Obtain AP pelvis and lateral hip radiographs to rule out bony abnormalities and intra-articular hip pathology. Occasionally, rounded or irregular calcific deposits may be seen above the trochanter at the attachment of the gluteus medius. There may be irregular bony spurs at the lateral trochanter.

Bone scans and MRI scans are rarely needed to make the diagnosis but may occasionally be helpful to rule out uncommon conditions such as occult fractures, tumors, or osteonecrosis of the femoral head.

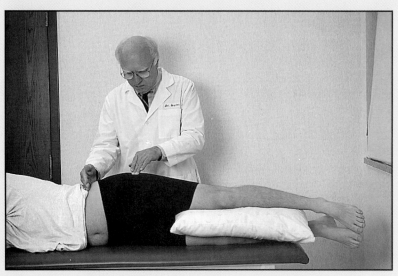

Figure 2
Palpate the greater trochanter with the patient in the lateral decubitus position

Differential diagnosis

Metastatic tumor (radiographic findings, weight loss, constitutional symptoms)

Osteoarthritis of the hip (painful internal rotation)

Sciatica (pain posteriorly or onto top of foot, reflex changes)

Snapping hip (an obvious snap of the iliotibial band)

Trochanteric fracture (radiograph, limp persists when walking, positive Trendelenburg's sign)

Adverse outcomes of the disease

Chronic pain, a limp, and/or complaints of sleep disturbances are possible.

Treatment

NSAIDs and an iliotibial band stretching program benefit many patients, especially in the younger age groups. The most direct treatment is injection of a local anesthetic and corticosteroid preparation into the greater trochanteric bursa (see Trochanteric Bursitis Insertion). Trochanteric injection relieves symptoms successfully in more than 90% of patients. Occasionally, repeat injections are required for symptomatic relief. Surgery is indicated only rarely for intransigent cases.

Adverse outcomes of treatment

NSAIDs may cause gastric, renal, or hepatic complications. In some patients, pain may persist, and while rare, infection from the injection may develop, but it can be largely avoided by use of careful sterile technique.

Referral decisions/Red flags

Failure of treatment, diagnostic uncertainty, and/or suspected fracture are indications for further evaluation.

procedure

Trochanteric Bursitis Injection

Materials

- Sterile gloves

- Skin preparation solution (iodinated soap or similar antiseptic solution)

- Sterile drape

- 12-mL syringe

- 20- or 22-gauge, 1½-inch needle (use a spinal needle in larger patients)

- 3 to 5 mL of 1% local anesthetic

- 40 to 80 mg of a corticosteroid preparation

- Adhesive dressing

Step 1

Wear protective gloves at all times during the procedure and use sterile technique.

Step 2

Ask the patient to lie in the lateral decubitus position with the affected hip turned upward. Place a pillow between the patient's knees to relax the iliotibial band and reduce the pressure required to inject the solution.

Step 3

Cleanse the skin with iodinated soap or similar antiseptic lotion.

Step 4

Draw lidocaine (short acting), bupivicane (long acting), or a combination of the two, into a 10-mL syringe.

Step 5

Draw the chosen dose of corticosteroid preparation into the same syringe and mix the two solutions.

Step 6

Palpate the greater trochanter and identify the point of maximum tenderness.

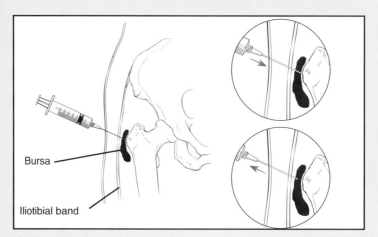

Bursa

Iliotibial band

Figure 1
Location for needle insertion

Trochanteric Bursitis Injection (continued)

Step 7

Insert the needle until it contacts bone, then withdraw it 1 or 2 mm so that the tip is in the bursa and not in the bone (Figure 1). Usually, a $1^1/_2$-inch needle is sufficient, but for larger patients a spinal needle may be needed to reach the trochanteric bursa. Do not withdraw the needle too far or it will be above the fascia lata and outside the trochanteric bursa.

Step 8

Aspirate to ensure that the needle is not in an intravascular position, then inject the steroid preparation/local anesthetic mixture in 1 to 2 mL aliquots.

Step 9

Partially withdraw the needle, then reinsert it and inject another aliquot. Continue this to infiltrate an area of several square centimeters around the point of maximal tenderness.

Step 10

Withdraw the needle completely and apply gentle pressure over the injection site.

Step 11

Dress the puncture wound with a sterile adhesive bandage.

Adverse outcomes

Although rare, infection or allergic reactions to the local anesthetic or corticosteroid preparation are possible. Always query the patient about medication allergies before the procedure. In some patients with diabetes, poor control of blood glucose levels may occur, but this is usually temporary. If the local anesthetic is injected near a major nerve (the sciatic or femoral nerve), transient nerve dysfunction may occur. Infiltration of the nerve can lead to permanent dysfunction. A minority of patients require more than one injection to achieve lasting pain relief; however, repetitive injection of the trochanteric bursa with corticosteroids should be avoided.

Aftercare/patient instructions

Advise the patient that as the local anesthetic wears off, pain may persist or become worse for a few days, until the corticosteroid takes effect. Instruct the patient to attempt weightbearing as tolerated and to contact you if symptoms recur or if redness, fever, immobilizing pain, or any other evidence of a local problem related to the injection occur.

Acknowledgements

Figure 1 is adapted with permission of the Mayo Foundation, Rochester, MN.

hip and thigh

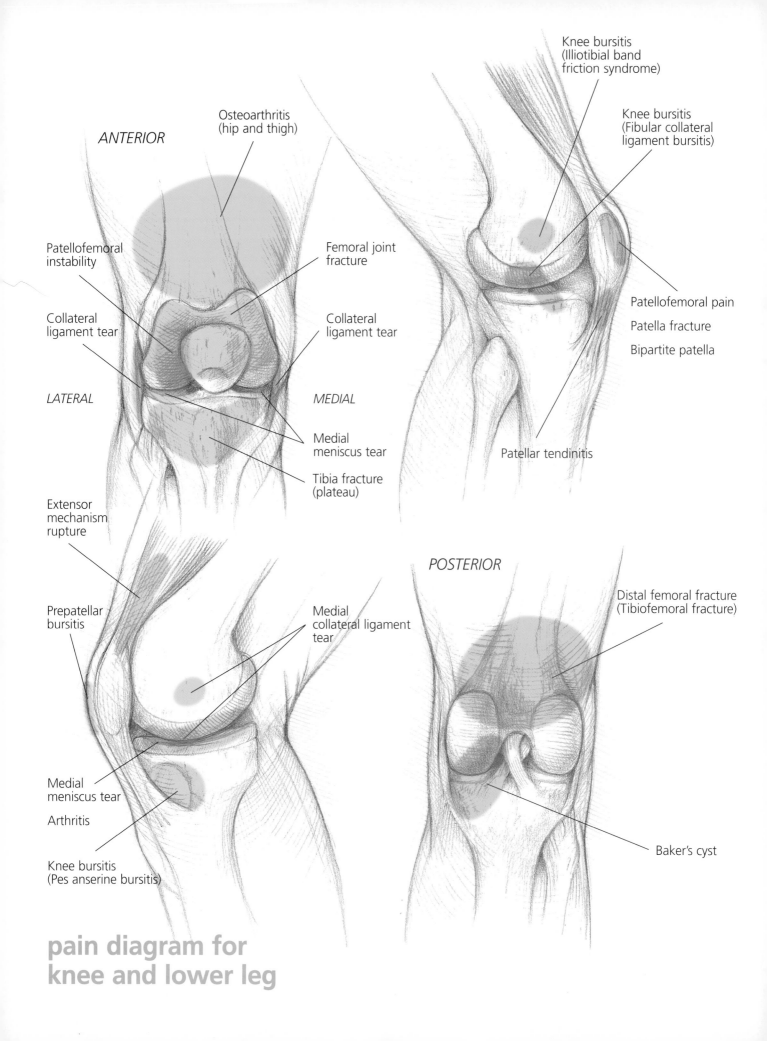

ANTERIOR

Osteoarthritis
(hip and thigh)

Patellofemoral
instability

Femoral joint
fracture

Collateral
ligament tear

Collateral
ligament tear

LATERAL

MEDIAL

Medial
meniscus tear

Tibia fracture
(plateau)

Knee bursitis
(Illiotibial band
friction syndrome)

Knee bursitis
(Fibular collateral
ligament bursitis)

Patellofemoral pain

Patella fracture

Bipartite patella

Patellar tendinitis

Extensor
mechanism
rupture

Prepatellar
bursitis

Medial
collateral ligament
tear

POSTERIOR

Distal femoral fracture
(Tibiofemoral fracture)

Medial
meniscus tear

Arthritis

Knee bursitis
(Pes anserine bursitis)

Baker's cyst

**pain diagram for
knee and lower leg**

knee and lower leg

Section Editor

Scott S. Kelley, MD
Assistant Professor of Orthopaedics
University of North Carolina, Chapel Hill
Chapel Hill, North Carolina

Joseph M. Erpelding, MD
Orthopedic Surgeons, P.S.C.
Billings, Montana

Jeffrey Kobs, MD
Raleigh Orthopaedic Clinic
Raleigh, North Carolina

Charles L. Nance, MD
Wilmington Orthopaedic Group
Wilmington, North Carolina

David A. Rendleman, III, MD
Bone and Joint Surgery Clinic
Raleigh, North Carolina

knee and lower leg— an overview

In this section of *Essentials of Musculoskeletal Care,* problems of the knee and lower leg are detailed, with specific focus on fractures and ligament problems. Most knee problems can be diagnosed by obtaining a careful history, physical examination, and radiographs. Patients with knee problems often report pain, instability, stiffness, swelling, locking, or weakness. These findings can occur in or around any aspect of the knee and are often localized even further proximally or distally, depending on the location of the structure involved. Examination of patients with knee problems should also always include evaluation of internal rotation of the hip joint, as patients with intrinsic hip problems may present with thigh pain and other symptoms that mimic knee disorders.

Radiographic examination of the knee should include AP and lateral radiographs. If the patient is able to stand, obtain a standing AP of both knees so that the injured knee can be compared with the opposite, uninjured knee. When symptoms are localized to the patellofemoral joint, a patellofemoral view is helpful. All radiographs should be evaluated for changes in bony architecture, including lytic and blastic lesions.

AP radiographs are best used to evaluate for medial and lateral compartment arthritis, fractures of the distal and proximal femur, and alignment of the femur to the tibia. A lateral radiograph is helpful in assessing the patellofemoral joint for fractures, degenerative changes, and relationship to the jointline. Patellofemoral views are best used to assess for subluxation of the patella and arthritis of the patellofemoral joint.

MRI scans play a role in surgical planning, but are rarely needed in diagnostic work-up.

Type of pain

Acute pain
The evaluation of the knee joint following a traumatic injury is a complex process; possible diagnoses fall into five basic categories—any of which can be painful: 1) fractures, 2) meniscal injuries, 3) ligamentous injuries, 4) extensor mechanism injuries, and 5) contusions.

Fractures can involve the distal femur, patella, proximal tibia, and fibula. The diagnosis can be confirmed following inspection for deformity, palpation for tenderness in the bone itself, and radiographic evaluation. Patellar fractures may result from indirect forces, such as a fall, but fractures of the tibia and femur at the knee usually result from major trauma. Dislocations are often reduced at the scene when a helper extends the patient's knee for transportation.

Obtaining a complete history is key in diagnosing meniscal injuries. A history of locking, tenderness along the jointline, and in the case of a bucket handle tear, a locked knee, are indicative. Some patients report that manipulating or pushing on the knee will stress and unlock it.

Patients with ligamentous injuries have acute pain, swelling, and instability. Patients with injuries to the extensor mechanism report sudden weakness or collapse.

Chronic pain

Nontraumatic conditions in which patients have pain include arthritis, tumors, sepsis, and overuse syndromes (including bursitis/tendinitis and anterior knee pain). Arthritis is relatively easy to diagnose because symptoms localize to the jointline and are associated with loss of motion and radiographic changes.

Tumors are characterized by night pain (relentless pain in which the patient is unable to sleep) and often can be palpated or identified on radiographs. Metastasis to the knee is rare; the most common malignant tumors are osteosarcoma (adolescence) and chondrosarcoma (adults). The most common benign tumor is a giant cell tumor, which typically occurs in adults age 20 to 30 years.

Sepsis in the knee joint is rare in adults; it is more commonly located in the prepatellar bursa. The exact location of the infection is easily determined by thorough inspection and palpation of the involved area. Sepsis in either region is characterized by enlargement, erythema, and loss of motion in the knee. With infection in the prepatellar bursa, there is swelling over the patella. With infection in the knee cavity, there is swelling around (medial, lateral, and above), but not over the patella.

Bursitis/tendinitis and anterior knee pain have similar characteristics; both are usually chronic, often secondary to overuse, and often bilateral. The pain is typically worse with rising or walking after sitting, worse at night, and worse with prolonged exercise or use.

Location of pain

Anterior knee pain

Tenderness at the upper pole of the patella indicates a tendinitis or partial tear of the quadriceps insertion on the patella. Pain at the upper and lateral poles of the patella suggests a bipartite patella. At the inferior pole of the patella or at the tibial tubercle, pain occurs with tendinitis or overuse injury of the patellar ligament (also called the infrapatellar tendon).

Medial knee pain

Pain at the medial jointline, midway between the front and back of the knee, is common with a torn meniscus, especially with degenerative tears that occur as a result of minor trauma such as twisting or rising from a squat. If the meniscal tear is sufficiently large or loose, patients may also report catching or locking. Tenderness in this area is also common with degenerative joint disease of the knee. Pain above and/or at the medial jointline, with localized swelling and a history of recent injury, usually indicates a sprain or tear of the medial collateral ligament origin from the femoral condyle.

Below the medial jointline lies the insertion of the pes anserine tendons (composed of the medial hamstring tendons, and so named because it looks like a goose foot) and the superficial portion of the medial collateral ligament. Pain in this area, in the absence of trauma, suggests a bursitis under the pes anserine tendons. Note that this condition often occurs in relation to an osteoarthritis of the medial knee joint.

Lateral knee pain

Pain over the lateral femoral condyle indicates an iliotibial band friction syndrome, usually associated with overuse or erratic exercise habits. Pain over the lateral jointline usually indicates a disorder of the lateral meniscus, or osteoarthritis of the lateral joint (more common in female or obese patients).

Posterior knee pain

Pain at the posteromedial corner of the knee may indicate a tear of the medial meniscus (at the jointline), a Baker's cyst, or both. Popliteal aneurysms may also be painful in the popliteal area, and many patients with knee effusions indicate popliteal pain from the distention of the joint capsule.

Instability

The knee joint is actually two joints: one between the tibia and the femur (tibiofemoral joint), and one between the patella and the femur (patellofemoral joint) (Figure 1). True instability means one bony component moves on another in an abnormal fashion, such as the patella sliding laterally on the femur in recurrent subluxation of the patella, or the tibia moving anteriorly on the femur in the anterior cruciate deficient knee. True instability is usually described as a "giving way" or "slippage" in the knee joint. Buckling, on the other hand, means a collapse of the knee, often secondary to pain or muscle weakness in the axis of the quadriceps mechanism.

Tibiofemoral instability

Differentiating instability from ligamentous injury is difficult in a painful, traumatized knee. During examination, it is important to assess the following four ligament complexes: 1) the anterior cruciate ligament, 2) the posterior cruciate ligament, 3) the posterolateral ligaments, and 4) the posteromedial ligaments. Although there are many different tests to evaluate stability, the following are most accurate:

- Lachman's test for the anterior cruciate ligament;

- posterior drawer for the posterior cruciate ligament;

- varus and valgus stress testing for the medial lateral structures.

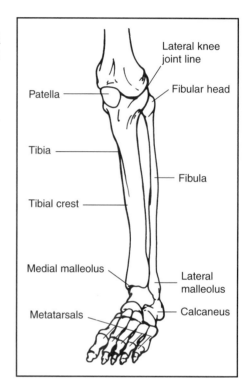

Figure 1
Anatomy of the knee joint

Complete descriptions and illustrations of these tests are included in the Physical Exam portion of this section. Varus and valgus stress testing should be performed in full extension to assess the more posterior structures, and in slight flexion to isolate the medial and lateral collateral ligaments. Any recurvatum (back knee) suggests injury to posterior structures as well.

Chronic instability of the knee occurs with severe arthritis as the ligaments are not at their full tension due to loss of articular cartilage and bone height as a result of the arthritis.

Patellofemoral instability

Instability in the extensor mechanism is usually due to a lateral dislocation or subluxation of the patella, and the diagnosis is made by palpation. If the patella is dislocated, the patient's knee will often be locked in approximately 45° of flexion. With a subluxation, the patient will be apprehensive when the patella is displaced laterally.

Stiffness

Stiffness at the knee is the most common complaint that accompanies an effusion. The distention of the knee cavity prevents full flexion, and the patient feels the knee is "stiff," but may not notice it is also swollen.

Arthritis is the other common cause of stiffness, and patients often report that the knee sticks or locks momentarily as they walk. This occurs because the articular surfaces are rough and act like two pieces of sandpaper rubbing on each other.

Stiffness can also result from any inflammatory condition at the knee, including arthritis, overuse syndromes, and traumatic effusion.

Swelling

When patients report swelling, they are most often talking about puffiness in the infrapatellar bursa. This bursa is located behind and to either side of the patellar ligament (infrapatellar tendon). Think of this structure as the knee's thermometer: it swells with a variety of knee disorders and is obvious to patients because it feels tense when they kneel, is prominent when they rub their knees, and is obvious when they look at their knees in the mirror.

With a true intra-articular effusion, there is distention around and above the patella in the knee cavity. Patients often notice this as stiffness; they cannot fully flex their knee since the knee cavity is filled with water or blood and cannot be compressed.

Locking

True locking occurs when the knee is stuck, usually in about 45° of flexion, and the patient is unable to unlock the knee without manipulating it in some fashion. Patients may indicate that pushing on the lateral knee will allow something to slip and unlock, or that rotating the tibia on the femur may allow the joint to become "unstuck."

Pseudolocking occurs with arthritis, when the adjacent rough surfaces stick momentarily as they glide onto one another. It may also occur with minor knee conditions such as a medial synovial plica, which becomes momentarily stuck under the patella as the knee extends.

Weakness

The onset of weakness of the muscles about the knee can occur acutely or gradually. Acute traumatic weakness usually occurs as a result of disruption of the extensor mechanism. There are three locations in which this can occur: a tear of the patella tendon below the patella, a tear of the quadriceps tendon above the patella, or a fracture through the patella. With partial ruptures, the extensor mechanism may continue to function, but without proper protection, the tear may become complete.

knee and lower leg—physical exam

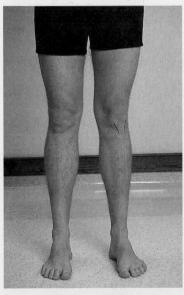

1

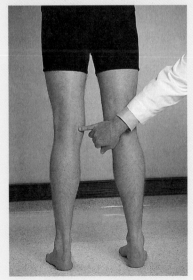

2

Figure 1
Knee, anterior

With the patient standing, look for valgus (knock-knee) or varus (bowleg) deformities, swelling below the patella (infrapatellar bursitis), or swelling above the patella (intra-articular effusion). Internal femoral torsion rotates the knee so the patellae point toward one another.

Figure 2
Knee, posterior

With the patient standing, look for swelling in the popliteal fossa, indicating a Baker's cyst, in comparison with the opposite knee. Palpate the popliteal fossa with the patient supine and the muscles relaxed.

knee and lower leg

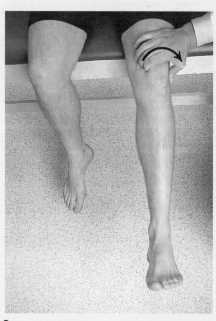

3

Figure 3
Patellar instability

Place the thumb of your outer hand on the patient's lateral femoral condyle, and pull the patella laterally with the fingers as the patient extends and flexes the knee. With patellar instability, the patient will often grasp your hand as the knee starts to subluxate, or the lateral movement (subluxation) of the patella may be obvious as the knee approaches about 30° of flexion. This should also be performed while supporting the lower leg to allow the quadriceps muscle to relax.

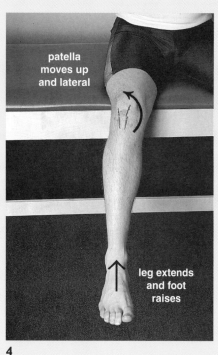

patella moves up and lateral

leg extends and foot raises

4

Figure 4
Patellar instability

Sit in front of the patient as the patient flexes and extends both knees. Watch for an inverted "J" shaped motion of the patella as it moves proximally, indicating susceptibility to patellar subluxation.

knee and lower leg

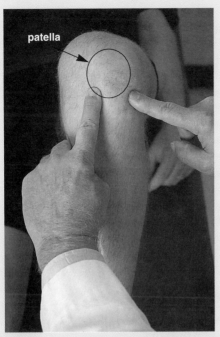

5

Figure 5
Infrapatellar bursa

Palpate below the patella, on either side of the infra-patellar tendon, for swelling. Often this is visible, with a dumbbell-like swelling on either side of the tendon. The asymmetry is easily seen with the patient seated or standing.

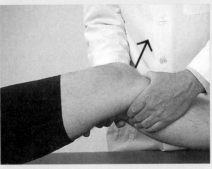

6

Figure 6
Lachman's test

With the patient supine, grasp the distal thigh above the patella with one hand to stabilize the thigh (thumb should wrap over the thigh just above the patella). Grasp the proximal tibia with the other hand and pull the tibia forward.

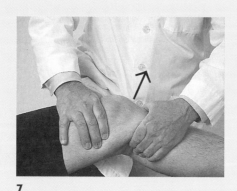

7

Figure 7
Lachman's test (alternate method)

With the patient supine, place your knee under the patient's knee and press the proximal thigh down on your knee to stabilize the femur. Grasp the proximal tibia with the other hand and pull it forward. This method of Lachman's test is especially useful if the patient's thigh is large and difficult to grasp. In both methods, normal motion is equal to the opposite knee (which should always be tested first). A torn cruciate ligament allows excessive anterior glide of the tibia, and there is no sharp end point to the anterior motion. Lachman's test is a more sensitive indicator of cruciate ligament injury than the drawer test.

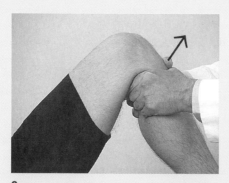

8

Figure 8
Drawer test

With the knee flexed to 90°, sit on the patient's foot and grasp the proximal tibia with both hands. With the upper edge of your index fingers, palpate the hamstring tendons to ensure that they are relaxed. Slide the tibia anteriorly for an anterior drawer test, or posteriorly for a posterior drawer test. Compare with the uninjured knee, which should always be examined first.

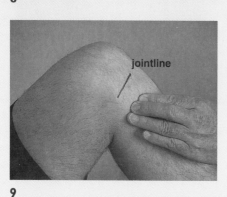

jointline

9

Figure 9
Pes anserine bursa

Find the jointline by palpating below and medial to the patella, feeling for the upper edge of the tibia. Palpate below the jointline on the proximal medial tibial flare to identify the pes anserine bursa, located beneath the conjoined medial hamstring tendons (pes anserinus).

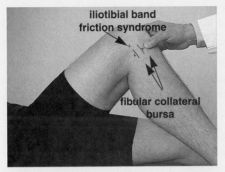

10

Figure 10
Iliotibial bursa

With the knee flexed 30° or so, palpate the lateral femoral condylar prominence lateral to the midpoint of the patella. Tenderness here is often associated with an iliotibial band friction syndrome seen in runners.

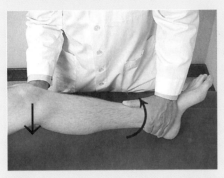

11

Figure 11
Medial collateral ligament in flexion

With the knee slightly flexed (20°), place the outer hand on the lateral side of the knee, grasp the medial foot or ankle with the opposite hand, and abduct the knee. Excessive motion, usually coupled with pain, indicates stretching or a tear of the medial joint capsule. Test the knee in extension next.

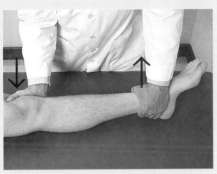

12

Figure 12
Medial collateral ligament in extension

With the knee extended, place the outer hand on the lateral side of the knee, grasp the medial foot or ankle with the opposite hand, and abduct the knee. Excessive motion, usually coupled with pain, indicates stretching or a tear of the medial joint capsule and posterior capsule. There may also be an associated tear of a cruciate ligament (see Lachman's test or drawer test).

knee and lower leg

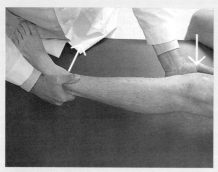

13

Figure 13
Lateral collateral ligament in flexion

With the knee slightly flexed (20°), place the inner hand on the medial side of the knee, grasp the foot or ankle with the opposite hand, and adduct the knee. Excessive motion, usually coupled with pain, indicates stretching or a tear of the lateral joint capsule. Test the knee in extension next.

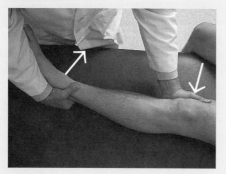

14

Figure 14
Lateral collateral ligament in extension

With the knee extended, place the inner hand on the medial side of the knee, grasp the foot or ankle with the opposite hand, and adduct the knee. Excessive motion, usually coupled with pain, indicates stretching or a tear of the lateral joint capsule and posterior capsule. There may also be an associated tear of a cruciate ligament (see Lachman's test or drawer test).

15

Figure 15
Straight leg raise, active

With the patient supine, ask the patient to stiffen the knee and raise the leg. If the patient is unable to keep the knee straight while raising the leg, suspect rupture of the extensor mechanism with rupture of the tendon proximal or distal to the patella, or fracture of the patella. With patellar fracture, there is usually a dome-shaped swelling over the patella itself.

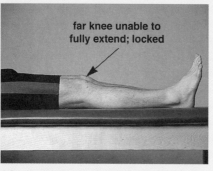

far knee unable to fully extend; locked

16

Figure 16
Knee, lateral

Observe for maximum extension of the knee. Inability to fully extend the knee in comparison to the uninjured side may indicate a tear of the anterior cruciate or a displaced or locked tear of a meniscus. (Note that the knee is unable to fully extend.)

knee and lower leg

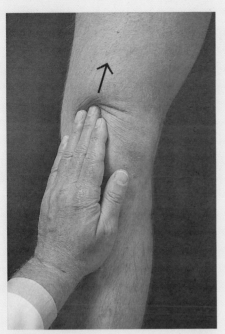

17A

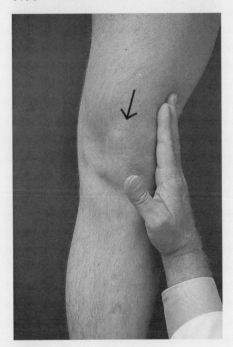

17B

Figure 17
Bulge sign

(A) With the patient's knee extended and resting flat on the examining table, milk the joint fluid up into the suprapatellar pouch by moving your hand proximally along the area medial to the patella. (B) Next, milk the fluid down from the suprapatellar pouch to the medial knee by moving your hand from above the lateral side of the patella along the lateral knee down to the tibia. Excessive joint fluid will create a bulge as the area medial to the patella distends with the displaced fluid. This test detects small intra-articular effusions.

knee and lower leg

The Essentials of Musculoskeletal Care

Anterior Cruciate Ligament Tear

Synonyms

Torn cruciate

Crucial ligament tear

ACL tear

Anterior cruciate insufficiency

Rotary instability of the knee

Anterolateral instability of the knee

Definition

Anterior cruciate ligament (ACL) tears result from a traumatic (complete or partial) rupture of the primary anterior and rotational stabilizer of the knee (Figure 1). The tear may occur alone or with other associated injuries, typically a meniscal or medial collateral ligament tear.

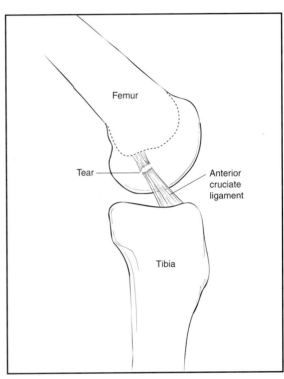

Figure 1
Complete ACL tear

Clinical symptoms

Patients usually report a history of a significant twisting injury to the knee. One third of patients also describe a "popping" sensation at the time of injury. Effusion may be significant and usually occurs rapidly following the injury. Acutely, most patients have painful range of motion and profound difficulty ambulating after the injury. If left untreated, patients may present late after one or more episodes of "giving way" of the knee, commonly followed by knee pain and swelling. Untreated instability may also lead to a future meniscal tear that may contribute to long-term degenerative arthritis.

knee and lower leg

Tests

Exam
A large effusion is commonly present for many days after the injury. An anterior drawer test with the knee flexed to 90° is often positive, but Lachman's test with the knee flexed only 15° to 20° is more sensitive and is easier to perform immediately after the injury (Figure 2).

Most patients are seen 1 to 2 days following the injury, at which time the knee is painful, making examination difficult. Repeat examinations are often needed to establish a diagnosis.

Diagnostic
AP, lateral, and Merchant radiographs are usually negative, but they may show an avulsion of either the tibial insertion of the ACL or the lateral capsular margin of the tibia.

Arthrocentesis can be performed to relieve pressure and pain and will generally return blood or serosanguinous fluid. If fat globules rise to the surface of the aspirate, a fracture is likely. After 24 hours, the blood may be clotted and difficult to aspirate.

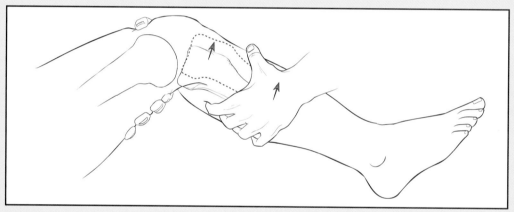

Figure 2
Correct position for Lachman's test

Differential diagnosis

Collateral ligament tear (varus/valgus laxity and/or local tenderness)

Meniscal tear (tenderness along the jointline)

Osteochondral fracture (positive radiographs)

Patellar tendon rupture (inability to straight leg raise)

Patellofemoral dislocation (palpable patellar dislocation or apprehension when stressing reduced patella)

Posterior cruciate ligament tear (posterior drawer or sag sign)

Tibial plateau fracture (positive radiographs and bony tenderness)

Adverse outcomes of the disease

Chronic knee instability is possible, resulting in subsequent repetitive knee injuries and eventual degenerative joint disease. Approximately one third of patients will not be able to work or participate in recreational activities without significant knee instability. Conservative treatment, including physical therapy and the use of a functional knee brace, may be sufficient for another third of patients. The remaining patients are able to work adequately, but they are unable to confidently participate in sports. Many of these patients have associated and surgically treatable tears of the meniscus, which may greatly decrease subsequent adverse outcomes such as osteoarthritis.

Treatment

Treatment of an acute ACL injury varies according to the patient's age, activity level, and the presence of associated injuries (collateral ligament or meniscal injuries). Initial treatment should include rest, ice compression, and elevation (RICE), along with a knee immobilizer, crutches, and NSAIDs. If the knee is tense and painful, aspiration of the hemarthrosis is appropriate (see Knee Joint Aspiration).

Definitive care of an ACL injury can only be determined after a thorough physical examination and diagnostic evaluation. Treatment may range from physical therapy and bracing to surgical reconstruction of the ligament and other associated injuries.

Adverse outcomes of treatment

NSAIDs may cause gastric, renal, or hepatic complications. Failed conservative treatment is often attended by persistent instability, pain, recurrent effusions, and meniscal tears.

Postoperative problems include failure of ligament reconstruction and persistent pain at the patellar graft donor site.

Referral decisions/Red flags

An acute ACL injury or the presence of a knee effusion secondary to trauma indicates the need for further evaluation. Associated collateral ligamentous injuries and meniscal tears are often difficult to diagnose, but may significantly increase adverse outcomes.

knee and lower leg

Acknowledgements

Figure 2 is adapted with permission from Feagin JA (ed): *The Crucial Ligaments: Diagnosis and Treatment of Ligamentous Injuries About the Knee.* New York, NY, Churchill Livingstone, 1988, p 10.

procedure

Knee Joint Aspiration

Materials

- Sterile gloves

- Skin preparation solution (iodinated soap or similar antiseptic solution)

- 25-gauge needle

- 30+ mL syringe with an 18-gauge, 1½" needle

- 5 mL of a local anesthetic

- 40 to 80 mg of corticosteroid preparation (optional)

- Adhesive dressing

Step 1

Wear protective gloves at all times during the procedure and use sterile technique.

Step 2

Cleanse the skin with an iodinated soap or similar antiseptic solution.

Step 3

Use the 25-gauge needle to infiltrate the skin with a local anesthetic, approximately 2 cm proximal and lateral to the patella (Figure 1).

Step 4

Withdraw this needle and insert the 18-gauge needle-syringe combination through the same tract into the suprapatellar pouch. There is usually a slight give as the needle penetrates the knee capsule. Ensure that approximately half of the 1½" needle remains outside the skin.

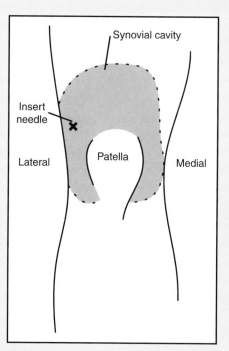

Figure 1
Location for needle insertion

Knee Joint Aspiration (continued)

Step 5

Aspirate synovial fluid, emptying the syringe as needed. When material no longer flows, use one hand to compress the suprapatellar area and force fluid into the lateral pouch. If the needle becomes clogged while there is still considerable effusion present, rotate the needle or inject a small amount of aspirate to see if the channel is clear. If it is not clear, introduce the needle another ¼" and try again. At times, hypertrophic synovial tissue may cover the needle inlet as the volume of joint fluid is reduced. Further aspiration may be impossible.

Step 6

After the aspiration is complete, inject any solutions, such as 40 to 80 mg of corticosteroid preparation or a local anesthetic, and withdraw the syringe.

Step 7

Dress the puncture wound with a sterile adhesive dressing.

Adverse outcomes

Aspiration of a hemarthrosis that is more than 1 day old may be impossible until the clot undergoes lysis and liquefies, which may take a week. Although rare, infection and hemarthrosis may occur.

Aftercare/patient instructions

Advise the patient to rest the knee as much as possible the first day following the injection. Instruct the patient to call you if there is increased heat, rapid recurrence of the fluid, or severe increase in pain.

knee and lower leg

Arthritis of the Knee

knee and lower leg

Synonyms

Osteoarthritis

Degenerative joint disease

Rheumatoid arthritis

"Wear and tear" arthritis

Gout

Pseudogout

Definition

There are three basic categories of arthritis: degenerative (osteoarthritis, degenerative joint disease), destructive (autoimmune, inflammatory, rheumatoid arthritis), and secondary arthritis (traumatic, septic). Osteoarthritis is the most common form of knee arthritis and can involve any or all three compartments in the knee: the medial compartment (medial tibial plateau and medial femoral condyle); the lateral compartment (lateral tibial plateau and lateral femoral condyle); or the patellofemoral compartment (patella and femoral trochlear notch). The medial compartment is most commonly involved, producing a bowleg (genu varum) deformity. Knock-knee (genu valgum) deformity comes from involvement of the lateral compartment. The patellofemoral compartment may be arthritic in association with tibiofemoral arthritis, or may be involved on its own.

Clinical symptoms

Degenerative arthritis most commonly affects patients age 55 years and older. It is associated with obesity and family history. Onset of pain is insidious, associated with activity in the early stages of the condition; and in the later stages, occurs even at rest.

Inflammatory arthritis can occur at any age: juvenile rheumatoid arthritis in patients younger than age 20 years, systemic lupus erythematosus in early adulthood (25 to 45 years), and gout and pseudogout in older adults (older than age 50 years).

Secondary arthritides are usually posttraumatic, including sequela of fractures, meniscal tears (and/or meniscectomy), and from chronic instability such as in ACL deficient knees.

Regardless of the initial cause, patients with end-stage arthritis have pain on weight bearing, sensations of buckling or giving way, stiffness, and joint swelling that can limit motion of the knee at both extremes of flexion and extension. Patients often report symptoms of locking or catching, similar to a meniscal tear due to the "sticking" of the rough joint surfaces and/or inflamed soft-tissue envelope.

Tests

Exam

Examination most commonly reveals a bowleg deformity that is sometimes subtle and must be viewed with the patient standing, using the opposite knee as a comparison. There is often mild effusion, tenderness along the medial joint line or pes anserine bursa, and palpable osteophytes along the medial distal femur. Knock-knee deformity is more common in women or patients with rheumatoid arthritis. Arthritis of the patellofemoral compartment is usually indicated by crepitation, a history of difficulty with stairs, and a sensation of buckling.

Diagnostic

Weightbearing AP radiographs with the knees in full extension and partial flexion show narrowing of the joint space. Hallmarks of degenerative arthritis include asymmetric joint narrowing, sclerosis, cysts, and osteophytes (Figure 1). Hallmarks of inflammatory arthritis include symmetric joint narrowing, osteoporosis, and periarticular bone erosion. Lateral and Merchant views are nonweightbearing and help define the status of the tibiofemoral and patellofemoral joint compartments. "Tunnel" or "notch" views are helpful to demonstrate or locate an osteochondral loose body.

Analysis of synovial fluid is often not helpful unless crystalline disease or sepsis is suspected.

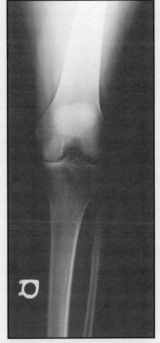

Figure 1
Early degenerative arthritis with medial compartment narrowing and sclerosis

Differential diagnosis

Bursitis (tenderness below the medial jointline)

Hip pathology (decreased internal rotation of the hip)

L3 or L4 lumbar radiculopathy (diminished knee reflex, numbness)

Meniscal tear (history of trauma with subsequent locking)

Osteonecrosis of the femur or tibia (in elderly women, use of steroids)

Septic arthritis (fever, malaise, abnormal joint fluid)

Tendinitis (tenderness directly over a tendon)

Adverse outcomes of the disease

A limp, episodes of falling, chronic pain, and a loss of knee function with weightbearing and walking may develop if the patient is not treated. There is often a general decline in the patient's overall physical condition as the patient becomes more sedentary due to a reduced activity level.

knee and lower leg

Treatment

Intermittent NSAIDs may help acute exacerbations of pain, especially in gouty arthritis, but many patients do just as well with acetaminophen. Elastic bandages, ice, heat, and liniments may temporarily relieve aching. Use of an exercise bicycle and a program of water aerobics provide a comfortable means of maintaining strength and motion if they do not increase effusions and pain. Patients with a history of falling or a limp should use a cane in the hand opposite the affected knee. Pain at rest or severe functional limitations indicate failure of conservative treatment.

Adverse outcomes of treatment

NSAIDs may cause gastric or hepatic complications. In addition, fluid retention and diminished renal function can develop, especially when NSAIDs are used in combination with antihypertensives. Repeated intra-articular injections of corticosteroid are contraindicated since relief is often temporary and is associated with occasional sepsis and accelerated destruction of cartilage.

Referral decisions/Red flags

Patients who report pain at rest or who have significant functional limitation need further evaluation.

Baker's Cyst

Synonyms

Popliteal cyst

Definition

A Baker's cyst is a fluid-filled sac located at the medial border of the popliteal fossa of the knee. This cystic sac may communicate with the knee cavity and be associated with degeneration of the posterior horn of the medial meniscus, with or without a tear of the meniscus. Most often it originates from the medial hamstring tendons. In rheumatoid arthritis, these cysts may become gigantic and dissect distally to the ankle.

Clinical symptoms

Swelling and tenderness are the common complaints. Patients often describe a feeling of "fullness" and aching in the popliteal area. Stiffness usually indicates an associated intra-articular effusion. Symptoms are worst as the cyst expands, and sometimes become tolerable as the cyst reaches a stable size.

Tests

Exam

Inspect the popliteal fossa with the patient standing and facing away from you. Prominence of the popliteal fossa, primarily on the medial side, may be apparent. Palpate both popliteal fossae, first with the patient prone, then supine with the knee flexed. The mass is most commonly located just lateral to the medial hamstring tendons.

Knee motion is usually normal, unless there is an associated effusion (possible meniscal tear or arthritis) or locking (displaced meniscal tear).

In patients with rheumatoid arthritis, the popliteal cyst may dissect distally, perhaps as far as the posterior aspect of the ankle. Enlargement of the calf and possible thrombophlebitis in these patients may be due to a Baker's cyst.

Diagnostic

Aspiration will confirm the cystic nature of the mass. AP and lateral radiographs may show calcification in the posterior meniscal area (meniscal degeneration, pseudogout), or calcification in a solid tumor (such as a synovial sarcoma).

Special diagnostic studies, such as an MRI scan, may be appropriate to confirm the diagnosis and rule out a solid tumor if aspiration is inconclusive or if surgery is planned (Figure 1).

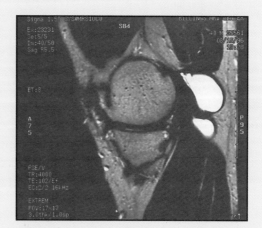

Figure 1
MRI scan of knee demonstrating cystic nature of Baker's cyst

Differential diagnosis

Rheumatoid arthritis or other inflammatory arthritides (other joint involvement, positive serologic studies)

Soft-tissue sarcoma (synovial sarcoma, fibrosarcoma, rhabdomyosarcoma)

Thrombophlebitis (pain and swelling in the calf)

Adverse outcomes of the disease

Compression of neurovascular structures may increase symptoms. Secondary thrombophlebitis is rare but may occur. Dissection of the cyst distally may lead to calf enlargement and increasing pain, especially in patients with rheumatoid arthritis.

Treatment

Aspiration may provide transient relief, especially if the cyst is enlarging and tense (see Knee Joint Aspiration). Note that the recurrence rate is high following aspiration. Surgery is appropriate if the symptoms become intolerable or threaten neurovascular function.

Adverse outcomes of treatment

Recurrence following aspiration is common, but it may be delayed and the patient's discomfort may not be as severe. Recurrence following surgery occurs in fewer than 5% of patients.

Referral decisions/Red flags

Patients with a suspected tumor, intolerable symptoms, or neurovascular symptoms need further evaluation.

Bipartite Patella

Synonyms

None

Definition

Bipartite patella is a common variant of patellar ossification, with a secondary center of ossification forming at the superolateral pole of the patella in 75% of patients, the lateral pole of the patella in 20% of patients, or the inferior pole of the patella in 5% of patients. The two (or sometimes three) bony pieces of the patella are joined by thick fibrous tissue that may be sprained and may cause recurring patellar pain associated with sports, stair climbing, or rising from a chair. Men are more commonly affected than women.

Clinical symptoms

Tenderness and pain at the superior and lateral regions of the patella are the common complaints. Patients may report increased discomfort climbing up or down stairs, difficulty in playing sports, especially those involving jumping, and they may report discomfort when rising from a chair to a standing position.

Tests

Exam
Examination reveals tenderness at the superolateral pole of the patella. While there is often a prominence in this area when compared with the opposite knee, the condition is not uncommonly bilateral.

Diagnostic
AP and lateral radiographs confirm the diagnosis. Bipartite patellar fragments have rounded edges and may vary in size from the size of a lima bean to almost half of the patella (Figure 1). The line dividing the two fragments is usually oblique. In contrast, fractures usually have sharp edges and are most often transverse.

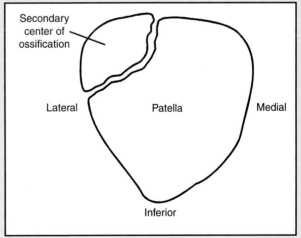

Figure 1
Bipartite patella

Differential diagnosis

Patellar fracture (sharp lines, associated bleeding into the prepatellar bursa)

Quadriceps tendinitis (local tenderness at insertion)

Adverse outcomes of the disease

Pain and limited function are common.

Treatment

Most adolescents with symptoms will improve as they limit and/or decrease jumping activities. Those who participate in competitive sports need a physical therapy program consisting of isometric quadriceps exercises, cycling, and sometimes rest from their sport.

If symptoms are acute and disabling, 2 weeks of rest with a knee immobilizer may be needed. Patients may remove the immobilizer intermittently to do range of motion exercises.

Surgery may be necessary if pain or the loss of function becomes unmanageable.

Adverse outcomes of treatment

Infection and/or rupture of the quadriceps tendon may occur postoperatively.

Referral decisions/Red flags

Failure of conservative treatment or confirmation of the diagnosis prior to treatment are both indications for further evaluation.

Bursitis of the Knee

Synonyms

Housemaid's knee

Pes anserine bursitis

Definition

Bursitis is inflammation of an adventitial bursa, whose purpose is to allow one structure (skin, tendon, or ligament) to glide smoothly over another. Bursae lie between the skin and bony prominences, between tendons and ligaments, and between tendoligamentous structures and bone. They are lined by synovial tissue, which produces a small amount of fluid and decreases friction between the adjacent structures. Chronic friction leads to inflammation and subsequent thickening of the bursa and surrounding structures.

Clinical symptoms

As with any overuse syndrome, pain with activity is common. Bursae that lie between tendinous or ligamentous structures produce pain that is worst when first rising, feels better with motion, and is worse at night. The patient often has a limp when first rising, and after a period of continued exercise such as walking or running, the patient reports aching.

The prepatellar bursa is superficial and lies between skin and the patella. When inflamed and filled with fluid, it forms a dome-shaped swelling over the patella and causes mild discomfort, but no night pain.

Tests

Exam

Frequently involved structures about the knee include the patellar tendon, pes anserine tendon, iliotibial band, and joint capsule. The pes anserine bursa lies under the conjoined insertion of the hamstrings on the medial flare of the tibia below the knee. It may become inflamed with overuse syndromes (running, walking, etc), but is often tender in patients who have osteoarthritis of the knee (Figure 1). The saphenous nerve exits beneath this structure and may become entrapped or stretched, producing tenderness over the posterior portion of the pes anserine tendon.

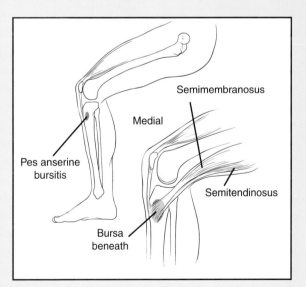

Figure 1
Region of pes anserine tenderness—medial view

The Essentials of Musculoskeletal Care   331

knee and lower leg

Iliotibial band friction syndrome seen in runners occurs over the lateral femoral condyle where the iliotibial band rubs back and forth, creating chronic inflammation and thickening (Figure 2).

Capsular bursae lie at the level of the joint and are often tender in conjunction with a tear of the meniscus (often degenerative in nature). Pain at the medial joint line is usually associated with meniscal pathology. Pain under the proximal portion of the lateral collateral ligament is best palpated with the affected foot resting on the opposite knee (male cross-leg position) and is often seen in association with lateral tibiofemoral arthritis but may be a primary condition from repetitive trauma.

Diagnostic

AP and lateral radiographs should be obtained to rule out a tumor, which is usually characterized by night pain.

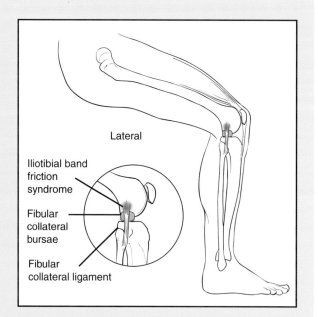

Figure 2
Region of iliotibial band bursitis—lateral view

Differential diagnosis

Medial meniscus tear (effusions, catching, and locking)

Osteoarthritis of the medial or lateral knee compartment (intra-articular effusion, osteophytes)

Patellar fracture (trauma, intra-articular effusion)

Rheumatoid arthritis (other joint involvement or bilateral involvement)

Saphenous nerve entrapment (numbness in medial shin, dysesthesias)

Septic knee (flexion contracture, swelling around but not over the patella)

Tumor

Adverse outcomes of the disease

Patients may continue to have pain and a limp. Occasionally the bursitis is associated with a partial tear of a tendon that may spontaneously rupture.

Treatment

Conservative treatment includes a short-term course of NSAIDs, ice, and a reduction or substitution of athletic activities. Modalities such as ultrasound and phonophoresis often help as well. Stretching of tight tendons (hamstrings, iliotibial band) can be performed through a home therapy program. An injection of a cortico-steroid preparation is appropriate if there is no response or night pain is intolerable (see Knee Joint Injection). Surgical treatment is rarely necessary. Recurring pes anserine bursitis is usually related to osteoarthritis of the knee.

Adverse outcomes of treatment

NSAIDs may cause gastric, renal, or hepatic complications. Spontaneous rupture may also occur following a corticosteroid injection.

Referral decisions/red flags

Failure of conservative treatment and/or tendon rupture indicate the need for further evaluation.

procedure

Knee Joint Injection

Materials

- Sterile gloves

- Skin preparation solution (iodinated soap or similar antiseptic solution)

- 3-mL syringe with a 25-gauge needle

- 3-mL syringe with a 22-gauge needle

- 2 mL of a local anesthetic

- Mixture of 80 mg of a corticosteroid preparation and 2 mL of a local anesthetic

- Adhesive dressing

Step 1

Wear protective gloves at all times during the procedure and use sterile technique.

Step 2

Cleanse the skin with an iodinated soap or similar antiseptic solution.

Step 3

Use the 25-gauge needle to infiltrate the skin with 2 mL of a local anesthetic, just posterior to the upper lateral pole of the patella.

Step 4

Withdraw this needle and insert the 22-gauge needle-syringe combination through the same tract into the suprapatellar pouch, posterior to the upper pole of the patella (Figure 1). There is usually a slight give as the needle penetrates the knee capsule. Ensure that approximately half of the 1½" needle remains outside the skin.

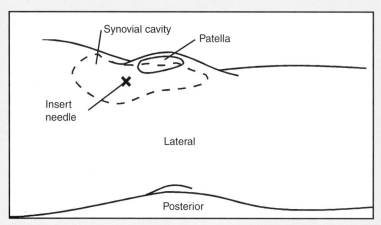

Figure 1
Location for needle insertion

knee and lower leg

Knee Joint Injection (continued)

Step 5

Inject 0.1 mL of the corticosteroid/local anesthetic mixture. If the fluid flows easily, complete the injection and withdraw the needle.

If the fluid does not flow easily, introduce the needle another ¼" and try again.

Step 6

Dress the puncture wound with a sterile adhesive dressing.

Adverse outcomes

Although rare, infection and hemarthrosis may occur.

Aftercare/patient instructions

Instruct the patient to call you if there is any sudden increase in knee stiffness, fever, or severe pain.

knee and lower leg

Collateral Ligament Tear

knee and lower leg

Synonyms

Medial or lateral capsular tear or sprain

Medial or lateral collateral ligament tear or sprain

Definition

A traumatic partial or complete tear of the primary medial or lateral stabilizer of the knee characterizes this condition. These injuries may occur alone or associated with a meniscal, anterior cruciate ligament, or posterior cruciate ligament tear.

Clinical symptoms

The mechanism of injury in a medial collateral ligament (MCL) tear is commonly a valgus (abduction) force without rotation, such as in a football clipping injury. The less common lateral collateral ligament (LCL) tear is the result of a pure varus (adduction) producing force to the knee. Most patients are able to ambulate after an acute injury and may continue to participate in athletics. Patients report localized swelling or stiffness, and medial or lateral pain and tenderness. Patient reports of instability and mechanical symptoms such as "locking" or a "popping" are infrequent after an isolated collateral ligament injury. After several days, localized ecchymosis may occur and a small effusion may develop.

Tests

Exam

Examine the normal knee first. This is important to understand what is normal for the patient and to reduce the patient's fears about examining the painful limb.

The MCL may be tender along its entire course from the medial femoral condyle to its tibial insertion (Figure 1). Isolated tenderness at its most proximal or distal extent may signify an avulsion-type injury. The LCL may be tender anywhere along its course from the lateral femoral epicondyle to its insertion on the fibular head. The MCL is best palpated in slight flexion, while the LCL is best examined in the "figure four" position. The degree

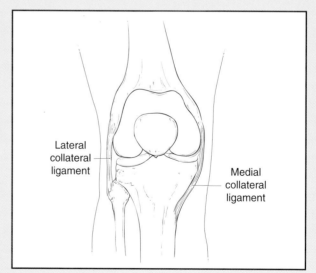

Figure 1
Medial and lateral knee structures

Lateral collateral ligament

Medial collateral ligament

of laxity following collateral ligament injury serves as a guide to treatment decisions.

The knee is examined in 25° of flexion to relax the posterior capsule while a varus and then valgus stress is applied to the knee. A joint space opening of less than 5 mm is considered a grade I (interstitial) tear, while an opening of greater than 10 mm is considered a grade III (complete) tear. A grade II (partial) tear falls between these extremes. If a patient has significant pain and guarding of the muscles bridging the knee, then instability (joint opening) may be masked. Be aware of the possibility of a false-negative exam.

Patients will often have swelling in and around the injured ligament, but the presence of a significant knee effusion might indicate an associated intra-articular injury.

Diagnostic
AP and lateral radiographs, while usually negative, may reveal an avulsion from the femoral origin of the MCL or the fibular insertion of the LCL.

Differential diagnosis

Anterior cruciate ligament tear (positive anterior drawer and Lachman's tests)

Meniscal tear (jointline tenderness)

Osteochondral fracture (radiographic evidence of loose osteochondral fragment)

Posterior cruciate ligament tear (positive posterior drawer test or sag sign)

Tibial plateau fracture (bony tenderness and radiographic evidence at fracture)

Adverse outcomes of the disease

While instability of the knee is rare, it does occur in association with disruption of the posterior lateral ligament.

Treatment

Treatment of all grades of isolated MCL tears is usually conservative, as the potential for MCL healing is great and late ligamentous laxity is uncommon. However, an injury of the anterior cruciate ligament or posterior cruciate ligament must be definitively ruled out. Grade I sprains with no effusion will likely resolve within a couple of weeks. Rest, ice, compression, and elevation (RICE) coupled with crutches and short-term NSAIDs are usually adequate. With grade II sprains, use a hinged brace and allow weightbearing as tolerated. For grade III injuries, use a hinged brace with gradual return to full weightbearing over the course of 4 weeks. Rehabilitation includes early range of motion (including cycling) and quadriceps strengthening exercises.

While grade I and II LCL tears should also be treated conservatively, grade III tears invariably involve a tear of the posterolateral capsular complex and are best treated surgically to avoid late instability. Grade I and II collateral ligament tears are typically braced for 4 to 6 weeks, while grade III tears frequently require up to 3 months of bracing and physical therapy prior to returning to unrestricted activity.

knee and lower leg

Adverse outcomes of treatment

NSAIDs are typically used for less than 1 month; therefore, gastric, renal, or hepatic complications are less likely to occur than when they are used for chronic conditions necessitating longer term treatment. Although frank instability is very uncommon following an isolated collateral ligament injury, chronic pain and a tendency for recurrent injury may occur. Missed associated diagnoses such as meniscal and ACL tears may complicate the course of conservative treatment and may ultimately require surgical intervention.

Referral decisions/Red flags

Patients with hemarthrosis or ligamentous instability need further evaluation. Failure to respond to conservative treatment may mean a missed diagnosis, such as an associated ligamentous rupture or meniscal tear.

Extensor Mechanism Rupture

Synonyms

None

Definition

Rupture of the extensor mechanism of the knee occurs in one of two ways: in younger patients as a result of a sudden or violent force (such as jumping, heavy lifting); and in older patients as a result of relatively trivial force. In either group, there may have been some prior aching. This condition affects older patients who have typically been somewhat sedentary and have suddenly increased their activity level, or patients who have had some preexisting or co-existing condition such as diabetes mellitus, rheumatoid arthritis, and other systemic inflammatory disorders, or prior knee surgery.

Clinical symptoms

Patients typically have pain and swelling. Many report hearing or feeling a pop, or they may describe the sensation as a rubber band breaking.

Tests

Exam

With partial ruptures, patients may (or may not) be able to extend the knee, but extensor strength is always poor. Most patients cannot extend their knee or do a straight leg raise, but they may be able to walk if they keep a hand on their thigh and maintain their knee in extension. There is often a palpable defect in the quadriceps tendon above the patella or the infrapatellar tendon below the patella (Figure 1). There is also often a hemarthrosis.

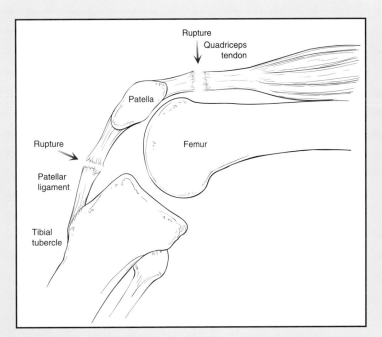

Figure 1
Areas where rupture can occur and should be palpated on exam

knee and lower leg

Diagnostic
Plain lateral radiographs may show that the patella is in a higher than usual location with rupture of the infrapatellar tendon, or a slightly lower than usual location with rupture of the quadriceps tendon. An MRI scan will confirm the rupture of the tendon, but this is rarely necessary in the presence of the strong clinical findings mentioned above.

Differential diagnosis

Fracture of the patella (dome-shaped swelling over patella, intact tendons)

Rupture of the anterior cruciate ligament (no palpable tendon defect)

Septic arthritis (fever, chills)

Adverse outcomes of the disease

Weakness of the knee extensors may be of great significance in young adults whose normal activities are strenuous. Late diagnosis may lead to contracture of the muscles and inability to repair the lesion.

Treatment

Surgical repair is almost always the treatment of choice.

Adverse outcomes of treatment

Infection and weakness of the extensor mechanism are possible following surgery.

Referral decisions/Red flags

Patients who are unable to extend the knee or have markedly reduced extensor strength need further evaluation.

Fracture of the Patella

Synonyms

None

Definition

Fracture of the patella is one of several potential injuries to the extensor mechanism of the knee. Most fractures are transverse in nature since the mechanism is one of distraction, or pulling apart of the bone. Many are comminuted, especially in the inferior fragment, since the other mechanism of injury is a direct blow to the bone while the quadriceps is under tension.

Clinical symptoms

Pain and inability to extend the knee are characteristic. Patients may report a fall directly on the patella or a fall from a height. The knee often strikes the dashboard in automobile accidents. There is rapid onset of swelling if the fracture fragments are displaced, because the posterior surface of the patella bleeds into the knee cavity and the anterior surface bleeds into the large anterior bursa.

Tests

Exam
Bleeding into the knee joint produces a large hemarthrosis with swelling above and beside the patella. Bleeding into the anterior bursa produces a characteristic dome-shaped swelling anterior to the patella, a hallmark of either patellar fracture or patellar bursitis.

The patient is unable to extend the knee against gravity if the extensor mechanism is torn in line with the patellar fracture. If the fracture is undisplaced and the extensor mechanism remains intact, the patient will be able to extend the knee, but not without pain.

Because many of these fractures are associated with a direct blow to the patella, there may be compromise of the skin as well, creating an open fracture. Any bleeding anterior wound associated with a patellar fracture should be classified as an open fracture.

Diagnostic
AP and lateral radiographs of the patella should be obtained to identify the type of fracture pattern (Figure 1).

The lateral view provides the best image to evaluate distraction of the fragments. If there are two main fragments and they are less than 6 mm apart, the extensor mechanism is often intact, which can be confirmed on examination.

Comminution of the distal fragment is evident in both the AP and lateral views.

Bipartite patella occurs when a secondary ossification center occurs in the patellar anlage. It produces a separate bony fragment located at the superior and lateral pole of the patella and may be quite large (30% of the patella). Features that distinguish it from a fracture include the following: rounded borders of the bone fragments; location at the superolateral pole; and lack of dome-shaped swelling anterior to the patella on the radiograph and physical examination. This condition is frequently seen bilaterally on comparison radiographs.

Aspiration of the knee with an 18-gauge needle will return blood. Marrow fat will float to the top of the blood, confirming a fracture.

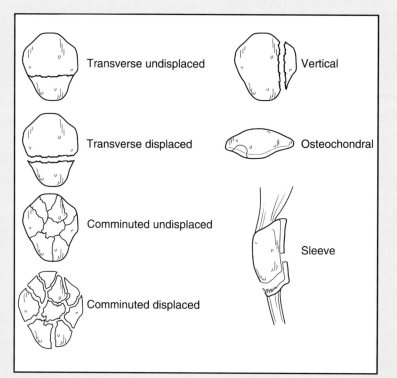

Figure 1
Common patellar fracture patterns

Differential diagnosis

Bipartite patella (rounded borders of bone fragments)

Patellar bursitis (normal radiographs)

Patellar dislocation or subluxation (Merchant view shows malalignment)

Adverse outcomes of the disease

Loss of knee extension, further displacement of bone fragments, patellar arthritis, and/or loose bodies within the knee joint are possible.

knee and lower leg

Treatment

Immobilize the knee in extension for 6 weeks if the fragments are separated by less than 5 mm (AP view) and less than 2 mm displaced (lateral view) and the patient retains extensor function. During that time (usually after 3 or 4 weeks), the patient may remove the splint for pain-free range of motion exercises as comfort improves.

Aspiration of the joint may relieve the pressure of a large hemarthrosis (see Knee Joint Aspiration), although the communication of the joint and anterior bursa through the fracture usually decompresses the knee enough to make aspiration unnecessary.

Adverse outcomes of treatment

Further displacement of fragments, failure of a fixation device, postoperative infection, and nonunion of fragments are all possible.

Referral decisions/Red flags

Any fracture with loss of extensor function, comminution of the patellar fragments, or an associated dislocation or subluxation requires surgical intervention. Patients with more than 1 mm of anteroposterior offset of the fracture fragments or more than 5 mm of separation also need additional evaluation. All open fractures require urgent evaluation.

knee and lower leg

Acknowledgements

Figure 1 is reproduced with permission from Carpenter WB, Kasman R, Matthews LS: Fractures of the patella, in Schafer M (ed): *Instructional Course Lectures 43*. Rosemont, IL, American Academy of Orthopaedic Surgeons, 1994, pp 97–108.

Fracture of the Tibia

Synonyms

None

Definition

Tibial diaphyseal fractures occur between the proximal and distal metaphyseal flares. These fractures are considered open if there is any break in the skin at or near the fracture site or closed if the skin is intact.

Pilon fractures are high-energy breaks that involve the distal tibia and the ankle joint. With these fractures, the articular surface of the distal tibia is in multiple pieces with fracture lines extending into the distal third of the tibia.

Clinical symptoms

Pain and inability to bear weight on the affected limb are the principal symptoms. Patients with these fractures often have other associated injuries.

Tests

Exam

Examination reveals swelling, ecchymosis, tenderness, and deformity. Instability of the limb is common, especially when both the tibia and fibula are fractured. Punctures or lacerations at or near the fracture site signal an open fracture until examination proves otherwise. Penetrating wounds with sharp objects can travel long distances under the skin. Sharp spikes of bone can also cut through the skin at some distance from the area where the fracture is seen radiographically, since the leg can "telescope" on itself after the bone is broken.

Absence of the posterior tibial arterial pulse may indicate serious arterial injury. Muscle paralysis indicates nerve compromise or compartment syndrome. This examination is important, not only to identify soft-tissue injuries, but also to document that such disruptions were caused by the original injury rather than the subsequent treatment. Repeat this examination at frequent intervals to monitor for evolving compartment syndrome. Pain with passive extension or flexion of the toes is the most sensitive sign of compartment syndrome.

Evaluate carefully for fractures in other regions, especially the spine. Other injuries may include liver laceration, ruptured spleen, perforated viscus, cardiac contusion, pneumothorax, pulmonary contusion, closed head injury, and other fractures.

Diagnostic

Obtain plain AP and lateral radiographs to visualize the tibia and an oblique view at the knee or ankle if a joint fracture is possible. Look for air in the soft-tissue shadow on the radiographs. Include the joint above and below the suspected fracture site to rule out other fractures or dislocations (Figure 1).

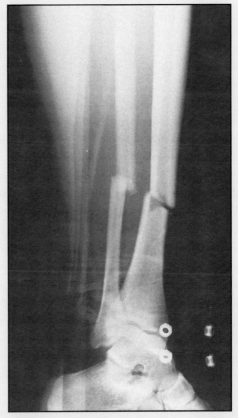

Figure 1
Fracture of the tibia and fibula in which a splint is in place

Differential diagnosis

Contusion

Sprain

Adverse outcomes of the disease

Nonunion (failure to heal) or malunion (healing crooked) may occur. Compartment syndrome may result in extensive muscle and nerve necrosis. Osteomyelitis can complicate open fractures, and posttraumatic arthritis commonly complicates pilon fractures.

Treatment

Since the tibia is a weightbearing structure, the goals of treatment are healing of the fracture with preservation of normal alignment, joint function, and length.

If there is an open wound, swab it for aerobic and anaerobic culture, irrigate, and cover it with saline-soaked gauze. Give a tetanus booster, if needed, or give immune globulin if the patient has a grossly contaminated wound and no record of appropriate prior immunizations.

Begin antibiotics as soon as possible after taking cultures; most experts favor a first-generation cephalosporin, such as 1 g of cephalothin, in combination with a loading dose of aminoglycoside, such as gentamicin, 1.5 mg/kg. Remember that a loading dose need not be reduced based on renal compromise.

The limb should be splinted in a well-padded, long leg splint. Apply ice and elevate the extremity, just above the level of the heart. Excessive elevation may reduce perfusion pressure, worsening or causing ischemia. Give nothing by mouth since emergency surgery is frequently necessary for these injuries.

Adverse outcomes of treatment

These are the same as for conservative treatment, but also include failure of fixation.

Referral decisions/Red flags

All tibial fractures are complex injuries that require further evaluation. Patients with open fractures and/or compartment syndrome require emergency care.

Acknowledgements

Figure 1 is reproduced with permission from Levine AM (ed): *Orthopaedic Knowledge Update: Trauma.* Rosemont, IL, American Academy of Orthopaedic Surgeons, 1996, pp 171–182.

Fracture of the Tibiofemoral Joint

Synonyms

Tibiofemoral fracture

Intra-articular fracture

Plateau fracture

Condylar fracture

Osteochondral fracture

Definition

Fractures of the tibiofemoral joint threaten the weightbearing and motion functions of the knee. They can involve the medial or lateral tibial plateau (flat articular surface of the tibia), the anterior tibial spine with its attached anterior cruciate ligament (ACL), and the medial or lateral femoral condyle (Figures 1-3).

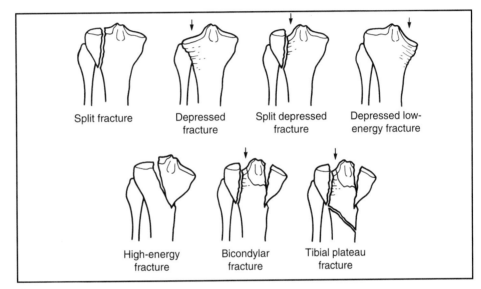

Figure 1

Common proximal tibial intra-articular knee fractures

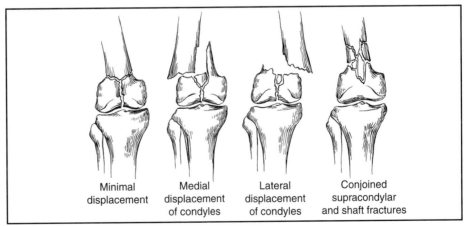

Figure 2

Supra- and intercondylar knee fractures (Neer classification)

Clinical symptoms

In the absence of osteoporosis, patients typically have a history of a significant injury, such as skiing, falling from a height, automobile trauma, etc. The knee swells rapidly because most fractures breach the articular surface and bleed directly into the joint. Pain comes from the fracture itself and the high pressure hemarthrosis within the knee cavity. The patient is unable to bear weight.

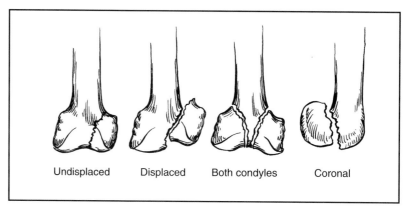

Undisplaced   Displaced   Both condyles   Coronal

Figure 3
Femoral condylar knee fractures

Tests

Exam

Examination typically reveals a hemarthrosis, with distention of the knee cavity above and to either side of the patella. Obtain radiographs if there is any hint of medial/lateral instability.

Lachman's test is positive if the anterior tibial spine is fractured (site of attachment of the ACL).

There may be a varus (adduction) or valgus (abduction) deformity if the tibial articular surface (tibial plateau) or femoral condyle are involved and displaced.

If marrow fat is floating on an aspirate of bloody fluid from the knee, there is a high likelihood of fracture.

Diagnostic

AP, lateral, and oblique radiographs demonstrate most knee fractures. At times, a tunnel view will help identify anterior tibial spine fractures or loose bodies. CT scans are used primarily for preoperative assessment.

Fractures of the tibial plateau (flat articular surface of the tibia), either medial or lateral, are sometimes difficult to visualize. If there is bony tenderness along the tibial plateau and a hemarthrosis with negative radiographs and a stable cruciate, a CT scan should be considered to identify a possible occult plateau fracture. Depressions in the central portion of a tibial plateau are difficult to identify; instability (varus or valgus) with increased density of the subarticular bone will increase suspicion.

Osteochondral fractures (slice fractures) from a femoral condyle show a loose body, often thin, within the joint.

Differential diagnosis

ACL tear (hemarthrosis, no fracture)

Loose body (osteochondritis dissecans; the edges are rounded, not sharp)

Patellar fracture (tenderness over patella and radiographic evidence)

Adverse outcomes of the disease

Malunion, nonunion, posttraumatic arthritis, and/or instability of the knee are possible. Compartment syndrome is uncommon, but may occur with severe displacement of the tibial plateau. Thrombophlebitis and pulmonary embolism commonly occur with these injuries. Posttraumatic infection or osteomyelitis is also possible.

Treatment

For patients seen within the first 6 hours following the injury, consider evacuation of tense, painful hemarthroses (see Knee Joint Aspiration); if the patient is seen after that, the blood often clots and cannot be aspirated easily. Look for marrow fat floating on top of the bloody aspirate to confirm an intra-articular fracture.

Initial treatment should include rest, ice, compression, and elevation (RICE) along with a knee immobilizer, crutches, and NSAIDs.

The definitive treatment of fractures around the knee is determined by a thorough assessment of the radiographs and a CT scan, if necessary. Surgical intervention is often required.

Adverse outcomes of treatment

Infection, nonunion, and malunion are all possible. Missed diagnosis of a fracture, or further displacement of a fracture, either by too-early weightbearing or by normal pull of the muscles crossing the knee joint can also occur.

Referral decisions/Red flags

Patients who have a tibial plateau fracture, a femoral condylar fracture, any instability, air in the tissues, any open fracture, and/or any associated circulatory compromise need further evaluation.

Acknowledgements

Figure 1 is adapted with permission from Perry CR: Fractures of the tibial plateau, in Schafer M (ed): *Instructional Course Lectures 43*. Rosemont, IL, American Academy of Orthopaedic Surgeons, 1994, pp 119–126.

Figure 2 is adapted with permission from Neer CS, Grantham SA, Shelton ML: Supracondylar fractures of the adult femur. *J Bone Joint Surg* 1967;49A:592.

Figure 3 is adapted with permission from Rockwood CA, Green DP (eds): *Rockwood and Green's Fractures in Adults,* ed 2. Philadelphia, PA, JB Lippincott, 1984, vol 2, p 1444.

Knee Joint and Bursa Infection

Definition

Synonyms

Septic prepatellar bursitis

Septic knee

The knee is the most commonly infected joint, either intra-articularly or in its adjacent patellar bursa. In most instances, the knee is the only joint involved. However, if the knee is one of several joints involved, the patient is usually immunocompromised and has an existing source of bacterial sepsis.

Clinical symptoms

Patients report sudden, severe pain, stiffness (effusion), and warmth associated with fever. Occasionally, there is a history of penetrating injury, intra-articular injection, or prior successful joint replacement. Chronic infection is unusual.

The prepatellar bursa may become infected as a result of trauma, such as a direct blow or chronic trauma as in "housemaid's knee." Infection may also be secondary to an adjacent distal infection spread by lymphatic drainage. The swelling is extra-articular. Nearby abrasions or pimples, or a foot infection, may also be potential sources.

Tests

Exam

With infections in the prepatellar bursa, there is usually erythema, edema, and a dome-shaped swelling over the patella that is easily identified by sighting across the normal knee to the infected bursa; the patella itself may not be easily palpable. The clinical picture of prepatellar bursitis is similar to that of a subcutaneous abscess (Figure 1). Surface findings are more dramatic with an infected bursa, in contrast to the systemic signs of a septic joint.

The cavity of the knee joint is large and extends above the patella 2" or 3"; swelling within the joint appears around but not over the patella (Figure 2). Patients often report stiffness rather than swelling since flexion is blocked due to elevated hydrostatic pressure. The position of greatest comfort is flexion to about 30° to maximize joint volume. Generally, patients with septic knee joints have extremely painful motion, if there is any motion at all. When the patient has a significant septic effusion, the patella is lifted away from the distal femur by the fluid; the patella can then be balloted on the anterior femur. The presence of heat, erythema, and enlarged regional nodes are further indications of infection.

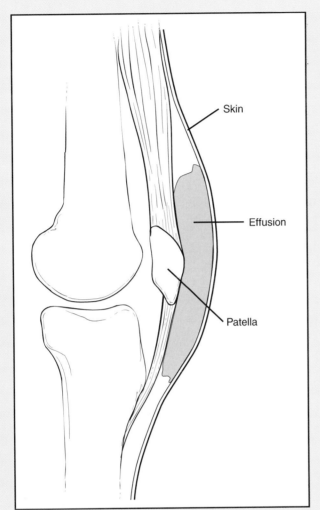

Figure 1
Involved region with septic prepatellar bursitis

Figure 2
Involved region with septic intra-articular knee effusion

Diagnostic

Radiographs are usually normal but an effusion may be noted. A penetrating injury to the knee will often show intra-articular air. Laboratory studies should include a CBC and a differential C-reactive protein or erythrocyte sedimentation rate. Joint aspirate should be examined for cell count (WBC > 25,000/mm³), crystals (pH drop in infection may precipitate gout), and then cultured. A Gram stain should be ordered to identify bacteria and also to provide early diagnosis so that definitive antibiotic therapy can be started.

Prepatellar bursitis is rarely associated with intra-articular infection; use care not to aspirate a joint through the infected tissue at the margin of a prepatellar bursal infection.

knee and lower leg

Differential diagnosis

Crystalline arthropathies (crystals on aspiration)

Gonorrhea (primarily in women, associated vaginitis, associated migratory polyarthralgias; culture genitourinary tract)

Immunodeficiency states (decreased WBC, decreased CD4)

Lyme disease (often 6 months after the infection; positive ELISA)

Prior joint replacement (1% to 2% chance)

Rheumatoid arthritis (in men at risk for polyarticular septic arthritis)

Subcutaneous abscess (superficial laceration in location other than anterior knee region)

Adverse outcomes of the disease

Delayed treatment of joint infections can destroy the joint cartilage. Joint replacements may then be contraindicated, leaving fusion (arthrodesis) the only option.

Treatment

Patients with blood or fluid in the bursa are at risk for a *Staphylococcus aureus* infection. Initial treatment consists of aspirating the fluid, obtaining a culture, and then prescribing oral antibiotics (see Knee Joint Aspiration). If the patient does not respond to initial treatment within 48 hours, inject a local anesthetic, and then incise, drain, and pack the area of infection.

Initial treatment of knee joint infections should consist of complete aspiration of the septic fluid, followed by bacterial culture and appropriate culture-based antibiotics. Immobilize the joint for rest. Surgical drainage of the joint via arthroscopy or arthrotomy is usually required. Antibiotic therapy, based on the organism and its sensitivities, should be continued until the erythrocyte sedimentation rate is normal, which may not occur in certain associated disease states, or until all obvious signs of infection are resolved. Gonoccocal infections respond well to IV penicillin.

Adverse outcomes of treatment

Despite appropriate diagnosis and treatment, joint destruction is still possible. Inoculation of a noninfected joint is possible if the joint is aspirated through the cellulitis or abscess of a prepatellar bursa.

Referral decisions/Red flags

Up to 75% of patients with septic bursitis require surgical incision and drainage. Septic arthritis is often a surgical emergency, especially with Staphylococcus organisms, which often loculate and are not cleared by repeated aspirations. A team approach is the best way to manage these potentially devastating infections.

Meniscal Tear

Synonyms

Torn cartilage

Locked knee

"Bucket handle" meniscal tear

Definition

A meniscal tear is characterized by a traumatic or degenerative tear of the medial or lateral meniscus. Meniscal tears may occur as isolated injuries or in association with a medial collateral or anterior cruciate ligament (ACL) tear.

Clinical symptoms

Patients typically report a significant twisting injury to the knee. Older patients with a degenerative tear, however, may have a history of minimal or no trauma, such as arising from a squatting position. Most patients can ambulate after an acute injury and may continue to participate in athletics. The acute event is typically followed by the insidious onset of swelling and stiffness. Untreated, the patient may present weeks after the initial injury with painful "clicking," "popping," or "locking," followed by recurrent knee stiffness. Most patients report stiffness rather than swelling; the effusion makes the knee feel tight.

Tests

Exam

A small to moderate effusion is commonly present for many days after the injury, although in patients presenting late, there may be no effusion. A torn meniscus will be tender along the joint line, usually near its adjacent collateral ligament (Figure 1). The McMurray test, with valgus stress on the knee, and the tibia externally or internally rotated as the knee is brought into extension from a fully flexed position, may elicit a painful click in the medial or lateral joint line. This test is very nonspecific.

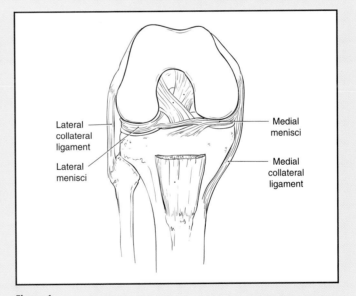

Lateral collateral ligament

Lateral menisci

Medial menisci

Medial collateral ligament

Figure 1
The relationship of the meniscus to the collateral ligaments

Diagnostic

In patients older than age 40 years, obtain the following radiographic views: AP in full extension and 30° of flexion; lateral; and bilateral Merchant views. These images will help rule out osteoarthritis, osteochondritis dissecans, patellar malalignment, and anterior tibial spine fractures.

If an effusion occurs within 3 hours of the injury, arthrocentesis may be indicated, since hemarthrosis is uncommon in an isolated meniscal injury and should raise suspicion of an ACL tear or intra-articular fracture. Most gradual-onset effusions are mild, but if the patient complains of stiffness and has an effusion, the aspirate will likely be clear and straw-colored, indicating meniscal tears or osteoarthritis.

Differential diagnosis

Anterior or posterior cruciate ligament tear (hemarthrosis)

Anterior tibial spine fracture (hemarthrosis, fracture evident on AP and lateral views)

Medial collateral ligament tear (instability with valgus stress and localized tenderness at bone insertion)

Osteoarthritis (joint space narrowing on standing views)

Osteochondral fracture (radiographic evidence of loose osteochondral fragment)

Osteochondritis dissecans, or intra-articular loose body (evident on radiograph)

Patellar malalignment (Merchant view will show lateral subluxation, tilt, or patellar arthritis)

Tibial plateau fracture (bony tenderness and radiographic evidence of fracture)

Adverse outcomes of the disease

Recurrent locking, damage to the articular cartilage, and subsequent osteoarthritis are possible, particularly with a delay in definitive treatment. While small, peripheral tears may heal, patients with recurrent stiffness, locking, or pain have a mechanically significant tear.

Treatment

In the absence of locking and instability, initial treatment should consist of rest, ice, compression, and elevation (RICE) for comfort, a short course (1 to 2 weeks) of NSAIDs, and gradual return to activity.

Recurrent catching, popping, or effusion usually indicates a mechanically significant lesion that may need surgical debridement.

Adverse outcomes of treatment

NSAIDs may cause gastric, renal, or hepatic complications. Postoperative failure of meniscal repair and infection are possible.

Referral decisions/Red flags

Persistent pain, locking, ligamentous instability, or fracture usually indicates a surgical lesion. Failure to properly diagnose or adequately treat a meniscal tear in an active patient results in osteoarthritis.

Acknowledgements

Figure 1 is adapted with permission from Mikosz RP, Andriacchi TP: Anatomy and biomechanics of the knee, in Callaghan JJ, Dennis DA, Paprosky WG, et al (eds): *Orthopaedic Knowledge Update: Hip and Knee Reconstruction.* Rosemont, IL, American Academy of Orthopaedic Surgeons, 1995, pp 227–239.

knee and lower leg

Patellar/Quadriceps Tendinitis

Synonyms

Jumper's knee

Osgood-Schlatter disease

Sinding-Larsen-Johannson disease

Definition

Extensor mechanism tendinitis is an overuse or overload syndrome involving either the quadriceps tendon at its insertion on the upper pole of the patella, or the infrapatellar tendon at its origin from the inferior pole of the patella or its insertion at the tibial tubercle. Younger patients with this condition are often engaged in jumping sports (jumper's knee) or have erratic exercise habits. This condition may also develop in older patients following a lifting strain or a significant change in their exercise level. Weight gain is sometimes a factor.

Clinical symptoms

Anterior knee pain is the hallmark. Patients often point to a tender spot where symptoms concentrate. They may report night pain, or pain with sitting, squatting, or kneeling. Climbing stairs often increases the pain.

Tests

Exam

Palpation reveals tender points near the bony attachment of the quadriceps tendon or the infrapatellar tendon. There may also be increased heat and mild swelling in the tender area. Examination in the area of the infrapatellar bursa (below the patella and behind the infrapatellar tendon) often reveals puffiness (Figure 1). Knee motion is normal, but painful at full extension and sometimes painful with resisted extension. In addition, there may be tenderness at the distal end of the quadriceps muscle, above the bony insertion on the patella. With the patient supine, assess the integrity of the anterior cruciate (ACL) and posterior cruciate (PCL) ligaments with a Lachman's test (anterior and posterior drawer test with the knee in 20° of flexion).

Symptoms are usually exacerbated by hyperflexion of the knee, which increases the stresses on the extensor mechanism. If the condition is long-standing, there may be more than 2 cm of quadriceps atrophy.

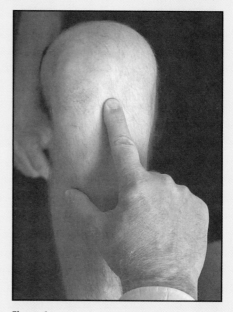

Figure 1
Region of tendinitis in a patient with patellar tendinitis

knee and lower leg

Diagnostic
AP, lateral, and bilateral Merchant radiographs may show some hyperostosis at the upper or lower pole of the patella, but these are usually normal. If a partial rupture (significant weakness in extension but no palpable defect) of the tendon is possible, an MRI scan may be of assistance in confirming the diagnosis.

Differential diagnosis

ACL or PCL injury (positive Lachman's test)

Inflammatory conditions (systemic disorders)

Knee joint infection (fever, warmth around the knee, elevated erythrocyte sedimentation rate or WBC)

Partial rupture of the extensor mechanism (weakness, palpable defect)

Adverse outcomes of the disease

Pain and sometimes spontaneous rupture of the tendon can occur.

Treatment

Treatment is primarily symptomatic and typically nonsurgical, but involves two critical aspects. First is a period of rest. Depending on the severity of the condition, this may vary from 3 or 5 days to as long as 3 weeks. In some instances, patients may need to use a knee immobilizer intermittently. NSAIDs may help control symptoms, but corticosteroids should be avoided around the extensor mechanism as the risk of rupture of the extensor mechanism is a strong possibility.

The second phase of treatment is focused on regaining strength and pain-free range of motion. Physical therapy focusing on quadriceps strengthening and extensor stretching should be initiated, often in conjunction with ultrasound and phonophoresis. Use of a knee sleeve brace with a patellar cutout is helpful.

Adverse outcomes of treatment

NSAIDs may cause gastric, renal, or hepatic complications. Pain is possible, and sometimes spontaneous rupture of the tendon can occur. Persistent functional impairment is the primary typical disability.

Referral decisions/Red flags

Patients with a possible rupture of the extensor mechanism need further evaluation.

Patellofemoral Instability and Malalignment

knee and lower leg

Synonyms

Patellar subluxation

Patellar instability

"Miserably malaligned patella"

Patellar dislocation

Definition

Patellofemoral instability is a continuum from malalignment to instability to dislocation. These conditions occur when the patella is not centralized and tracking correctly on the distal end of the femur. Malalignment may exist without subluxation, and subluxation may exist without malalignment, but the two are usually associated. The malaligned patella tracks too far laterally, resulting in excessive wear of its lateral face, and leads to chondromalacia and eventually osteoarthritis. These conditions are more common in women, in patients with ligamentous laxity, or patients of shorter stature. The origin of the subluxation is often complex and multivariate. Factors that increase the Q angle, such as internal torsion of the femur, lateral insertion of the infrapatellar tendon on the tibia, and knock-knee, are common. Other factors include hypoplasia of the lateral condyle of the femur, a high-riding patella (patella alta), and back-knee (genu recurvatum). These conditions are often bilateral.

Clinical symptoms

The patient usually reports that the knee buckles or that the kneecap slips off (instability). Pain, with or without instability, may be the primary complaint. Patients often indicate they are unable to participate in sports. Those who are able to compete may notice crepitation and stiffness. Symptoms are worse in activities that require pivoting away from the fixed foot, such as basketball or volleyball.

Tests

Exam

Patients with recurring subluxations of the patella often become apprehensive about any activity. These patients often cry out or grab your hand as you touch their patella. They have had so many episodes of subluxation, most of which are painful and transiently disabling, that even a physical examination is threatening. For those reasons, direct examination of the patella should come at the end of the exam.

First examine the patient standing. Watch the patient walk to see if the patellae point toward each other (a sign of internal femoral torsion); look for genu valgum (knock-knee) and inadequate development of the vastus medialis obliquus (VMO) muscle at the distal and medial thigh. Measure the Q angle, which is an angle constructed by a line drawn from the anterior superior iliac spine to the center of the patella, and from there to the center of the tibial tubercle. In women, it should be less than 9° with the knee in 90° of flexion. In men, it should be less than 8° with the knee in 90° of flexion.

With the patient seated, check for excessive internal femoral torsion (internal rotation exceeds external rotation by more than 30°). Have the patient move the knee back and forth through flexion and extension; as the knee nears full extension, the patella moves laterally, scribing the course of an "upside down J." Palpate for crepitation as the patient extends the knee. Finally, as the patient extends the knee, attempt to subluxate the patella laterally by pushing against its medial side; this may increase discomfort or cause the patella to subluxate and the patient to grab your hand (Figure 1). As the patient extends the knee, push or pull the patella to the lateral side to demonstrate instability. This can also be done with the patient supine, the knee relaxed, and the patella pushed laterally.

With the patient supine and the knee extended and relaxed, palpate the undersurface of the patella; there may be significant tenderness in the inferior medial portion. Measure the thigh circumference bilaterally, looking for a difference of more than 2 cm.

Patients with complete disclocations usually present with the knee locked in flexion. What at first may be confused with a meniscal tear or knee dislocation can easily be diagnosed by careful palpation of the patella (Figure 2).

Foot moves out and up

Figure 1
Testing for patellar subluxation

Diagnostic
Appropriate radiographs include AP, lateral, and bilateral Merchant views. Lateral displacement of the patella and tilting of the patella are seen in the Merchant view (Figure 3).

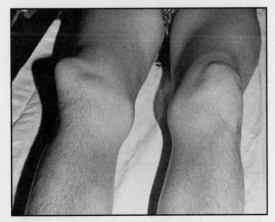

Figure 2
Lateral dislocation of the patella

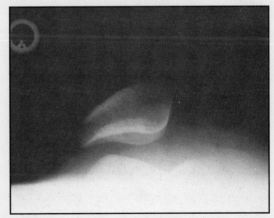

Figure 3
Merchant view of the knee with bilateral patellar tilt and subluxation

knee and lower leg

Differential diagnosis

Chondromalacia (arthroscopic diagnosis)

Quadriceps weakness (right to left differences, consider EMG/nerve conduction velocity studies)

Adverse outcomes of the disease

Persistent abnormal tracking of the patella usually leads to patellofemoral arthritis. Recurrent instability causes many patients to lead a more sedentary lifestyle.

Treatment

Initial treatment is directed toward improving the size and function of the VMO, which pulls the patella medially during terminal extension of the knee. Physical therapy should focus on short-arc ($0°$ to $15°$) activities. Cycling is beneficial.

Use of a patella stabilizing brace may be of some benefit, especially if the malalignment is only of a mild variety. The goal is to return the patient to normal activity patterns within 4 to 12 weeks.

For patients with a "miserably malaligned patella" (internal femoral torsion, increased Q angle, and, sometimes, genu valgum), an exercise regimen may need to be augmented by surgery.

Adverse outcomes of treatment

Persistent subluxation and postoperative infection are both possible.

Referral decisions/Red flags

Patients with a "miserably malaligned patella" and those with recurrent effusions (stiffness) need further evaluation.

Acknowledgements

Figure 2 is reproduced with permission from Crosby LA, Lewallen DG (eds): *Emergency Care and Transportation of the Sick and Injured,* ed 6. Rosemont, IL, American Academy of Orthopaedic Surgeons, 1995, p 555.

Patellofemoral Pain

Synonyms

Patellofemoral pain syndrome

Anterior knee pain

Chondromalacia

Definition

As a pathology term, chondromalacia refers to softening or fissuring of the articular surface of a joint; it may or may not be associated with symptoms. With fraying of the patellar articular surface, audible crepitation is a common symptom, especially as patients rise from a chair or climb stairs. Forces on the articular surface of the patella in a typical 200-lb man may vary from 600 to 3,000 lb per square inch in activities, ranging from walking to running. The most common causes are cumulative trauma, traumatic dislocations of the patella, patellar malalignment, increased patellar compression (associated with obesity and/or weight lifting), and primary osteoarthritis. The ultimate outcome is osteoarthritis of the patellofemoral joint, affecting women more often than men and hastened by obesity. Associated effusions are rare. In the absence of a clear history of trauma, the condition is often bilateral.

Clinical symptoms

Patients most commonly report anterior knee pain in the region of the patella that is worse after prolonged sitting (theater sign), climbing stairs, jumping, or squatting. In most instances, there is no preexisting trauma, although patients may have a history of a direct blow on the patella. Some patients report buckling of the knee, or sometimes a sticking sensation of the patella.

Tests

Exam

It is important to reserve all patellar manipulation until the end of the examination; otherwise, the patient will likely resist the entire examination. First, examine the patient standing. Watch the patient walk to see if the patellae point toward each other (a sign of internal femoral torsion), look for genu valgum (knock-knee) and for inadequate development of the vastus medialis obliquus (VMO) muscle at the distal and medial thigh.

Measure the Q angle, which is an angle constructed by a line drawn from the anterior superior iliac spine to the center of the patella, and from there to the center of the tibial tubercle. In women, it should be less than 22° with the knee in extension, and less than 9° with the knee in 90° of flexion. In men, it should be less than 18° with the knee in extension, and less than 8° with the knee in 90° of flexion.

knee and lower leg

With the patient seated, check for excessive internal femoral torsion (internal rotation of the hip exceeds external rotation by more than 30°). There is often palpable (and sometimes audible) crepitation, which may exist in asymptomatic knees. Have the patient move the knee back and forth through flexion and extension; as the knee nears full extension, the patella may move laterally more than a centimeter, or may even subluxate (instability). As the patient is extending the knee, attempt to subluxate the patella laterally by pushing against its medial side; this may increase the patient's discomfort or cause the patella to subluxate. The patient usually grabs the examiner's hand or expresses apprehension if there is patellar instability.

With the patient supine and the knee relaxed, palpate the undersurface of the patella; there may be significant tenderness under the medial or lateral surface. Measure the thigh circumference bilaterally looking for a difference of more than 2 cm or specific atrophy of the VMO. Have the patient stiffen the knee and palpate the quadriceps muscle, especially the VMO, for firm tone. Pain associated with chondromalacia is often increased by hyperflexion of the knee, which increases pressure on the patella.

Effusions are uncommon and usually indicate osteoarthritis or meniscal tears.

Diagnostic

Appropriate radiographs include the AP, lateral, and bilateral Merchant views. A Merchant view helps to rule out malalignment and arthritis (Figures 1 and 2).

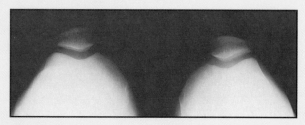

Figure 1
Normal Merchant view, patella well-aligned in femoral groove

Figure 2
Merchant view showing bilateral patellar subluxation

Differential diagnosis

Meniscal tear (with or without locking; jointline tenderness)

Patellar malalignment (clinical and radiographic malalignment)

Patellar osteoarthritis (in older patient, effusion; crepitus and radiographic evidence on Merchant view)

Patellar tendinitis (jumper's knee, inferior pole tenderness; local tenderness at patellar tendon)

Quadriceps tendinitis (local tenderness at insertion)

Adverse outcomes of the disease

Pain and dysfunction are the principal problems. Quadriceps atrophy, weakness, and buckling are all possible. Tendon strains and overload may be associated with weakness. Patellar osteoarthritis is also a possibility.

Treatment

All patients should be started on a program of quadriceps (especially VMO) strengthening and flexibility. Initially, full-arc quadriceps exercises (0° to 90°) should be avoided and short-arc activities (0° to 15°) emphasized. Once quadriceps strength and flexibility has been recovered, then greater range of motion activities, such as cycling, can be initiated. Use of a simple knee sleeve with a patellar cutout or strap may help. Some patients benefit from intermittent, short-term use of NSAIDs. Weight loss is recommended if there is concomitant obesity.

Adverse outcomes of treatment

NSAIDs may cause gastric, renal, or hepatic complications. In addition, aggressive full-arc quadriceps exercises may aggravate the symptoms.

Referral decisions/Red flags

Patients with persistent symptoms, including pain, recurrent effusions, or jerky or painful patellar motion indicates the need for further evaluation to consider surgical intervention.

Posterior Cruciate Ligament Tear

Synonyms

PCL tear

Definition

The posterior cruciate ligament (PCL) runs obliquely from the femur anteriorly to the tibia posteriorly. This injury, which is characterized by a traumatic rupture of the primary posterior stabilizer of the knee, may occur alone, but more often it occurs in association with a tear of the anterior cruciate ligament (ACL), collateral ligament, or meniscus.

Clinical symptoms

Patients typically report that they have fallen on a flexed knee or sustained a blow to the anterior aspect of their bent knee (dashboard injury). Swelling is marked and usually immediate (within 3 hours). Most patients have a painful range of motion and profound difficulty ambulating after the injury. If untreated, patients may present late with knee pain and weakness, but reports of giving way or instability are rare.

Tests

Exam

Examination reveals a large effusion, which persists for several days following the injury. Knee aspiration almost always reveals a hemarthrosis, indicating a fracture or tear of an intra-articular structure. The posterior drawer test with the knee flexed to 90° and the foot externally rotated is usually positive. Visual inspection, with the patient relaxed in a supine position and both knees flexed to 90°, often demonstrates a posterior sag of the injured knee (Figure 1). This is best seen looking from the lateral side of the affected knee across to the other knee. Posterior cruciate tears are often part of a knee dislocation, with serious multidirectional instability and associated vascular injury.

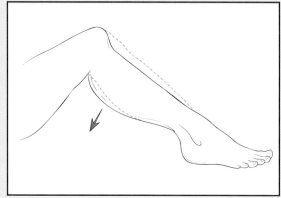

Figure 1
Posterior sag of the proximal tibia

Diagnostic

AP and lateral radiographs rarely show an avulsion of the posterior tibial insertion of the PCL. The "tunnel" view is best to visualize the intercondylar notch to identify the occasional bony avulsion.

knee and lower leg

Differential diagnosis

ACL tear (positive anterior drawer and Lachman's test)

Collateral ligament tear (varus/valgus laxity and/or local tenderness)

Meniscal tear (jointline tenderness)

Osteochondral fracture (radiographic evidence of loose osteochondral fragment)

Patellar tendon rupture (inability to straight leg raise)

Patellofemoral dislocation (palpable patellar dislocation or apprehension when stressing reduced patella)

Tibial plateau fracture (bony tenderness and radiographic evidence of fracture)

Adverse outcomes of the disease

Occasional instability is possible, but some patients are able to return to their previous level of sports activity. Chronic pain is disabling when it occurs. Meniscal tears and osteoarthritis develop in some patients.

Treatment

Initial treatment consists of rest, ice, compression, and elevation (RICE), crutches, and short-term NSAIDs. Once the swelling subsides and the knee is comfortable, a program of physical therapy should be initiated to regain range of motion and establish quadriceps dominance, as this will help substitute for the lost function of the PCL. Some patients require surgery. The need for surgical reconstruction of the PCL is assessed based on the patient's response to conservative treatment.

Adverse outcomes of treatment

NSAIDs may cause gastric, renal, or hepatic complications. Chronic instability, subsequent meniscal tears, and osteoarthritis are all possible. Postoperative graft failure, recurrent instability, persistent pain, or infection can also develop.

Referral decisions/Red flags

Patients with arterial injury and/or multiple ligament injuries need further evaluation to confirm the diagnosis. Diagnostic dilemma, including the need to rule out associated pathology, and treatment of persistent pain, instability, or loss of function indicate the need for additional evaluation.

Acknowledgements

Figure 1 is adapted with permission from Rockwood CA, Green DP (eds): *Rockwood and Green's Fractures in Adults,* ed 2. Philadelphia, PA, JB Lippincott, 1984, vol 2, p 1525.

knee and lower leg

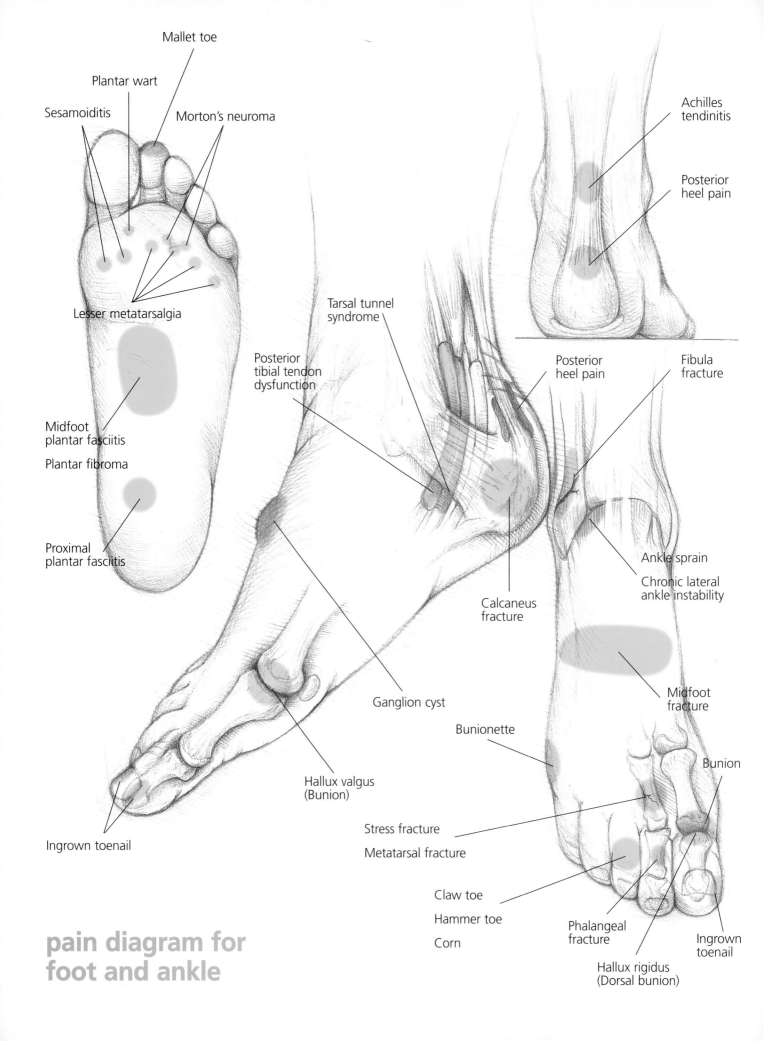

Mallet toe

Plantar wart

Sesamoiditis

Morton's neuroma

Lesser metatarsalgia

Midfoot
plantar fasciitis

Plantar fibroma

Proximal
plantar fasciitis

Ingrown toenail

Tarsal tunnel
syndrome

Posterior
tibial tendon
dysfunction

Ganglion cyst

Hallux valgus
(Bunion)

Achilles
tendinitis

Posterior
heel pain

Posterior
heel pain

Fibula
fracture

Ankle sprain

Chronic lateral
ankle instability

Calcaneus
fracture

Midfoot
fracture

Bunionette

Bunion

Stress fracture

Metatarsal fracture

Claw toe

Hammer toe

Corn

Phalangeal
fracture

Ingrown
toenail

Hallux rigidus
(Dorsal bunion)

pain diagram for
foot and ankle

Section Editor

Glenn B. Pfeffer, MD
California Pacific Medical Center
San Francisco, California
Assistant Clinical Professor
Department of Orthopaedics
University of California, San Francisco

Michael R. Clain, MD
Attending Orthopaedic Surgeon
Greenwich Hospital
Greenwich, Connecticut
Clinical Instructor, Orthopaedics
Yale University
New Haven, Connecticut

Carol Frey, MD
Associate Clinical Professor of Orthopaedic Surgery
University of Southern California
Director, Orthopaedic Foot and Ankle Center
Manhattan Beach, California

Mark J. Geppert, MD
Orthopaedic and Trauma Specialists
Somersworth, New Hampshire

Mark E. Petrik, MD
Clinical Assistant Professor
University of Louisville
School of Medicine
Louisville, Kentucky

foot and ankle— an overview

One mile of walking generates more than 60 tons of stress on each foot. Therefore, it is not surprising that more than 20% of musculoskeletal problems affect the foot and ankle. Most of these problems can be easily treated in the office setting. This section of *Essentials of Musculoskeletal Care* describes basic foot and ankle problems and details comprehensive care that can be provided in an efficient, cost-effective manner.

The key to the successful diagnosis of a foot problem is to determine the exact location and duration of symptoms. A pertinent medical history should also be obtained. Various systemic illnesses increase the foot's susceptibility to injury. Diabetes, peripheral vascular disease, neuropathy, and inflammatory arthritis all directly affect the foot. A history of bilateral foot pain should always trigger a careful search for a possible systemic etiology. A spinal etiology should also be considered in patients with bilateral symptoms.

Most patients with foot problems report pain. Acute pain (of less than 2 weeks duration) is an unusual presentation unless a fracture is present. Always consider a stress fracture if a patient reports recent onset of pain over the metatarsals (especially in the distal aspects of the second or third metatarsals). Chronic pain is much more common; the key to successful diagnosis in these patients is to determine the exact location of the pain.

This section also contains basic foot procedures that can be done in the office. A variety of specific devices (pads, inserts, heel cups, etc) should be stocked in your office; however, the two devices that are indispensable are the metatarsal/neuroma pad and the heel cup.

Forefoot problems

Bunions, hammer toes, claw toes, ingrown toenails, metatarsalgia, and interdigital neuromas account for most instances of forefoot pain. All these problems are detailed in chapters in this section.

Forefoot problems occur in women nine times more often than in men and are directly attributable to wearing tight, fashionable shoes. Shoe modification (lower heels, wider shoes, etc) is always the first line of treatment. In retractory cases, radiographs may be helpful. When necessary, always obtain standing AP and lateral radiographs because standing views show the weightbearing foot, which is a source of critical clinical information. Other common problems in the forefoot are hallux rigidus (arthritis of the metatarsophalangeal joint of the great toe) and stress fractures. Painful limited extension (dorsiflexion) of the great toe is consistent with hallux rigidus; pain and tenderness directly over the second or third metatarsals indicate a stress fracture until proven otherwise.

Midfoot problems

Chronic dorsal pain at the midfoot most commonly occurs secondary to degenerative arthritis involving one or more of the midfoot joints. Patients are often able to pinpoint the exact location of the pain and will have point tenderness in that area. In addition to standing AP and lateral radiographs, nonweightbearing oblique views may be helpful in demonstrating the involved joints. Pain on the plantar aspect of the midfoot is unusual and occurs with midfoot plantar fasciitis or plantar fibromas.

Hindfoot problems

Plantar heel pain is the most common problem in the foot and ankle. Usually, the cause is plantar heel pain/proximal plantar fasciitis. The pain associated with this condition is often worst with the first few steps out of bed and resolves with rest. Patients have focal tenderness directly over the plantar medial heel. On examination, considerable pressure often has to be applied in this area to duplicate the patient's symptoms.

Posterior heel pain is directly related to irritation from the shoe counter. There is often an associated superficial irritation (bursitis) of the posterior heel. Plantar and posterior heel pain can progress to the point where a patient is walking on tiptoes to avoid irritating the painful areas. When evaluating a patient with posterior heel pain, make sure that the problem is not more proximal within the Achilles tendon, as a partial or even complete rupture of the tendon can be missed.

Ankle problems

More than 25,000 individuals sprain an ankle each day in the United States. Acute anterolateral ankle pain, swelling, and frequently ecchymosis are the hallmarks of this condition. Patients have tenderness over the lateral ankle ligaments, although it may be difficult to localize this tenderness. Radiographs should be ordered if there is any focal bony tenderness over the fibula, base of the fifth metatarsal, talus, or calcaneus. Always obtain a radiograph if there is any question about the diagnosis.

Chronic ankle pain usually occurs at the anterolateral aspect of the ankle. For these patients, document a history of intermittent giving way, which may be consistent with a diagnosis of ankle instability. Chronic low-grade pain and swelling are consistent with injury to the peroneal tendons or an occult fracture (osteochondral lesion) of the ankle joint. Subtalar synovitis or arthritis may also present in a similar fashion.

A commonly overlooked problem in the hindfoot is posterior tibial tendinitis/dysfunction. This condition is characterized by pain and tenderness over the medial ankle posterior and distal to the medial malleolus. Advanced dysfunction will cause an acquired flatfoot that can cause additional pain on the lateral side of the ankle as the tip of the fibula abuts the collapsing foot. Tarsal tunnel syndrome can also cause chronic medial ankle pain, but will almost always be associated with neurogenic symptoms into the plantar aspect of the foot.

foot and ankle

foot and ankle— physical exam

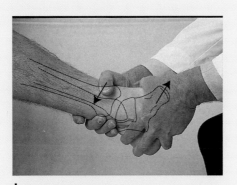

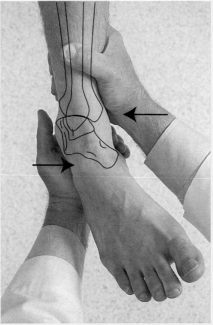

1

Figure 1
Anterior drawer test (ankle)

With the patient's tibia stabilized, grasp the calcaneus and pull forward to demonstrate anterior instability of the talus in the ankle mortise. There will be asymmetrical or excessive motion with chronic ankle ligament laxity and severe acute ankle ligament tears. This view is from the lateral side.

Figure 2
Varus stress test (ankle)

With the tibia stabilized, grasp the calcaneus and talus and invert the hindfoot. There will be excessive or asymmetrical motion with chronic ankle ligament laxity and acute severe ligament tears.

2

foot and ankle

3

Figure 3
Claw toe

Grasp the patient's foot and examine the toes for an extended metatarsophalangeal joint and flexed proximal interphalangeal (PIP) joint. Multiple toes tend to be involved. There is often a hard callus over the PIP joint dorsally. Claw toe commonly occurs in patients with cavus foot deformities, Charcot-Marie-Tooth disease, or rheumatoid arthritis.

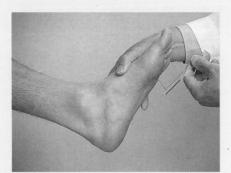

4

Figure 4
Sensitivity test

Use a 0.10-g, 5.07 diameter filament to determine if protective sensation exists in a patient with peripheral neuropathy of any origin.

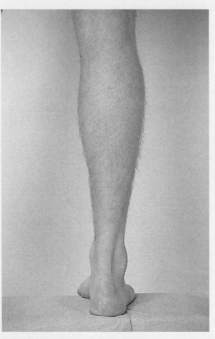

5

Figure 5
Posterior calf and foot

With the patient standing facing away, observe whether the Achilles tendon inserts into a neutral or slightly valgus calcaneus (turned out heel) with no more than one or two lateral toes visible from behind. In a patient with an acquired flatfoot from posterior tibial tendon dysfunction, look for unilateral increased valgus of the calcaneus and more than two visible toes ("too many toes sign"). A varus calcaneus (turned in heel) occurs with a cavus foot. An inflamed prominence of the posterior heel is called a "pump bump."

foot and ankle

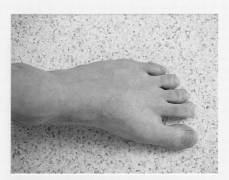

6

Figure 6
Top of the foot

Observe the top of the patient's foot for corns at the proximal interphalangeal joint of the toes, a bunion at the great toe metatarsophalangeal (MP) joint, tailor's bunion or bunionette at the fifth MP joint, in addition to evidence of poor nail care. Palpate the top of the foot for tenderness at the MP joint, which may be present with metatarsalgia, arthritis, Freiberg's infraction, osteonecrosis, or synovitis.

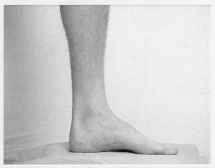

7

Figure 7
Medial aspect of the foot and ankle

With the patient standing, look for a high arch (cavus foot), flatfoot (pes planus), or undue prominence of the medial midfoot (accessory navicular). The arches of both feet should be symmetric.

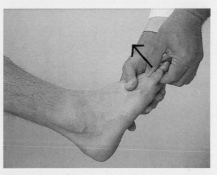

8

Figure 8
Assessing instability

With the patient sitting, stabilize the foot, then grasp the proximal phalanx of the toe and move the joint in a dorsal and plantar direction. Instability is often present following chronic synovitis or long-standing claw toe deformity.

foot and ankle

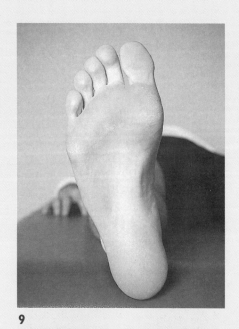

9

Figure 9
Plantar surface

With the patient supine, observe the bottom of the foot for plantar warts (which usually do not occur beneath the metatarsal head), a plantar callus (which does occur beneath the metatarsal head), or ulceration (especially in diabetic feet).

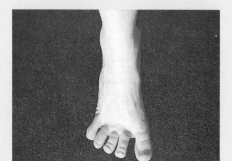

10

Figure 10
Spread toes

Ask the patient to spread or fan the toes as widely as possible. Look for corns or ulceration between the toes. Inability to actively spread the toes may indicate loss of intrinsic function.

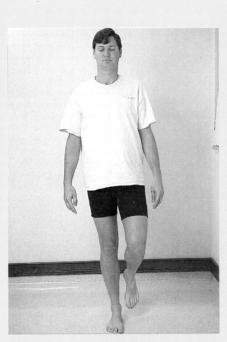

11

Figure 11
Romberg's test

Ask the patient to stand on one foot with eyes closed. Poor balance indicates altered proprioception or cerebellar dysfunction.

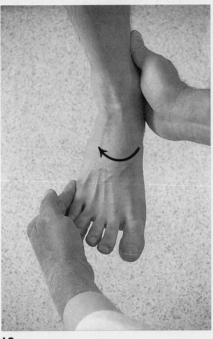

12

Figure 12
Resisted eversion

To test the strength of the peroneal muscles, stabilize the patient's tibia, and resist the patient's attempt to evert the foot. Weakness indicates injury to the peroneal tendons or a lesion involving the peroneal nerve or S1 nerve root.

foot and ankle

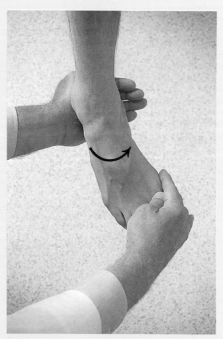

13

Figure 13
Resisted inversion

To test the strength of the posterior tibial tendon, stabilize the patient's tibia with the foot flexed, and resist the patient's attempt to invert the foot. To neutralize the inversion force of the tibialis anterior muscle, it is important to keep the foot flexed downward while performing this test. Weakness indicates injury or rupture of the posterior tibial tendon, or a lesion involving the posterior tibial nerve or L5 nerve root.

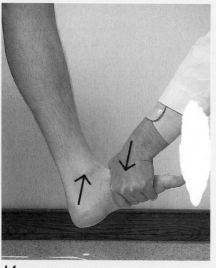

14

Figure 14
Resisted foot dorsiflexion

To test the strength of the anterior tibial muscle and toe extensors, ask the patient to resist as you push down on the patient's foot. Weakness indicates injury to the anterior tibial tendon or muscle, the toe extensors, or a lesion involving the peroneal nerve or L5 nerve root.

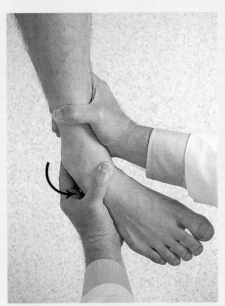

15A

Figure 15
Passive heel inversion

To test inversion of the subtalar joint (talocalcaneal joint) with the patient seated (A), grasp the patient's foot and invert the foot. Restricted motion may be seen in patients with subtalar arthritis, end-stage posterior tendon dysfunction, or tarsal coalition (bony connection between talus and calcaneus). View B shows the examination with the patient prone.

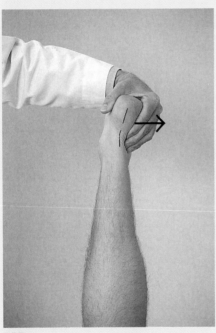

15B

foot and ankle

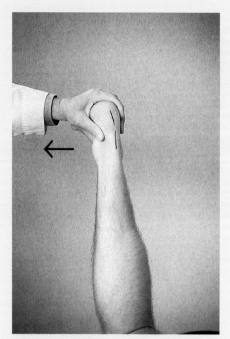

16A

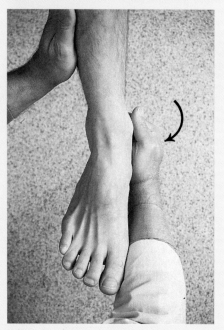

16B

Figure 16
Passive heel eversion

To test eversion of the subtalar joint (talocalcaneal joint) with the patient prone (A), grasp the patient's foot and evert the foot. Restricted motion is seen in similar conditions as those that cause restricted inversion. View B shows the examination with the patient seated.

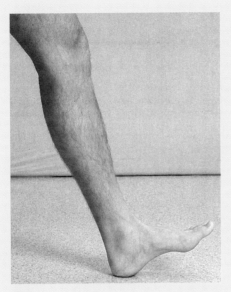

17A

Figure 17
The phases of gait

(A) Heel strike; (B) with the foot flat, a slight high arch or mild cavus is present; and (C) toe off.

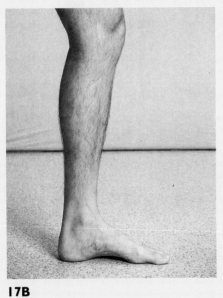

17B

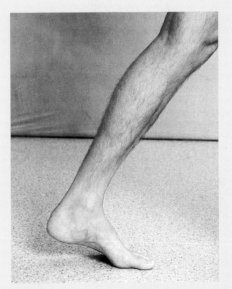

17C

foot and ankle

18

Figure 18
Dorsiflexion of the great toe

While palpating the plantar surface of the first metatarsophalangeal joint, dorsiflex the great toe. Increased pain and tenderness beneath the first metatarsal head may indicate sesamoiditis or a sesamoid fracture. Reduced motion may indicate hallux rigidus.

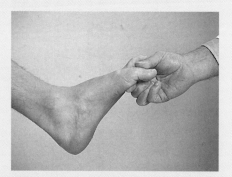

19

Figure 19
Plantar flexion of the great toe

While stabilizing the first metatarsal, plantar flex the great toe. Flexion may be decreased with hallux rigidus.

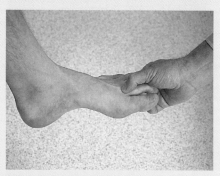

20

Figure 20
Palpation of the metatarsophalangeal (MP) joint

Palpate the dorsal and plantar surfaces of the MP joint. Tenderness indicates subluxation, dislocation, synovitis, or arthritis of the MP joint. Presence of a plantar callus may indicate excessive prominence of a metatarsal head, claw toe, or altered sensation on the plantar aspect of the foot.

foot and ankle

21A

Figure 21
Interdigital neuroma test

(A) With upward pressure between adjacent metatarsal heads, compress the metatarsals from side to side with the free hand. (B) The upward pressure places the neuroma between the metatarsal heads, allowing it to be compressed during side-to-side compression. Interdigital neuromas are almost always located between the second and third or third and fourth metatarsal heads.

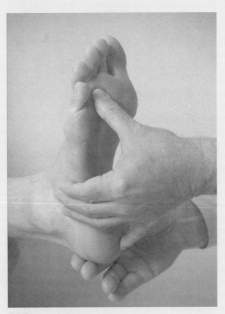

21B

foot and ankle

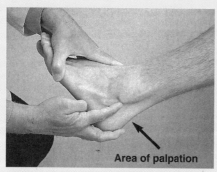

22

Figure 22
Palpation of peroneal tendons

Palpate behind and below the fibular malleolus for tenderness or crepitation involving the peroneal tendons, or for subluxation of the tendons during active rotation of the foot and ankle.

Area of palpation

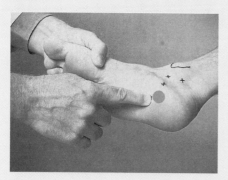

23

Figure 23
Proximal plantar fasciitis

Palpate the area of maximal tenderness in a patient with proximal plantar fasciitis.

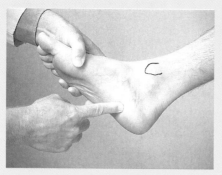

24

Figure 24
Entrapment of first branch nerve

Palpate the area of maximal tenderness in a patient with entrapment of the first branch of the lateral plantar nerve (Baxter's nerve). Tenderness is often associated with chronic plantar fasciitis.

foot and ankle

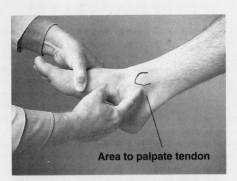

Area to palpate tendon

25

Figure 25
Posterior tibial tendon palpation

As the patient inverts and everts the foot, palpate posterior to the medial malleolus along the course of the tendon for tenderness and crepitance, which indicates inflammation or rupture of the posterior tibial tendon. The most common site of injury to the tendon is 1 to 2 cm distal to the medial malleolus.

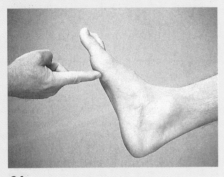

26

Figure 26
Sesamoid palpation

Palpate the area beneath the first metatarsal head for tenderness. The tender spot will move as the toe is dorsiflexed and plantar flexed. The medial sesamoid is the most commonly injured site.

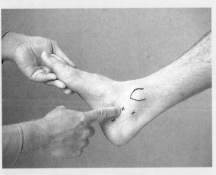

27

Figure 27
Tarsal tunnel

Palpate the area inferior to the posterior margin of the medial malleolus for tenderness. Percussion over the nerve should reproduce a patient's symptoms, indicating tarsal tunnel syndrome.

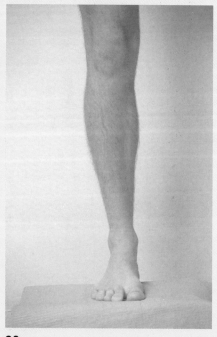

28

Figure 28
Anterior view of the foot and ankle

With the patient standing, view the foot and ankle from the front to look for medial curvature of the foot, abnormal rotation of the limb (placing the foot in either external or internal rotation), bunions, bunionettes, hammer toes, or similar foot deformities.

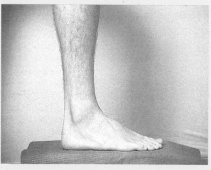

29

Figure 29
Lateral view of the foot and ankle

With the patient standing, view the foot from the lateral side to look for callosities, anterior ankle swelling, or prominence of the posterior calcaneus.

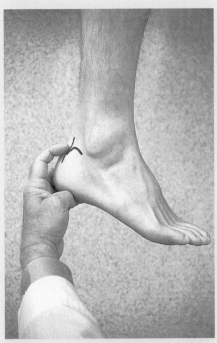

30A

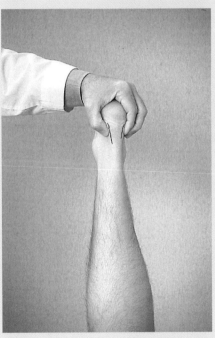

30B

Figure 30
Posterior heel pain

Palpate either side of the Achilles tendon insertion on the calcaneus (A). Tenderness or swelling on either side of the Achilles tendon indicates retrocalcaneal bursitis, which is often associated with a "pump bump" from a prominence of the posterior superior calcaneus (Haglund's deformity). Tenderness directly over the tendon insertion indicates insertional tendinitis of the Achilles tendon. Swelling superficial to the Achilles tendon insertion indicates a superficial bursitis. View B shows the examination with the patient prone.

foot and ankle

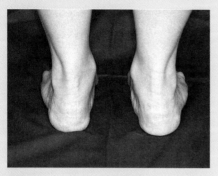

31

Figure 31
Weight bearing feet

In the normal physiologic position for weight bearing feet, the heels are in a slight valgus position.

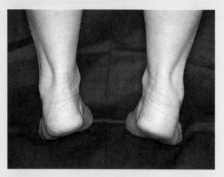

32

Figure 32
Standing on toes

When the patient stands on her toes, the heels move into a normal varus position.

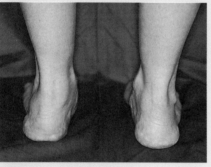

33A

Figure 33
"Too many toes" sign

A patient with an aquired flatfoot from posterior tibial tendon dysfunction will appear to have too many toes in the affected foot, as this patient does in the left foot, when the feet are viewed from behind in normal weight bearing position (A). When the same patient attempts to raise up on her toes, her left heel does not move into a normal varus position, as her right heel does (B).

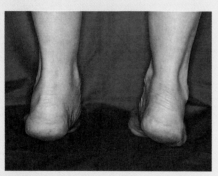

33B

foot and ankle

Achilles Tendinitis and Rupture

Synonyms

Heel cord

Tendoachilles

Gastrocsoleus complex

Definition

There are two principal groups of problems associated with the Achilles tendon. The first is a rupture of the tendon. The second is Achilles tendinitis, which can be divided into two groups—insertional, occurring at the bone tendon interface of the calcaneus, and noninsertional, occurring 4 to 5 cm proximal to the insertion into the calcaneus.

Clinical symptoms

Achilles rupture: Sudden, severe calf pain is described as a "gunshot wound" or as a "direct hit from a racquet." This condition typically affects middle-aged men who are weekend athletes, playing tennis, squash, basketball, etc. Pain may resolve quickly and be thought to be an ankle sprain. If neglected, the major disability will be weakness during ambulation.

Achilles tendinitis: Patients often have insidious pain in the area of the Achilles tendon that becomes worse with exercise. In some patients, the pain may actually improve with exercise. This condition typically develops following a change in training habits. Pain may vary in intensity from an annoyance to a significant disability.

Tests

Exam

Achilles rupture: Swelling and ecchymosis from the calf into the heel is common. Often the patient is unable to bear weight and may have a palpable defect in the tendon a few centimeters above the calcaneus. Perform the Thompson test by placing the patient prone with the knee and ankle at 90°. The test can also be done with the patient kneeling on a chair. A squeeze of the calf normally results in passive plantar flexion of the ankle (Figure 1); a positive test is the absence of plantar flexion. The test is most reliable within 48 hours of the rupture.

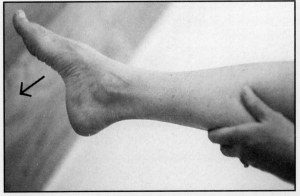

Figure 1
Squeezing the calf normally results in passive plantar flexion of the ankle

foot and ankle

Achilles tendinitis: Noninsertional tendinitis is often, but not always, associated with a discreet thickening of the tendon above the calcaneus. The patient may feel a squeaking sensation during plantar flexion of the ankle, and the tendon is tender. Insertional tendinitis is often associated with a protuberant posterolateral bony process of the calcaneus. Patients with insertional tendinitis are usually more comfortable in heels, boots, or an open back shoe.

Diagnostic

Lateral radiographs of the ankle rarely demonstrate a bony avulsion of the posterior calcaneus or ossification within the insertion of the tendon. Enlargement of the posterosuperior tuberosity of the calcaneus, called Haglund's syndrome, is commonly associated with soft-tissue swelling of the posterior heel at the Achilles insertion; this is often called a "pump bump."

Differential diagnosis

Medial gastrocnemius tear

Partial Achilles tear (probably quite uncommon)

Plantaris rupture (little evidence that this exists as an isolated entity)

Stress fracture of the tibia

Adverse outcomes of the disease

With Achilles rupture, there is weakness with push-off, which patients describe as feeling like walking on soft beach sand, and decreased athletic function. With Achilles tendinitis, there is chronic pain and decreased athletic function.

Treatment

Achilles rupture: Treatment options include casting in plantar flexion or surgical repair and casting. The specific decision about whether to choose conservative or surgical treatment is based on the patient's level of activity, age, medical condition, etc. If treatment is delayed beyond a few days, then the problem becomes more complicated to treat.

Achilles tendinitis: NSAIDs, a gentle Achilles stretching program, modification in training habits (shoe wear and mileage) are all indicated. In refractory cases, a simple heel lift (wearing a boot or inserting a $1/4''$ to $1/2''$ lift to be effective) and a course of physical therapy (emphasizing stretching, flexibility, and local massage and ultrasound) two to three times a week for 6 weeks may be helpful. Sometimes, a 4- to 6-week period of cast or cast boot immobilization, followed by the above treatment, is indicated. Corticosteroid injections increase the risk of tendon rupture; therefore, injection is not indicated.

foot and ankle

Adverse outcomes of treatment

NSAIDs may cause gastric, renal, or hepatic complications. Postoperative infection or skin slough because of tenuous blood supply to the posterior calf is another possibility.

Referral decisions/Red flags

History of a sudden "pop" means probable rupture in which the patient needs further evaluation immediately. Weakness and persistent tendinitis pain following conservative treatment also indicate the need for further evaluation.

Acknowledgements

Figure 1 is reproduced with permission from Lutter LD: Hindfoot problems, in Heckman JD (ed): *Instructional Course Lectures 42.* Rosemont, IL, American Academy of Orthopaedic Surgeons, 1993, pp 195–200.

foot and ankle

Ankle Sprain

Synonyms

Inversion injury

Lateral collateral ligament tear

Definition

Approximately 25,000 people sprain an ankle every day. Ankle sprains are not always simple injuries and can result in residual symptoms in up to 40% of patients. Although the lateral collateral ligaments (the anterior talofibular and calcaneofibular) are injured 85% of the time, other injuries can occur with an inversion injury, including peroneal tendon tear or subluxation, neurapraxia of the superficial and deep peroneal nerve, sprain of the subtalar joint, fracture of the base of the fifth metatarsal or talar dome, avulsion fracture of the calcaneus or talus, and injury to the calcaneocuboid joint (Figure 1).

Injury to the syndesmosis, the thick ligaments connecting the distal tibia and fibula, occurs in approximately 5% of all ankle sprains. With an injury to the syndesmosis, often referred to as a "high" ankle sprain, the recovery time from the injury is increased. Less common are injuries to the medial structures, including the deltoid ligament and posterior tibial tendon.

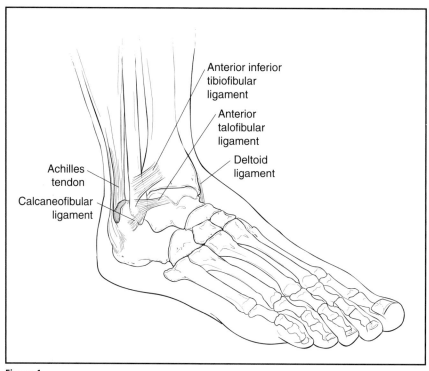

Figure 1
Ligaments of the ankle

Clinical symptoms

Pain over the injured ligaments, swelling, and loss of function are common. Patients who report feeling a "pop," followed by immediate swelling and the inability to walk after an inversion injury likely have a severe sprain. Determine if the patient has a history of ankle sprains and giving way, as this information will help differentiate an acute injury superimposed on chronic ankle instability from acute ligament injury.

Tests

Exam
Examination often reveals ecchymosis and swelling around the entire ankle joint, not just the lateral side. Tenderness on palpation over the anterior talofibular, posterior talofibular, and calcaneofibular ligaments may help identify which ligaments are injured. Palpate the lateral and medial malleoli and the base of the fifth metatarsal for crepitation or tenderness caused by a fracture.

The squeeze test, which is performed by compressing the tibia and fibula at the midcalf, and the external rotation of the foot, which is performed by placing the ankle in dorsiflexion and then externally rotating the foot, help identify injury to the syndesmosis. A positive result is indicated by the presence of pain over the distal tibiofibular junction (syndesmosis).

Diagnostic
Radiographs of the ankle are not required for an isolated ankle sprain. However, if there is any tenderness over the lateral malleolus, ankle joint, syndesmosis, or other bony structure, then radiographs are needed to rule out a fracture.

Differential diagnosis

Avulsion fracture of the calcaneus, talus, lateral malleolus, or base of the fifth metatarsal

Fracture of the lateral process of the talus

Fracture of the proximal fibula (Maisonneuve fracture)

Neurapraxia of the superficial or deep peroneal nerve

Osteochondral fracture of the talar dome

Peroneal tendon tear or subluxation

Subtalar joint sprain

Syndesmosis injury

Adverse outcomes of the disease

A missed fracture may have serious consequences. Following a severe sprain, there may be chronic pain, instability, and the possibility of arthritis. Chronic instability is usually secondary to incomplete rehabilitation and occurs in up to 40% of all ankle sprains.

Treatment

The goal of treatment is to prevent chronic pain and instability. Phase one consists of NSAIDs, ice, compression, and elevation. Use a brace or air stirrup for protection and to promote soft-tissue healing (Figure 2). For severe sprains, the use of a cast or cast boot for 2 weeks is acceptable but usually not required. Encourage weightbearing as tolerated, with use of crutches as needed. Forty-eight hours following the injury, contrast baths are most helpful to decrease swelling (see Contrast Baths).

Phase two begins when the patient can bear weight without increased pain or swelling, usually 2 to 4 weeks after the injury. Continue use of the air stirrup or brace. Begin exercises to increase peroneal and dorsiflexor strength; stretching the Achilles tendon is also essential. Continue this phase until the patient has full range of motion and 80% of normal ankle strength. Plantar flexion exercises are not included as they place the ankle in a position of least stability.

Phase three, which is usually 4 to 6 weeks after injury, consists of functional conditioning with proprioception, agility, and endurance training. Exercises that are helpful for proprioception include standing on the sprained ankle for several minutes with the opposite foot elevated and the eyes closed (Figure 3). Running in progressively smaller figures of 8 is excellent for agility and peroneal strength. During this time, the patient should be weaned from the air stirrup or ankle brace.

Figure 2
An air stirrup-type ankle brace

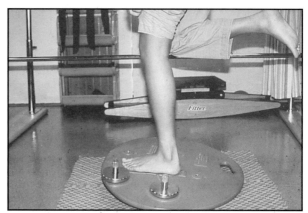

Figure 3
Exercise following an ankle sprain

This three-phase treatment program may take only 2 weeks to complete for minor sprains, or up to 6 to 8 weeks for severe injuries. For athletes with severe injuries who are returning to sports, the use of a functional brace or air stirrup, or taping on a long-term basis will help prevent recurrent injury. The use of a brace is particularly indicated for athletes in sports associated with a high risk for ankle sprains, such as basketball, volleyball, and soccer. Exercises during this phase should include peroneal strengthening in both dorsiflexion and plantar flexion, as well as continued Achilles tendon stretching.

Adverse outcomes of treatment

NSAIDs may cause gastric, renal, or hepatic complications. Casting and ankle immobilization for more than 3 weeks may cause stiffness and a slower return to sports or work. Therefore, these modalities should be reserved for only the most severe cases. Most ankle sprains can be treated with functional immobilization such as an air stirrup.

Referral decisions/Red flags

Fractures of the foot and ankle, tears or subluxation of the peroneal tendons, nerve injury, a history of repeated giving way (chronic instability), and failure to improve in 6 weeks with appropriate treatment all indicate serious injury.

Acknowledgements

Figure 1 is adapted with permission from The Physician and Sportsmedicine. McGraw-Hill Companies, 1992.

foot and ankle

procedure

Contrast Baths

Materials
2 buckets
Ice
Warm water

Step 1

To decrease inflammation in the foot, ankle, or hand, first soak the affected area in a cold water bath. The water temperature should be as cold as can be tolerated; the addition of several small ice cubes to the water is usually the best method to cool it. Soak for 30 seconds.

Step 2

Immediately place the affected area into a second bucket filled with water that is as warm as can be tolerated; generally around 104°F (40°C). Soak for 30 seconds.

Step 3

Place the affected area back into the cold water for 30 seconds. Continue to rotate between the cold and hot water for a total of 5 minutes. The first and last soaks should be in the cold water.

Aftercare/patient instructions

Advise the patient that, ideally, the contrast bath should be repeated three times a day. If that is not possible, then once in the morning and once in the evening is usually sufficient. Patients with a peripheral neuropathy and decreased sensibility should be careful to avoid injury from water of extreme temperature.

Bunion

Synonyms

Hallux valgus

Metatarsus primus varus

Definition

A bunion is a lateral deviation of the great toe at the metatarsophalangeal (MP) joint that may lead to a painful prominence of the medial aspect of this joint (medial eminence) (Figure 1). There is a familial tendency, with a female to male ratio of approximately 10:1.

Clinical symptoms

Pain and swelling, aggravated by shoe wear, are the principal complaints. There is usually a large, hypertrophic bursa over the medial eminence of the first metatarsal, and a pronated (rotated inward) great toe with subsequent callus on the medial aspect of the toe joint. Irritation of the medial plantar sensory nerve may cause numbness or tingling over the medial aspect of the great toe.

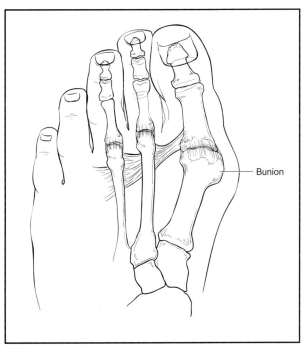

Figure 1
Anatomy of a bunion

Tests

Exam

Extension of the MP joint (normal = 70°) should be assessed. Valgus should occur at the MP joint, not the interphalangeal joint (hallux interphalangeus). Evaluate the lesser toes for associated deformities. A second toe that overrides the laterally deviated great toe is also a frequent problem. Other lesser toe problems include corns, calluses, hammer toes, and bunionette (a bunion-like prominence on the lateral side of the fifth MP joint).

Diagnostic

Standing AP radiographs of the foot and measured angles are used to grade the severity of a bunion deformity. The normal hallux valgus angle is up to 15°, and the normal intermetatarsal angle is no greater than 9° (Figure 2). Standing AP radiographs also assess lateral subluxation of the sesamoids, the shape of the metatarsal head, degenerative changes in the MP joint, hallux interphalangeus, and lesser toe abnormalities. A standing lateral radiograph of the foot is less helpful, but is important for evaluating any lesser toe subluxation or arthritic changes of the great toe MP joint.

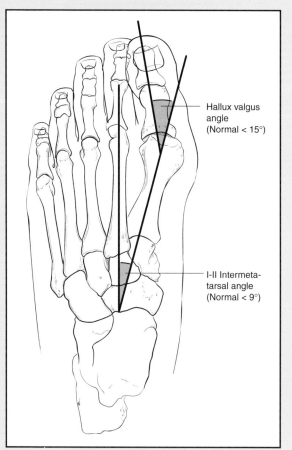

Hallux valgus angle
(Normal < 15°)

I-II Intermeta-tarsal angle
(Normal < 9°)

Figure 2
Measure the above angles from the patient's standing AP radiograph

Differential diagnosis

Gout

Hallux extensus (cock-up toe)

Hallux rigidus (dorsal bunion)

Hallux varus

Adverse outcomes of the disease

Chronic pain is a primary symptom, requiring activity modification and shoe restrictions.

foot and ankle

Treatment

Initial treatment of a bunion is patient education and shoe wear modification. Shoe wear modification is usually sufficient for mild to moderate deformities. Trace the outline of the foot and the shoe, measure the greatest width, and keep the shoes within 1/2″ of the foot measurement. Shoes should be constructed of soft uppers, with no thick stitching over the medial eminence. A shoe should be stretched out professionally, directly over the bunion prominence. Patients should avoid shoes with heels higher than 2 1/4″, as these place undue pressure on the forefoot and bunion prominence. Physical therapy, splints, and bracing are not helpful, although occasionally a medial longitudinal arch support can decrease the pressure on the bunion of a patient with a flatfoot.

For patients who have continued disability despite conservative treatment, several well-established surgical procedures are available. For mild to moderate deformities (hallux valgus < 25° to 30°, intermetatarsal angle < 14°), a distal osteotomy of the first metatarsal may be used, and for more severe deformities (hallux valgus > 35°, intermetatarsal angle > 14°), a proximal osteotomy is usually indicated. Occasionally, fusion or resection arthroplasty (Keller procedure) may be used. Joint replacement is rarely used because of the high complication rate.

Adverse outcomes of treatment

With surgical treatment of bunions, there is the possibility of recurrence, undercorrection, overcorrection (hallux varus), decreased function, stiffness, pain, loss of toe push-off, arthritis, osteonecrosis, malunion or nonunion of osteotomies, transfer lesions (metatarsalgia, corns), or iatrogenic neuromas.

Referral decisions/Red flags

Patients with persistent pain may require surgical correction.

Acknowledgements

Figure 2 is adapted with permission from Pedowitz W: Bunion deformity, in Pfeffer G, Frey C (eds): *Current Practice in Foot and Ankle Surgery.* New York, NY, McGraw Hill, 1993, pp 219–242.

The Essentials of Musculoskeletal Care 397

foot and ankle

Bunionette

Synonyms

Tailor's bunion

Definition

Bunionette is a deformity of the fifth metatarsophalangeal (MP) joint that is analogous to a bunion deformity of the great toe. It is characterized by prominence of the lateral aspect of the fifth metatarsal head and medial deviation of the small toe (Figure 1). Often, there is an associated overlying hard corn and painful bursitis. The usual cause is frequent wearing of tight, narrow, pointed-toe shoes.

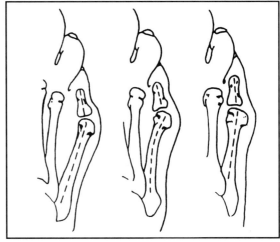

Figure 1
Variations of a bunionette

Clinical symptoms

Patients report pain, problems with finding comfortable shoes, and an overlying corn.

Tests

Exam
Deformity is evident on physical examination. When this deformity is associated with hallux valgus (bunion), it is called a splayfoot.

Diagnostic
Standing AP radiographs will show medial deviation of the fifth metatarsal, as well as lateral deviation of the fifth metatarsal shaft and/or a prominence on the lateral aspect of the fifth metatarsal head. The joint is usually normal.

Differential diagnosis

None

Adverse outcomes of the disease

The most common problem is persistent pain that is aggravated by shoe wear. Patients may also have associated hard and soft corns, with possible ulceration and infection.

Treatment

Patients should be advised to select proper shoe wear with a soft upper and a roomy toe box. A shoe repair shop can stretch the shoe over the bunionette. Custom shoe wear is an expensive option, but can be effective in patients who are not candidates for surgery. A metatarsal pad can help shift pressure off the fifth metatarsal head if it is cut lengthwise on a diagonal, then applied to the lateral aspect of the toe box, just proximal to the bunionette prominence (see Application of a Metatarsal Pad). A medial longitudinal arch support may help a patient who has a flexible flatfoot. This orthotic device rotates the forefoot slightly, which decreases direct pressure over the bunionette prominence. With continued symptoms, surgical excision of the bunionette or realignment osteotomy of the fifth metatarsal may be required.

Adverse outcomes of treatment

Treatment may fail to relieve pain or an ulceration may develop.

Referral decisions/Red flags

Failure of conservative treatment indicates the need for further evaluation.

Acknowledgements

Figure 1 is reproduced with permission from Coughlin MJ: Etiology and treatment of the bunionette deformity, in Greene WB (ed): *Instructional Course Lectures XXXIX*. Park Ridge, IL, American Academy of Orthopaedic Surgeons, 1990, pp 37–48.

foot and ankle

procedure

Application of a Metatarsal Pad

Materials

Felt or gel metatarsal pad

Temporary marker (lipstick, eyeliner)

Step 1

Ask the patient to mark the painful spot on the bottom of the foot with a material that transfers easily, such as lipstick or eyeliner (Figure 1). The mark should be approximately 0.50 cm square.

Step 2

Instruct the patient to stand in a shoe, without socks, to transfer the mark to the inside of the shoe.

Step 3

Place a metatarsal pad into the shoe just proximal to (toward the heel), not directly under, the mark (Figure 2). This ensures that the painful area is suspended by the proximally placed pad. Off-the-shelf felt or gel pads are easy to use, inexpensive, and effective, and come in different sizes to accommodate different sized lesions and feet.

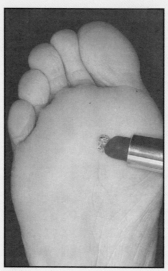

Figure 1
Marking the painful area

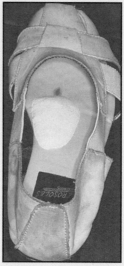

Figure 2
Placing the metatarsal pad

foot and ankle

Application of a Metatarsal Pad (continued)

Aftercare/patient instructions

Advise the patient that if the pad is effective, but the patient needs a more permanent device that can be transferred from shoe to shoe (Figure 3), a custom orthosis can be fabricated by a certified pedorthist.

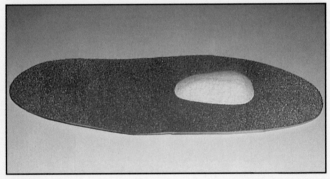

Figure 3
Pad can be moved from shoe to shoe

Chronic Lateral Ankle Instability

Synonyms

Trick ankle

Weak ankle

Definition

Chronic lateral ankle instability is classified as either functional or mechanical. Functional instability is an involuntary sensation of giving way that does not exceed the anatomic or physiologic range of motion of the ankle; causes include proprioception defects, pain, and muscle weakness. Mechanical instability is motion beyond anatomic or physiologic restraints and is demonstrable by clinical and radiographic examination. Mechanical instability follows an ankle sprain in which the lateral ligaments healed in a lengthened position. The most common cause of chronic ankle instability is incomplete rehabilitation following a sprain.

Clinical symptoms

A history of recurrent sprains and giving way is common. Pain is intermittent and usually occurs after each sprain. Patients may avoid specific activities that predispose them to injury.

Tests

Exam

After a recurrent sprain, swelling, ecchymosis, and tenderness on palpation over the injured ligaments are common. Clinical and radiographic stress testing is painful and may require injection of a local anesthetic into the lateral capsule for valid results. Perform the anterior drawer test with the knee flexed to 90° and the ankle in slight plantar flexion; grasp the heel and pull forward and inward while exerting a posterior force on the anterior tibia (Figure 1). Compare the distance the foot travels forward in the affected ankle with that in the asymptomatic ankle. Remember that a patient can have bilateral laxity.

Figure 1
Anterior drawer test

Occasionally, there will be a "suction sign" over the anterolateral aspect of the ankle joint, indicating a defect of the anterior talofibular ligament. In the talar tilt test, one hand supports the medial tibia and the other hand is placed on the lateral side of the heel with the foot in neutral position; the heel is inverted and the amount of varus tilt in the affected ankle is compared with that of the asymptomatic ankle (Figure 2). A Romberg test evaluates balance and functional instability. With this test, the patient is asked to stand on one leg and then the other, with the eyes opened and closed. Any difference in sway is noted. Another important part of the exam is to compare peroneal muscle strength. Place the patient's foot in plantar flexion and inversion and resist as the patient moves the foot laterally with peroneal contraction. Patients with recurrent ankle sprains often have weakness of the peroneals.

Diagnostic
Plain radiographs of the ankle are usually normal. Ossification in the syndesmosis or small fragments off the tip of the malleolus may indicate an old injury; fragments with sharp edges represent an acute fracture. Ossification inferior to the lateral malleolus is a strong indication of a significant sprain in the past. Osteochondral lesions of the talar dome can be associated with instability and repeated sprains.

The anterior drawer and talar tilt tests can be documented on radiographs. Compare both ankles; any subluxation greater than 4 mm is abnormal in the anterior drawer test. A difference of more than 6° of talar tilt between the two ankles is considered abnormal (Figure 3).

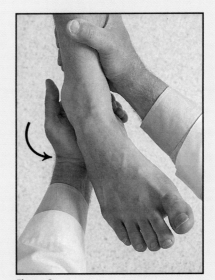
Figure 2
Varus stress test

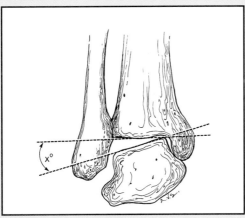
Figure 3A
When measuring talar tilt, a difference of more than 6° from the uninjured ankle is considered abnormal

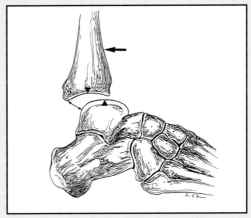
Figure 3B
An anterior translation of more than 4 mm is usually considered abnormal

foot and ankle

Differential diagnosis

Anterolateral soft-tissue impingement in the ankle joint

Cavovarus foot (inverted heel predisposes to repeated sprains)

Chronic entrapment of the superficial or deep peroneal nerves (pain with percussion over the nerve)

Herniated lumbar disk (causing peroneal weakness)

Loose bodies in the ankle joint

Osteochondral lesion of the ankle (pain with no history of giving way)

Peroneal tendon tear or subluxation

Subtalar instability

Syndesmosis injury

Adverse outcomes of the disease

Osteoarthritis, chronic instability, and pain are all possible.

Treatment

Conservative treatment involves strengthening the peroneal, the ankle dorsiflexor, and plantar flexor muscles, stretching the Achilles tendon, and practicing one-legged balance and other proprioception exercises. Use of a functional brace or ankle air stirrup during sports activities may be helpful on a long-term basis to prevent recurrent injury. Surgical reconstruction of the ankle ligaments should be considered for patients who have completed a full course of physical therapy and rehabilitation and continue to have recurrent sprains.

Adverse outcomes of treatment

Incomplete rehabilitation is the most common cause of chronic ankle instability and recurrent injury.

Referral decisions/Red flags

Fractures of the foot and ankle, tears or subluxation of the peroneal tendons, nerve injury, and failure to improve within 12 weeks with appropriate treatment are all indications for further evaluation.

Acknowledgements

Figure 3 is reproduced with permission from Pfeffer GB (ed): *Current Practice in Foot and Ankle Surgery.* New York, NY, McGraw-Hill HPD, 1994, p118.

Chronic Leg Pain

Synonyms

Periostitis

Shin splints

Stress fracture

Acute compartment syndrome

Chronic compartment syndrome

Peripheral nerve entrapment

Medial tibial stress syndrome (due to periostitis of the soleus origin)

Definition

A variety of specific anatomic lesions can cause chronic leg pain in the athlete. Stress fracture, chronic compartment syndrome, medial tibial stress syndrome (an enthesopathy or inflammation at the muscle origin from bone), and peripheral nerve entrapment are the most likely causes.

Clinical symptoms

The precise pain pattern of symptoms varies with the etiology. With stress fractures, pain occurs immediately upon weightbearing. Walking is uncomfortable, but running and jumping are worse. The pain has a deep, aching quality and is localized anteriorly and posteriorly over the bone.

With exercise-induced compartment syndrome, patients typically report pain over the muscles of the involved compartment (anterior or posterior) after 5 to 10 minutes of activity. The pain then gradually increases and can last anywhere from a few minutes to several hours following exercise.

The pain associated with medial tibial stress syndrome occurs with exercise over the posteromedial surface of the tibia. This area remains tender even after exercise.

The superficial peroneal nerve may become entrapped as it exits the fascia of the anterior compartment in the distal lateral leg, resulting in burning pain and dysesthesias over the anterior ankle and top of the foot.

Tests

Exam
Palpation of the crest of the tibia will usually identify a specific painful spot with a stress fracture. Compression of the tibia and fibula will result in pain at the site of the stress fracture.

Tenderness at the posteromedial border of the tibia near the junction of the middle and distal thirds usually indicates medial tibial stress syndrome, primarily at the origin of the posterior tibial muscle.

Localized nerve entrapment (superficial peroneal is most common) occurs at the site where the sensory nerve exits the fascia. There may be a small bulge of muscle through this fascial hole. The most common site is at the junction of the middle and distal thirds of the leg over the anterior compartment. There may also be a positive percussion sign (Tinel's sign) over this area, with tingling paresthesias produced by gentle percussion of the nerve at the site of the entrapment.

foot and ankle

Diagnostic

AP and lateral radiographs may reveal a stress fracture, either by showing an undisplaced crack (lucency) in the cortex, a smudge in the medullary bone, or subperiosteal new bone. If plain radiographs are normal, and symptoms have been present for at least 5 days, a bone scan will be positive with dense uptake. Medial tibial stress syndrome may also cause a positive bone scan along the medial tibia, but the uptake is less dense and appears as a thin long line of increased uptake along the medial tibia.

Exercise-induced compartment syndrome can be diagnosed only by demonstrating a pathologic elevation of compartment pressure during exercise and a delayed return to normal levels. This is done using intracompartmental pressure transducers attached to a wick or slit catheter during exercise.

The diagnosis of a peripheral sensory nerve entrapment is done on the basis of a clinical exam. Nerve conduction studies may be helpful, but these are often unable to detect a dynamic compression of a peripheral sensory nerve.

Differential diagnosis

None

Adverse outcomes of the disease

Persistent pain is the only adverse outcome generally arising from untreated exercise-induced compartment syndrome, medial tibial stress syndrome, or peripheral nerve entrapment.

Stress fractures, if unrecognized and untreated, can occasionally evolve into complete displaced fractures of the tibia or fibula.

Treatment

Reduction in activity is the first step in treatment of these various causes of leg pain. Athletes must return to an asymptomatic level of training and can gradually increase activities as long as they remain pain free. Ice massage after activity is helpful. NSAIDs may help medial tibial stress syndrome.

Crutches are usually not necessary for stress fractures unless simple walking is painful. A removable cast may be used during ambulation. Athletic activities should be avoided for 4 to 6 weeks, after which time a progressive training schedule can be resumed, increasing no more than 10% per week. Low-impact activities, such as "running" in a pool while supported by flotation devices, swimming, or cycling can be tried. In recalcitrant cases, casting and/or electrical stimulation may be helpful.

Peripheral nerve entrapments will sometimes respond to measures used for medial tibial stress syndrome. Compressive wraps or tight taping should be avoided. An injection of corticosteroid at the site of the entrapment may also be beneficial.

Surgical treatment may be necessary for recalcitrant compartment syndrome or peripheral entrapment neuropathy.

Adverse outcomes of treatment

NSAIDs may cause gastric, renal, or hepatic complications.

Referral decisions/Red flags

Patients who fail to respond to conservative treatment need further evaluation.

Claw Toe

foot and ankle

Synonyms

None

Definition

Claw toe deformity is characterized by extension at the metatarsophalangeal (MP) joint, and flexion at the proximal interphalangeal (PIP) joint and often the distal interphalangeal (DIP) joint (Figure 1). The deformity is usually the result of incompetence of the foot intrinsic muscles, secondary to neurologic disorders affecting the strength of these muscles (diabetes, alcoholism, or other peripheral neuropathies, Charcot-Marie-Tooth disease, spinal cord tumors). All toes, particularly the lesser toes, tend to be affected, rather than an isolated deformity in one toe. Painful calluses and difficulty with shoe wear are common (Figure 2).

Clinical symptoms

Pain is the principal symptom. A history of a changing deformity, especially rapid progression, suggests a related neurologic condition as the etiology of this condition.

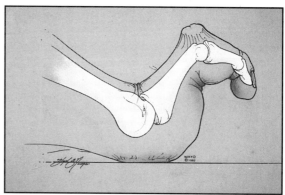

Figure 1
Clinical appearance of a claw toe

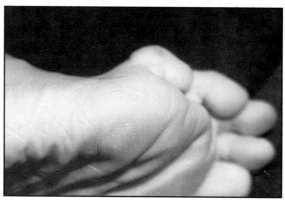

Figure 2
Plantar callus associated with claw toe

Tests

Exam

When the patient stands barefoot, the toes "claw" by flexing at the interphalangeal (IP) joints and extending at the MP joint. These deformities may be flexible or, in more advanced cases, fixed contractures. Dorsal dislocation of the proximal phalanx onto the metatarsal head may occur in advanced cases and is easily palpable. There are often calluses over the dorsal aspect of the IP joints, and sometimes on the tip of the toe. With advanced cases, the metatarsal head and resultant calluses become prominent on the plantar surface. A careful neurologic exam is warranted to rule out various causes of peripheral neuropathy, especially with bilateral involvement of multiple toes and a high arched cavus foot.

Diagnostic

Radiographs of the foot confirm the diagnosis, but are useful only in planning surgery. An MRI scan of the lumbar spine may be indicated if spinal pathology is suspected.

Consider electromyography and/or nerve conduction studies, serologic testing for vitamin deficiencies, and other laboratory studies to evaluate for peripheral neuropathy.

Differential diagnosis

Charcot-Marie-Tooth disease

Diabetes mellitus

Hammer toe

Mallet toe

Peripheral neuropathy

Spinal cord pathology

Adverse outcomes of the disease

Patients may have a persistent or progressively painful deformity. Claw toes may also contribute to the development of metatarsalgia by causing distal displacement of the fat pad, which is normally situated beneath the metatarsal heads and by the deformed toes causing downward pressure on the metatarsal heads. In advanced cases, dislocation of the MP joint is possible. Other possible problems include painful corns, ulceration, and infection.

foot and ankle

Treatment

Shoes with soft, roomy toe boxes help to accommodate the deformity. Patients should be advised to avoid constricting, high-heeled shoes. Early in the process, a home exercise program of passive manual stretching of the toe, and foot intrinsic exercises, such as picking up marbles with the toes, or laying a towel flat on the floor and crumpling it under the foot using the toes, may be useful. Passive stretching of the contracted joints is also helpful.

Commercially available straps and cushions sometimes provide symptomatic relief (Figure 3). The toe may be taped into flexion at the MP joint with a narrow strip of tape. Begin on the plantar aspect of the foot under the metatarsal head, looping over the dorsum of the proximal phalanx (while holding the toe plantar flexed at the MP joint) and crossing back over the starting point on the plantar aspect of the foot. Patients can do this at home.

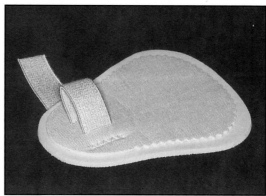

Figure 3
Type of splint used for claw toe

"In depth" shoes, which have an extra $3/8''$ depth in the toe box, accommodate fixed claw toes and are available in specialty shoe stores. Shoe repair shops can also help by stretching a small pocket in the toe box using a cobbler's swan, a ball-and-ring device that will crimp an outward bulge in the shoe.

Surgical correction may be necessary and is generally effective in patients for whom conservative treatment has failed. Correction of more than two toes on one foot is usually unnecessary, unless the problem is due to rheumatoid changes.

Adverse outcomes of treatment

Patients may still have persistent pain and/or calluses. Taping, strapping, and appliances should be avoided in patients with insensate feet as these methods may compromise circulation to the affected toe. Likewise, patients with vascular disease are susceptible to skin breakdown associated with taping and appliances.

Referral decisions/Red flags

Failure of conservative treatment or the presence of ulceration with suspected septic arthritis or osteomyelitis all indicate the need for further evaluation. Remember the need for possible neurologic consultation.

Acknowledgements

Figure 1 is reproduced with permission from The Mayo Foundation, Rochester, MN.
Figure 2 is reproduced with permission from California Pacific Medical Center.

Corns and Metatarsalgia

Synonyms

Clavus

Callosity

Heloma

Heloma molle

Definition

Callus is a hyperkeratotic lesion that forms in response to excessive pressure over a bony prominence. If the callus forms on a toe, it is called a corn. If it forms elsewhere (as under a metatarsal head) it is called a callus. Callosities on the sole of the foot are termed intractable plantar keratoses (if diffuse), or seed calluses (if descrete).

Intractable plantar keratoses and seed calluses usually occur beneath the metatarsal heads. They are often associated with metatarsalgia, a general term for pain about one or several of the metatarsal heads. Such pain may be caused by the callosity itself, or by some other manifestation of chronic pressure overload of the metatarsal head, such as overuse synovitis of the joint, attritional tearing of the joint ligaments, or flexor tendinitis. Such pressure overload is often caused by a cavus foot, high heels, claw toe deformity, or overuse (as in increased running). Osteonecrosis of the metatarsal head, known as Freiberg's infraction, will also cause metatarsalgia.

Corns (Figure 1) usually occur from inappropriately tight shoe wear with subsequent development of toe deformities (hammer toe, bunionette, claw toe). Hard corns occur over bony prominences while soft corns develop between the fouth and fifth toes in the web space (Figure 2). Periungual corns are small but painful lesions occuring at the edge of a nail.

Clinical symptoms

Patients with corns and calluses typically complain of painful cutaneous lesions. Patients will with metatarsalgia report pain beneath one or more metatarsal heads. The second metatarsal is typically invoved.

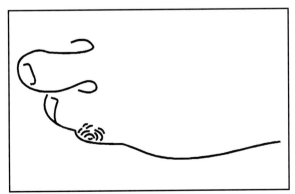

Figure 1
Hard corn at the lateral aspect of the PIP joint of the small toe

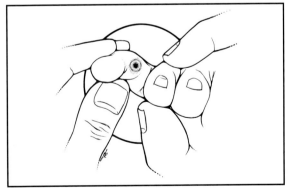

Figure 2
Soft corn on the medial aspect of the tip of the small toe

foot and ankle

Tests

Exam

The pared surface of a callus will have a uniform waxy appearance and occur over areas of pressure concentration or bony prominence (Figure 3). Warts are generally tender when pinched side to side, whereas corns and callosities are not. Presence of a tender callus under a metatarsal head indicates metatarsalgia. Corns occur over or between the toes. Subungual corns occur at the edge of a nail, often in association with a mallet toe or secondary to improper shoe fit.

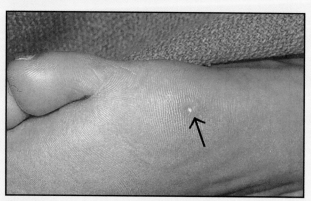

With tenderness over the dorsal surface of the metatarsophalangeal (MP) joint, perform the drawer test to assess for joint instability.

Figure 3
Clinical appearance of a seed callus beneath the fifth metatarsal head

Differential diagnosis

Interdigital neuroma (between the third and fourth metatarsal head or second and third metatarsal head, no callus)

Plantar warts (nonweightbearing areas, may have a nearby satellite lesion, and pared surface has multiple tiny points of hemorrhage near its base)

Synovitis of the MP joint (tenderness over the dorsal MP joint, no plantar callus, painful drawer test)

Adverse outcomes of the disease

Persistent pain, ulceration, or infection can develop. Metatarsalgia, if due to ligament tearing or joint synovitis, may result in gradual dislocation of the joint with persistent pain and/or deformity of the toe.

Treatment

Paring and pressure relief are the principal treatment of corns (see Trimming a Corn or Callus). Paring involves shaving the lesion layer by layer with a scalpel after the skin is prepared with alcohol or iodine. The goal is to remove enough of the avascular keratin to restore a more normal contour to the skin without drawing blood. This can be accomplished using a number 15 blade, without anesthetic, if performed gradually and with care. Paring also gives excellent short-term pain relief. Patients should then be instructed in self-care, using a pumice stone or callus file to regularly debride these lesions after soaking the foot or following a shower.

Treatment of metatarsalgia includes use of a metatarsal pad, trimming of the callus/corn, and correction of associated problems such as improper shoe fit and claw toe deformities.

Pressure is relieved by wearing roomier shoes, commercially available silicone cushions, or small foam donut pads to shift pressure from the lesion to the surrounding normal skin. For soft corns, a small amount of lamb's wool taped between the toes can wick away moisture and help to cushion the area.

If conservative measures fail, then surgical treatment to remove the underlying bony prominences is indicated. With soft corns, "syndactylization" (a partial webbing of the involved toes) may be required.

Adverse outcomes of treatment

Infection and bleeding can occur from excessively deep paring. Paring a soft corn can be especially difficult because of its awkward location in the web space. Medicated keratolytic corn pads often cause maceration and may result in infection; therefore, these pads are probably best avoided.

Referral decisions/Red flags

Failure to respond to conservative treatment, presence of ulceration, or infection are all indications for further evaluation of hyperkeratotic lesions. Deformity or persistent metatarsalgia may need further evaluation.

foot and ankle

Acknowledgements

Figure 1 is adapted with permission from Coughlin MJ, Mann RA: Lesser toe deformities, in Mann RA (ed): *Surgery of the Foot.* St. Louis, MO, CV Mosby, 1986, pp 148–170.

Figure 2 is reproduced with permission from Alexander IJ (ed): *The Foot: Examination and Diagnosis,* ed 1. New York, NY, Churchill Livingstone, 1990, p 77.

procedure

Trimming a Corn or Callus

Materials

Sterile gloves

Scalpel with #10 or
#15 blade, or
paragon blade

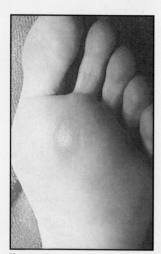

Figure 1A
A diffuse callus

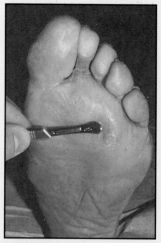

Figure 1B
Placement of the blade

Step 1

Wear protective gloves at all times during the procedure and use sterile technique.

Step 2

Anesthesia is not required to trim a corn or callus; however, the patient may soak the feet in water for several minutes to soften the skin prior to the procedure.

Step 3

Place the blade tangential to the lesion to shave it down (Figure 1). Bleeding should not occur.

Step 4

Take special care with a corn on the toe because the skin is thin and can be fragile. Shell out several millimeters of the hard central core of a plantar callus with the sharp tip of the #15 blade.

foot and ankle

Trimming a Corn or Callus (continued)

Adverse outcomes

Paring down a corn or callus too deeply may expose subcutaneous tissue. If bleeding occurs, the corn, wart, or callus has been trimmed too deeply.

Aftercare/patient instructions

Instruct the patient to continue to pare down the lesion daily after a shower or bath with a pumice stone or nail file.

foot and ankle

The Diabetic Foot

Synonyms

Charcot foot

Insensate foot

Neuropathic foot

Definition

Some diabetic foot problems are caused by vascular insufficiency, but most have a neuropathic basis, possibly worsened by vascular impairment. The neuropathy can appear as a sensory, autonomic, or motor disorder and is irreversible. The insensate foot fails to provide sensory feedback, causing the skin to break down due to unperceived repetitive trauma. Intrinsic bony forces and extrinsic forces, such as shoe wear, contribute to the breakdown. The three major clinical problems are diabetic ulceration, deep infection, and Charcot joints. Charcot joints are joints damaged and ultimately disintegrated by multiple nonhealing fractures in the insensate foot.

Clinical symptoms

Patients may not report symptoms, but often have pain at night, burning, and tingling. Onset of any new pain in a patient with diabetes should be investigated and its cause determined. Autonomic disorders present with dry, scaly, and cracking skin. Motor neuropathy causes contracture of the intrinsic foot musculature, resulting in clawing of the toes and formation of calluses under the metatarsal heads. Skin breakdown follows and leads to a painless ulcer. Deep infection is associated with long-standing ulceration and usually presents with a sudden increase in swelling, redness, drainage, and sometimes pain. With Charcot changes, an alteration in the shape of the foot, swelling, warmth, and redness are common (Figure 1). Some degree of pain may be present. In very early Charcot changes, patients may have swelling and erythema, but normal radiographs.

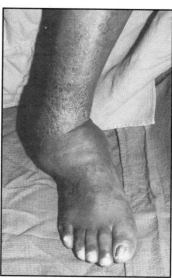

Figure 1

Severe deformity from Charcot breakdown of the ankle joint

Tests

Exam

When examining diabetic patients, it is essential to include a thorough evaluation of their feet. A significant number of amputations can be avoided simply by asking these patients to remove their shoes to allow for a brief inspection of their toenails, skin, and sensibility. Test sensibility with a sensory testing nylon filament. Patients who cannot feel the 10-g, 5.07 diameter filament have lost protective sensibility and must pay great attention to protective foot care. A pin may

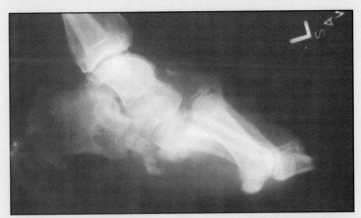

Figure 2
Lateral radiograph showing Charcot degeneration of the midfoot

also be used to detect decreased sensibility by comparing pin prick sensation on the dorsum of the foot to that on the medial side of the proximal leg. Consider customized, well-cushioned shoe wear for these patients.

Diabetic ulcers are insensate and can be easily inspected and probed to determine depth and size. With deep infection, the foot is swollen, hot, and red. The ulceration is likely to be long-standing and may communicate directly to bone or a joint. If bone can be palpated, osteomyelitis is likely present. Examination of a Charcot joint is likely to reveal a hot, red, swollen foot with intact skin. Pulses are usually strong in patients with Charcot joints. By elevating the foot above the heart for 5 minutes, the Charcot foot will lose its redness, whereas an infected foot will not. Evaluation of patients with a Charcot foot must include checking for cellulitis, osteomyelitis, and gout.

Diagnostic

Plain radiographs are necessary to help rule out osteomyelitis and Charcot fractures (Figure 2). An MRI scan may help confirm a deep abscess, but is rarely indicated. Vascular studies are appropriate if pulses are absent or if the patient has a nonhealing ulcer. The vascular workup may help determine the ability of the ulcer to heal.

There is no noninvasive study that differentiates Charcot radiographic changes from osteomyelitis. As a general rule, however, osteomyelitis will develop in an adult with diabetes only if the overlying skin has been violated (eg, ulceration or puncture). In rare instances, a biopsy of the bone may be required.

foot and ankle

Differential diagnosis

Cellulitis

Gout (may be confused with early Charcot changes)

Osteomyelitis (usually beneath an open skin ulcer)

Other neuropathies

Adverse outcomes of the disease

Skin ulceration, chronic osteomyelitis, Charcot joint, and gangrene all occur in the diabetic foot. Amputation is often necessary.

Treatment

The neuropathic foot is a major health problem and a common cause of hospitalization and amputation (see Care of a Diabetic Foot). Any neuropathy that has occurred is irreversible. The goal of treatment is prevention and patient education. When a problem exists, aggressive treatment is needed to avoid a more serious and debilitating situation. The appearance of a callus is the first phase of a diabetic ulcer and signals the need for adaptive shoe wear (contoured and molded insoles) and close follow-up.

Treatment of a diabetic ulcer requires removing the pressure causing the ulcer, allowing the ulcer to heal, and prescribing optimal shoe wear to prevent recurrence. Shoe wear changes, modification of an orthotic device, and total contact casting may be used for superficial ulcerations. For deeper ulcerations, these measures may be inadequate, and surgery may be required.

Treatment of a deep infection must be aggressive and prompt. Any abscess collection should be considered an emergency and surgically drained. Sometimes, an osteomyelitis that is limited, such as the distal phalanx, can be treated surgically. The infection is often polymicrobial and requires antibiotics. Prior to any elective surgery, the patient should undergo vascular evaluation to ensure that the surgical site has healing potential.

In the initial stage of a Charcot joint, the foot and ankle need to be unweighted and stabilized, usually with a cast. After the acute swelling and erythema have subsided, the patient can begin weightbearing with continued use of a cast or customized brace. The patient needs to be advised that the period of immobilization can be lengthy, often up to 12 months, and a permanent brace may be required for ambulation. When properly recognized and treated, acceptable limb salvage can be achieved. Occasionally, surgical reconstruction by arthrodesis is needed for severe deformity that is unbraceable.

Adverse outcomes of treatment

The adverse outcomes of treatment are the same as for the disease, but include surgical complications of infection, ischemia, and death.

Referral decisions/Red flags

Unexplained pain in a diabetic foot, sudden onset of swelling and pain, and nonhealing ulcerations all signal the need for further evaluation.

Acknowledgements

Figures 1 and 2 are reproduced with permission from Harrelson JM: The diabetic foot: Charcot arthropathy, in Heckman JD (ed): *Instructional Course Lectures 42.* Rosemont, IL, American Academy of Orthopaedic Surgeons, 1993, pp 141–146.

Information regarding the sensory testing nylon filament is provided by the Filament Project, 5445 Point Clair Road, Carville, LA 70721.

procedure

Care of a Diabetic Foot

Care of the feet

1. Never walk barefoot—always wear shoes or slippers.

2. Wash feet daily with mild soap and water.

   - Always test the water temperature with your hands or elbows before putting your feet in the water.

   - After washing, pat your feet dry; do not rub vigorously.

   - Use only one thickness of towel to dry your feet, especially between the toes.

   - Use a lanolin, coconut oil, or petroleum based lotion to keep skin soft and prevent skin from getting dry and cracked; however, do not use lotions on open wounds.

3. Inspect your feet daily for puncture wounds, bruises, pressure areas and redness, and blisters.

   - Puncture wounds—Have you stepped on any nails, glass, or tacks?

   - Bruises—Feel for swelling.

   - Pressure areas and redness—Check the six major locations for pressure on the bottom of the foot:

     a. Tip of the big toe

     b. Base of the little toes

     c. Base of the middle toes

     d. Heel

     e. Outside edge of foot

     f. Across the ball of the foot (metatarsal heads)

   - Blisters—Check the six major locations on the bottom of the foot for blisters, plus the tops of the toes and the back of the heel. *Never* pop a blister!

4. Seek treatment by a physician for any foot injuries or athlete's foot.

5. Do not use Lysol, iodine, cresol, carbolic acid, kerosene, or other irritating antiseptic solutions to treat cuts or abrasions on your feet. These products will damage soft tissue.

6. Do not use sharp instruments, drugstore medications, or corn plasters on your feet. Always seek the advice of your physician for any condition that needs such care.

Care of a Diabetic Foot (continued)

7. Always keep your feet warm.

   - Wear loose bed socks while sleeping.

   - Avoid frostbite by wearing warm socks and shoes during cold weather.

   It is important that you follow these additional guidelines:

   - Do not use a heating pad on your feet.

   - Do not place your feet on radiators, furnaces, furnace grills, or hot water pipes.

   - Do not hold your feet in front of the fireplace, circulators, or heaters.

   - Do not use a hair dryer on your feet.

8. Place thin pieces of cotton or lamb's wool between your toes if your toes overlap.

9. Do not sit cross-legged; it can decrease circulation to your feet.

10. Take care of your toenails in the following manner:

    - Soak or bathe feet prior to trimming.

    - Trim toenails straight across.

    - Never trim toenails into the corner.

    - Use a nail file or emery board for trimming; if toenails are thick, see your physician.

    - Make sure that you trim your nails under good lighting.

    - Consult your physician if there are any signs of an ingrown toenail. Do not treat an ingrown toenail with drugstore medications; however, you can place a thin piece of cotton under the corners of the toenail.

Socks and stockings

1. Wear clean, dry socks daily. Make certain that there are no holes or wrinkles in your socks or stockings.

2. Wear thin, white, cotton socks in the summer; they are more absorbent and porous. Change them if your feet sweat excessively.

3. Wear square-toe socks; they will not squeeze your toes.

Care of a Diabetic Foot (continued)

4. Wear pantyhose or stockings with a garter belt. It is important that you do not wear or use the following:

   • Elastic top socks or stockings, or knee-high stockings

   • Circular elastic garters

   • String tied around the tops of stockings

   • Stockings that are rolled or knotted at the top

Shoe wear

1. Always wear proper shoes. Check the following components daily to ensure that your shoes fit properly and will not damage your feet:

   • Shoe width—Make sure that the shoes are wide and deep enough to give the joints of your toes breathing room. Shoes that are too narrow will cause pressure bruises and blisters on the inside and outside edges of your foot at the base of the toes.

   • Shoe length—Shoes that are too short will cause pressure and blisters on the tops of your toes.

   • Back of shoe—Looseness at the heel will cause blisters at your heels.

   • Bottom of heel—Make sure there are no nails. The presence of holes indicates that there are nails in the heels.

   • Sole—Make sure that the sole is not broken. A break in the sole will allow nails or other sharp objects to puncture the skin.

2. Be careful what type of new shoes you purchase. Use the following guidelines when you look for new shoes:

   • Buy new shoes in the evening to allow for swelling in your foot.

   • Inspect your feet once an hour for the first few days. Look for red areas, bruises, and blisters.

   • Do not wear your new shoes for more than a half day for the first few days.

   • The following components in a shoe are desirable:

     a. Lace or adjustable closure

     b. Soft leather top

Care of a Diabetic Foot (continued)

      c. Leather soles (to allow foot to breathe; they mold to foot)

      d. Crepe soles (to provide a good cushion for walking)

- Avoid the following components in a shoe:

      a. Elastic across the top of the shoe

      b. Pointed-toe style (they constrict the toes)

      c. High heels

3. Put your shoes on properly.

- Inspect the inside of each shoe before putting it on. Make sure to remove any small stones or debris. Be certain that the inside of the shoe is smooth.

- Loosen the laces before putting on or taking off your shoes.

- Make sure that the tongue is flat, with no wrinkles.

- Be certain that you do not tie your laces either too tight or too loose.

Fracture of the Ankle

Synonyms

Pott's fracture

Bimalleolar fracture

Trimalleolar fracture

Maisonneuve fracture

Definition

Ankle fractures involve the medial or lateral malleolus, the posterior lip of the tibia (posterior malleolus), the collateral ligamentous structures, and/or the talar dome. Stable ankle fractures involve one malleolus but no ligament structures (Figure 1). Unstable ankle fractures involve both sides of the ankle joint, with either both malleoli fractured or a fracture of the distal fibula with disruption of the deltoid ligament (Figure 2). In these cases, the ankle is vulnerable for displacement, instability, and posttraumatic arthritis.

Clinical symptoms

Patients usually report acute pain following trauma. The etiologies are as varied as the circumstances, but there is usually some element of rotation or twisting.

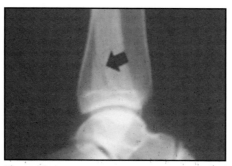

Figure 1A
Nondisplaced fracture of the lateral malleolus

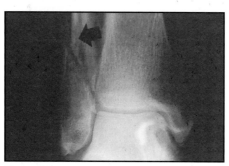

Figure 1B
Minimally displaced fracture of the lateral malleolus

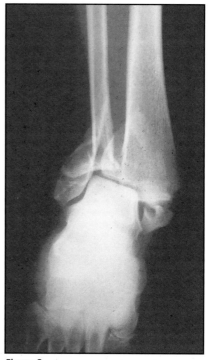

Figure 2
Displaced bimalleolar ankle fracture that requires immediate reduction

Tests

Exam

Swelling medially, laterally, and/or posteriorly accompanies most ankle fractures. There is extreme tenderness at the fracture site. There is often a palpable gap on the medial side, and the ankle is painful to all manipulation. Displacement may be rotational (foot rolled externally) or lateral offset of the foot from the tibia.

An isolated fracture of the lateral malleolus with tenderness over the medial malleolus is presumed to be a bimalleolar injury with a tear of the deltoid ligament from the medial malleolus.

Palpate proximally to assess for tenderness at the proximal fibula, as this coupled with swelling of the medial ankle may indicate a Maisonneuve fracture, which is a fracture of the proximal fibula, a tear of the medial deltoid ligament, and a disruption of the ankle mortise.

Examination should include evaluation of the posterior tibial pulse, and plantar sensation (posterior tibial nerve).

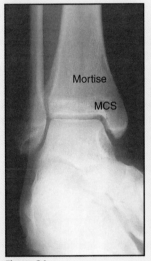

Figure 3A
Internal rotation mortise view of the ankle (MCS = medial clear space)

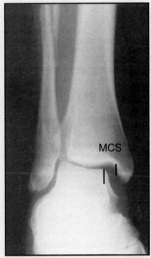

Figure 3B
Lateral fibular fracture with deltoid disruption and widening of the MCS

Diagnostic

AP, lateral, and oblique (slightly internally rotated AP) radiographs will show most fractures. The oblique view is often called the mortise view, since the relationships of the tibia, fibula, and talus are clearest in this projection (Figure 3). AP and lateral views should include the proximal fibula and tibia if there is tenderness in that area.

Minimally displaced fractures may not appear on initial radiographs; therefore, radiographs should be repeated in 10 to 14 days, if there is a high suspicion of a fracture.

With a rotational injury, a portion of the talus can be sliced off the top or lateral border, creating a free fragment or osteochondral fracture. This is best seen in the mortise or internal rotation oblique view (Figure 4). A CT scan is often required for complex fractures with articular surface involvement or involvement of the lateral portion of the distal tibia.

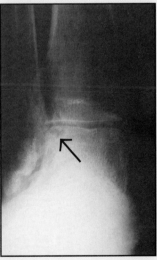

Figure 4
Lateral talar dome fracture

foot and ankle

Differential diagnosis

Fracture of the base of the fifth metatarsal

Osteochondral fracture of the talar dome, lateral process of the talus, and anterior process of the calcaneus

Adverse outcomes of the disease

Instability, deformity, reflex sympathetic dystrophy, posttraumatic arthritis, posterior tibial nerve injury, and tarsal tunnel syndrome are possible, along with posterior or anterior compartment syndrome.

Treatment

Stable unimalleolar fractures (no ligamentous involvement) may be treated with a weightbearing cast or brace for 6 weeks, followed by a strengthening program. A nonweightbearing cast and prolonged immobilization for an additional 3 to 4 weeks should be considered in patients with peripheral neuropathy and decreased sensibility due to the possibility of a Charcot deformity. Surgical treatment should be considered for unstable fractures or those with a loose fragment.

Adverse outcomes of treatment

Infection, nonunion, malunion, posttraumatic arthritis, and reflex sympathetic dystrophy may develop following treatment.

Referral decisions/Red flags

Patients with signs of Maisonneuve fracture, bimalleolar and trimalleolar injuries, and osteochondral fragments need further evaluation.

Acknowledgements

Figure 2 is reproduced with permission from Grantham SA: Trimalleolar ankle fractures and open ankle fractures, in Greene WB (ed): *Instructional Course Lectures XXXIX*. Park Ridge, IL, American Academy of Orthopaedic Surgeons, 1990, pp 105–111.

Figures 3A and 3B are reproduced with permission from Stiehl JB: Ankle fractures with diastasis, in Greene WB (ed): *Instructional Course Lectures XXXIX*. Park Ridge, IL, American Academy of Orthopaedic Surgeons, 1990, pp 95–103.

Fracture of the Hindfoot

Synonyms

Talar fracture

Heel fracture

Calcaneal fracture

Definition

The two principal bones of the hindfoot, the talus and calcaneus, are usually fractured only as a result of severe trauma such as a motor vehicle accident or fall from a height. The adverse outcomes are potentially severe and disabling, and treatment is difficult. Fractures of the talus often interrupt the blood supply to the body of the talus and lead to osteonecrosis.

Clinical symptoms

Patients often report an inability to bear weight. In addition to the bony injury, there is usually involvement of the adjacent joint surface, either from penetration of the fracture, or from compression of the articular cartilage.

Tests

Exam
Examination reveals tenderness over the talo-navicular joint anterior to the medial malleolus, tenderness with side to side compression of the heel, and swelling in the heel and at the ankle. Check plantar sensation in the ball of the foot to evaluate posterior tibial nerve function. Compartment syndrome is difficult to evaluate with these injuries; however, significant swelling in the area of the arch is highly suggestive of plantar compartment syndrome.

Diagnostic
Obtain AP and lateral radiographs of the hindfoot, along with AP and mortise views of the ankle (Figure 1). A CT scan may be necessary if further evaluation is needed.

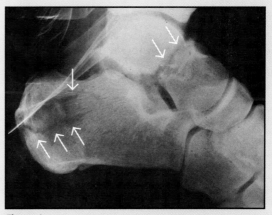

Figure 1
Fracture of the talus and calcaneus

foot and ankle

Differential diagnosis

Associated lumbar spine fracture (in 10% of patients)

Malleolar fracture (evident on radiographs, ankle swelling or deformity)

Medial or lateral ankle ligament injury (ankle swelling, instability to external or internal rotation)

Adverse outcomes of the disease

Chronic pain, reflex sympathetic dystrophy, osteonecrosis of the talus, posttraumatic arthritis, plantar compartment syndrome, or tarsal tunnel syndrome may result.

Treatment

Immediate treatment consists of splinting with a well-padded posterior splint from the toe to the upper calf. The extremity should be elevated above the level of the heart and then ice applied for 20 minutes after 1 to 2 hours. Note that many of these fractures often require immediate surgical reduction and fixation to minimize later complications.

Adverse outcomes of treatment

Infection, reflex sympathetic dystrophy, and nonunion are possible following surgery.

Referral decisions/Red flags

These patients need further evaluation immediately upon diagnosis.

Acknowledgements

Figure 1 is reproduced with permission from Levine AM (ed): *Orthopaedic Knowledge Update: Trauma.* Rosemont, IL, American Academy of Orthopaedic Surgeons, 1996, pp 191–209.

Fracture of the Metatarsal

Synonyms

Arch fracture

Forefoot fracture

Definition

Fractures of the forefoot involving the metatarsal bones, either single or multiple, usually heal with nonoperative treatment. A fracture in the proximal metaphysis of the fifth metatarsal (Jones' fracture) may fail to heal without prolonged nonweightbearing casting or surgery (Figure 1).

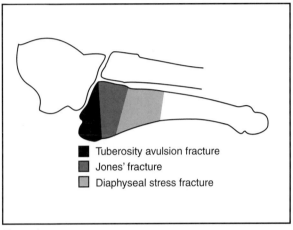

■ Tuberosity avulsion fracture
■ Jones' fracture
■ Diaphyseal stress fracture

Figure 1
Schematic representation of fracture zones for proximal fifth metatarsal

Clinical symptoms

Pain on weightbearing and swelling is common. Stress fractures usually occur after a sudden increase in activity, such as a new training regimen (increase in intensity or distance), change in running surface, or even prolonged walking.

Tests

Exam
Examination reveals tenderness and swelling over the metatarsal, along with ecchymosis and pain on rotation of the metatarsal (when twisting a single toe).

Diagnostic
Obtain AP, lateral, and oblique radiographs of the foot.

foot and ankle

Differential diagnosis

Lisfranc dislocation or sprain

Lisfranc fracture (tarsometatarsal joint)

Metatarsalgia (plantar pain over the metatarsal head)

Morton's neuroma

Adverse outcomes of the disease

Both malunion and nonunion are possible; however, nonunion is not common except in association with Jones' fracture. Metatarsalgia with a plantar callus may develop as a result of a dorsiflexed malunion. Compartment syndrome can develop following severe trauma and multiple metatarsal fractures.

Treatment

Conservative treatment of undisplaced metatarsal neck and shaft fractures should include use of a short leg cast, fracture brace, or wooden-soled shoe. Select the device that uses the minimum amount of immobilization while providing adequate comfort. Weightbearing is permitted as tolerated. Repeat radiographs over the next 3 weeks to observe for displacement, then at 6 weeks to confirm healing. Tenderness at the fracture site will diminish as the fracture heals. Fractures of the first metatarsal are often the result of a high-impact incident and often require surgical intervention (Figure 2).

Multiple metatarsal fractures and those with more than 4 mm of displacement or an apical angulation of more than 10° (in the lateral view) may require open or closed reduction to reestablish a physiologic weightbearing position of the metatarsal head.

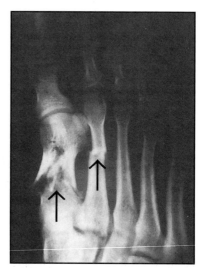

Figure 2
Radiograph showing fractures of the first and second metatarsals

Jones' fracture occurs through the proximal metaphysis of the fifth metatarsal and has a propensity for nonunion or delayed union. Initial treatment consists of immobilization in a nonweightbearing cast for 6 weeks, followed by a weightbearing cast until healing occurs.

Avulsion fractures of the base of the fifth metatarsal often fail to heal, but are rarely symptomatic. Initial conservative treatment includes use of wooden-soled shoes or an air stirrup or brace until symptoms subside.

Adverse outcomes of treatment

Malunion with painful plantar callosities under the metatarsal heads or dorsal corns due to friction over a prominent metatarsal head can occur.

Referral decisions/Red flags

Metatarsal fractures with more than 4 mm of displacement, more than 10° of angulation, multiple fractures, possible compartment syndrome, and Jones' fractures are all indications for further evaluation. Displaced or comminuted fractures of the first metatarsal also indicate the need for further evaluation.

Acknowledgements

Figure 1 is adapted with permission from Lawrence SJ, Botte MJ: Jones fractures and related fractures of the proximal fifth metatarsal. *Foot Ankle* 1993;14:358–365.

Figure 2 is reproduced with permission from Shereff MJ: Fractures of the forefoot, in Greene WB (ed): *Instructional Course Lectures XXXIX*. Park Ridge, IL, American Academy of Orthopaedic Surgeons, 1990, pp 133–140.

foot and ankle

Fracture of the Midfoot

Synonyms

Lisfranc fracture-dislocation

Definition

Fracture of the midfoot is an easy-to-miss, traumatic disruption of the tarsometatarsal joints involving fracture, dislocation, or both. Injury to this joint may occur as a result of significant trauma, such as a fall from a height, an automobile accident, or from an indirect mechanism as commonly occurs in athletes. This seemingly innocuous injury often has serious adverse effects. The critical injury is to the second tarsometatarsal joint, which is the stabilizing apex for the other tarsometatarsal joints since it "keys" into a slot in the cuneiforms.

Clinical symptoms

Patients often report an ankle sprain; however, pain is localized to the dorsum of the midfoot.

Tests

Exam

This injury is easily missed and often misdiagnosed as a sprain, since the swelling of the foot is very similar to that seen with a grade III ankle sprain. Careful examination will reveal that the maximum area of tenderness and swelling is more distal on the foot, over the tarsometatarsal joint rather than over the ankle ligaments. During examination, stabilize the hindfoot (calcaneus) with one hand, and rotate and/or abduct the forefoot with the other (Figure 1). This produces severe pain with a Lisfranc injury, but only minimal pain with an ankle sprain.

Diagnostic

Obtain AP, lateral, and oblique radiographs of the foot with the patient standing if possible. A common error is to obtain only ankle radiographs. Radiographs of the entire foot are important, as joint anatomy in the midfoot is complex and easy to misdiagnose. One key to diagnosis is to look for the normal alignment of the medial aspect of the middle cuneiform and the medial aspect of the base of the second metatarsal (Figure 2). The oblique view should show similar colinearity of the medial aspect of the fourth

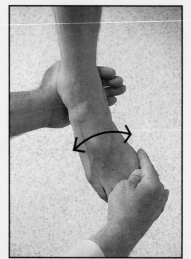

Figure 1
Stabilize the hindfoot with one hand and rotate and/or abduct the forefoot with the other

foot and ankle

metatarsal base and the medial aspect of the cuboid (Figure 3). Comparison views with the normal foot can be helpful.

If the AP radiograph shows that the second metatarsal base has shifted laterally, even by only a few millimeters, then a Lisfranc dislocation has occurred. A small avulsion fracture between the bases of the first and second metatarsals (called a Lisfranc fragment) also indicates significant injury to the tarsometatarsal joint (Figure 4).

If radiographs are normal but physical examination suggests injury to this joint complex, stress radiographs of the midfoot under local anesthetic or sedation may be indicated. Spontaneous reduction following complete dislocation is not uncommon. If confusion still exists, a CT or MRI scan may be helpful.

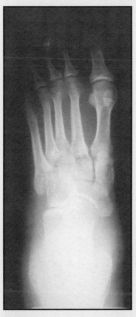
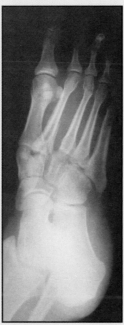
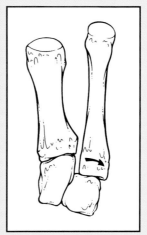

Figure 4
Lisfranc fragment

Figure 2
Normal alignment of the medial aspect of the middle cuneiform and the medial aspect of the base of the second metatarsal

Figure 3
Lisfranc fracture-dislocation with clear offset of the second and third metatarsal-cuneiform articulation

Differential diagnosis

Ankle fracture

Ankle sprain (focal pain over the lateral ankle ligament)

Metatarsal fracture

Navicular fracture

foot and ankle

Adverse outcomes of the disease

Lisfranc injuries have significant morbidity. Adverse outcomes include compartment syndrome, ischemic contracture of muscle with claw toes and sensory impairment, midfoot instability and arthritis, flatfoot, a limp, and pain.

Treatment

These injuries are usually associated with significant swelling. The foot should be elevated with no weight-bearing until the swelling subsides. A short leg, nonweightbearing splint or cast is appropriate. After the swelling decreases, nondisplaced injuries are treated by 6 to 8 weeks of nonweightbearing cast immobilization, followed by use of a rigid arch support for 3 months.

Compartment syndrome is difficult to diagnose in the foot. The only way to determine pressure in the foot is to obtain an intracompartmental pressure measurement. If there is any clinical suspicion, pressure measurements or surgical fasciotomy is indicated.

A fracture or fracture-dislocation with any displacement will usually require surgical stabilization.

Adverse outcomes of treatment

A high percentage of patients will have chronic pain in the midfoot even after appropriate cast immobilization or surgical treatment.

Referral decisions/Red flags

Any possibility of compartment syndrome requires immediate surgical evaluation. Often, even a minimally displaced fracture-dislocation requires surgical reduction.

Acknowledgements

Figures 2 and 3 are reproduced with permission from Levine AM (ed): *Orthopaedic Knowledge Update: Trauma.* Rosemont, IL, American Academy of Orthopaedic Surgeons, 1996, pp 191–209.

Figure 4 is reproduced with permission from Alexander IJ (ed): *The Foot: Examination and Diagnosis,* ed 1. New York, NY, Churchill Livingstone, 1990, p 131.

Fracture of the Phalanges

Synonyms

Broken toe

Definition

Phalangeal fractures usually involve the proximal phalanx of the toe and are caused by direct trauma or by stubbing. They rarely result in major disability. The fifth toe is most commonly affected.

Clinical symptoms

Patients have pain, swelling, or ecchymosis.

Tests

Exam
Examination may reveal deformity of the toe, but local bony tenderness, swelling, and ecchymosis are often the only principal findings.

Diagnostic
AP radiographs usually confirm the diagnosis.

Differential diagnosis

Freiberg's infraction (osteonecrosis of the metatarsal head)

Ingrown toe nail/paronychia

Metatarsalgia (plantar tenderness over the metatarsal head)

Metatarsophalangeal synovitis (tenderness over the MP joint)

Adverse outcomes of the disease

Without treatment, the patient will have pain and a limp until the fracture heals. Permanent deformity is also a possibility.

foot and ankle

Treatment

Phalangeal fractures are treated by "buddy taping" the fractured toe to an adjacent toe, usually the toe medial to the fractured one. Place a gauze pad between the toes to absorb and wick moisture and prevent maceration of the skin from sweating. The tape and gauze should be changed as often as needed.

If the fracture is displaced, perform a closed reduction under a digital block (see Digital Anesthetic Block [Foot]), then buddy tape the toe to its neighbor. Fractures involving the joint rarely require open reduction and pinning.

Adverse outcomes of treatment

Chronic swelling and deformity of the toe are possible.

Referral decisions/Red flags

Patients with an open fracture or a displaced articular fracture at the MP joint need further evaluation.

procedure

Digital Anesthetic Block (Foot)

The least painful and most reliable method of performing an anesthetic block to the toe is with an injection at the level of the metatarsal heads (Figure 1).

Materials

Sterile gloves

Skin preparation solution (iodinated soap or similar antiseptic solution)

Ethyl chloride spray

10-mL syringe

18-gauge needle

21-gauge, 1½ -inch needle

10 mL of 1% lidocaine or 0.5% bupivacaine, both *without* epinephrine

Adhesive dressing

Step 1

Wear protective gloves at all times during the procedure and use sterile technique.

Step 2

Use the 18-gauge needle to draw 10 mL of the local anesthetic into the syringe, then switch to the 21-gauge needle to preserve sterility.

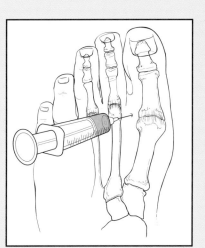

Figure 1
Insert the needle on either side of the metatarsal head

Step 3

Cleanse the dorsal surface of the foot with an antiseptic solution on either side of the metatarsal heads.

Step 4

Freeze the dorsal skin with an ethyl chloride spray.

Step 5

Insert the 21-gauge needle into the soft-tissue space on either side of the metatarsal head, until it just begins to tent the plantar skin. Remember that the sensory nerves travel along the plantar side of the metatarsal.

Step 6

Withdraw the needle 1 cm or until the tip rests at the level of the plantar aspect of the metatarsal head.

foot and ankle

Digital Anesthetic Block (Foot) (continued)

Step 7

Inject 3 mL of anesthetic, then an additional 2 mL as the needle is withdrawn. Make certain that some of the anesthetic is deposited subcutaneously around the dorsal sensory nerves.

Step 8

Repeat the procedure on the other side of the metatarsal of the toe that is to be numbed. Within 1 to 2 minutes, the involved toe should become numb.

Step 9

Dress the puncture wound with a sterile adhesive dressing.

Adverse outcomes

Although rare, infection is possible. Necrosis of a digit is possible if epinephrine is used in the anesthetic solution.

Aftercare/patient instructions

Advise the patient that a collection of fluid on the plantar aspect of the foot may appear, but that the fluid will dissipate within several hours after the block.

Fracture of the Sesamoid and Sesamoiditis

Synonyms

Turf toe

Dancer's toe

Definition

Sesamoid disorders include inflammation, fracture, or arthritis. The sesamoid bones are embedded in the flexor hallucis brevis tendon beneath the first metatarsal head (plantar surface). Sesamoiditis occurs from repeated stress and subsequent inflammation of the sesamoid secondary to repeated stress. It is a condition that is equivalent to metatarsalgia that can occur beneath the lesser metatarsals. Fractures may be acute or stress related.

Clinical symptoms

Pain under the first metatarsal head, specifically involving the sesamoid bones (medial and lateral), with or without swelling and ecchymosis, is common (Figure 1). The usual stresses involve dancing or running, but can occur with trauma from forced flexion or dorsiflexion (turf toe) and from falls. There may be a history of a push-off injury with acute onset of pain.

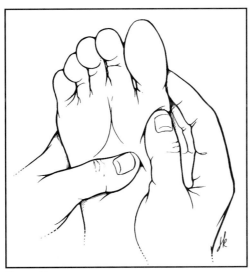

Figure 1A
Location of the medial sesamoid

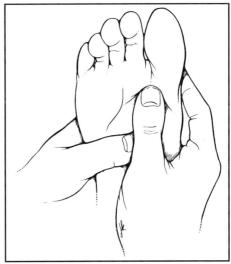

Figure 1B
Location of the lateral sesamoid

Tests

Exam

In both sesamoiditis and sesamoid fractures, examination reveals tenderness at the sesamoid bone directly beneath the metatarsal head. The tender spot will move with the sesamoid as the patient flexes and extends the great toe. Forced dorsiflexion of the toe is likely to be painful.

Diagnostic

Obtain AP, lateral, and axial views of the sesamoid bones. An oblique view of the sesamoid may also be helpful in visualizing a fracture line. Multipartite or bipartite sesamoid bones are common (25% of the population) and have rounded margins on the radiographs. These should not be confused with fractured sesamoids whose margins are sharp. A comparison radiograph with the opposite foot or a bone scan can help to identify an acute fracture or stress fracture.

Differential diagnosis

Hallux rigidus (limited dorsiflexion, dorsal spurs)

Hallux valgus (pain medially)

Joplin's neuroma (pain and tenderness over the medial sensory nerve)

Metatarsalgia (tenderness over the plantar aspect of the lesser metatarsal)

Adverse outcomes of the disease

Without treatment, the patient may have pain and a limp, and possibly nonunion.

Treatment

An acute sesamoid fracture is best treated with a stiff-soled postoperative shoe or cast. Patients may change to a well-cushioned shoe with a wide toe box as symptoms permit, usually about 4 weeks after the injury. A J-shaped pad can help relieve pressure under the first metatarsal head once the fracture has healed and the patient begins to ambulate in everyday shoes. Patients with sesamoiditis should be advised to avoid wearing high-heeled shoes. They should also be advised to tape the toe in plantar flexion, use sesamoid pads, or wear a short leg cast for 4 to 6 weeks to relieve pressure on the sesamoids and decrease inflammation. If these measures fail, use of selective injection of corticosteroid into the sesamoid-metatarsal joint may be needed.

Adverse outcomes of treatment

None

foot and ankle

Referral decisions/Red flags

Persistent pain or failure of conservative treatment are indications for further evaluation.

Acknowledgements

Figures 1A and 1B are reproduced with permission from Alexander IJ: *The Foot: Examination and Diagnosis,* ed 1. New York, NY, Churchill Livingstone, 1990, p 65.

foot and ankle

Ganglion Cysts and Plantar Fibromas

Synonyms

Mucoid cyst

Fibromatosis

Ledderhose disease

Definition

Both the plantar fibroma and the ganglion cyst appear as a single mass on the foot. A ganglion is usually a small, 2- to 3-cm cystic tumor that contains clear, jelly-like viscous fluid, which arises from a synovial sheath or joint capsule. A plantar fibroma is a benign thickening of the plantar fascia that can vary in size from 1 to 6 cm.

Clinical symptoms

A ganglion cyst is usually a painless nodule, although it often causes problems with shoe wear. A plantar fibroma appears as a mass on the bottom of the foot that may be painful and is always located within the plantar fascia. A fibroma tends to occur on the plantar surface of the foot, while a ganglion tends to appear on the top and sides.

Tests

Exam

A ganglion cyst is a discrete mass with a gelatinous core. The ganglion is usually movable with side-to-side pressure. A plantar fibroma may be focal or have multiple, discrete masses that are hard (rubbery) and are part of the plantar facial band.

Diagnostic

Plain radiographs are typically normal. Aspiration of a ganglion with a large-gauge needle will return a straw-colored gelatinous material. Sophisticated diagnostic imaging is generally not necessary for neither a ganglion cyst nor a plantar fibroma.

foot and ankle

Differential diagnosis

Giant cell tumor of the tendon sheath

Lipoma

Malignant sarcoma (synovial sarcoma, fibrosarcoma)

Neurofibroma

Adverse outcomes of the disease

Patients may report persistent discomfort and difficulty with shoe wear.

Treatment

The wall of the ganglion should be pierced three or four times with an 18-gauge needle to release the gelatinous core and promote complete collapse of the cyst. If the ganglion recurs and continues to be symptomatic, it should be excised surgically. The efficacy of corticosteroid injection into a ganglion has never been proven.

A plantar fibroma is best treated with shoe modifications and orthotic devices. Surgical excision is indicated in patients with persistent symptoms or from the shoe wear problems associated with the mass.

Adverse outcomes of treatment

Surgical excision of a plantar fibroma should be avoided, if possible, due to the high rate of recurrence. Malignant degeneration of a plantar fibroma rarely occurs.

Referral decisions/Red flags

If the diagnosis of the mass is not clear based on anatomy, examination, radiographs, and aspiration, further evaluation is needed to rule out other etiologies.

foot and ankle

Hallux Rigidus

Synonyms

Hallux limitus

Great toe arthritis

Definition

Hallux rigidus is a degenerative arthritis of the metatarsophalangeal (MP) joint of the great toe. The principal symptoms are pain and stiffness, especially in dorsiflexion. It is the second most common malady of the great toe, following bunion deformity, and affects approximately 2% of the population between the ages of 30 and 60 years.

Clinical symptoms

Patients have pain in the great toe joint with activity, especially in the toe-off phase of gait as they transfer weight to the opposite foot. The bony mass on the dorsum of the great toe metatarsal becomes red and irritated with shoe wear (Figure 1). The dorsal sensory nerve of the great toe may be irritated by the bony spur.

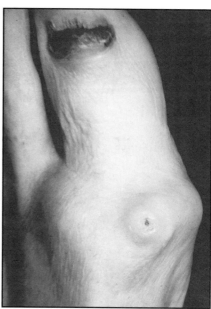

Figure 1A
Top view of dorsal prominence from underlying bony osteophyte

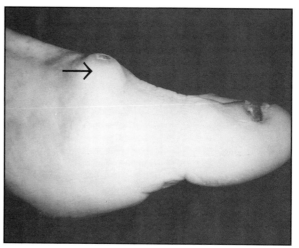

Figure 1B
Lateral view of same

Tests

Exam

Stiffness of the great toe with loss of extension at the MP joint is the hallmark. There is often a "dorsal bunion" caused by the osteoarthritic spurs located on the dorsal portion of the first metatarsal head (Figure 2). The toe is in normal alignment, unlike hallux valgus in which the great toe drifts laterally.

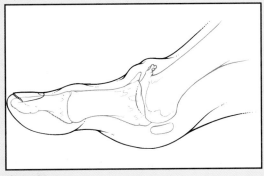

Figure 2A
Hallux rigidus joint line osteophytes

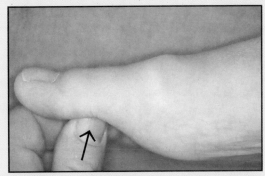

Figure 2B
Loss of extension is the hallmark

Diagnostic

AP and lateral radiographs show narrowing of the MP joint of the great toe and spurs predominantly on the dorsal and lateral aspects of the great toe (Figure 3).

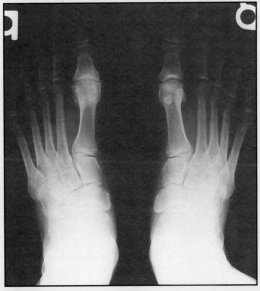

Figure 3
Standing AP radiograph of the feet. Left, foot with advanced arthritic changes. Right, foot with small medial and lateral spurs

foot and ankle

Differential diagnosis

Gout

Hallux valgus with bunion (medial prominence)

Adverse outcomes of the disease

Pain aggravated by walking is common.

Treatment

Use of a shoe with a large toe box helps diminish symptoms. A stiff-soled shoe modified with a steel shank limits dorsiflexion of the great toe and further decreases the pain that is caused by motion in the arthritic joint. This type of shoe modification can be done by a certified pedorthist. Patients should be advised to avoid wearing high-heeled shoes. NSAIDs, ice, and contrast baths may also help control symptoms for a short period of time (see Contrast Baths).

Surgical treatment consists of either excision of the spur (cheilectomy) or fusion of the joint (arthrodesis). Resection of the joint may be indicated in some patients. An artificial joint implant is not recommended.

Adverse outcomes of treatment

NSAIDs may cause gastric, renal, or hepatic complications. Otherwise, there are no known adverse outcomes associated with treatment, other than the usual surgical complications.

Referral decisions/Red flags

Failure of conservative treatment is an indication for further evaluation.

Acknowledgements

Figures 1A and 1B are reproduced with permission from Mann RA: Hallux rigidus, in Greene WB (ed): *Instructional Course Lectures XXXIX.* Park Ridge, IL, American Academy of Orthopaedic Surgeons, 1990, pp 15–21.

Figure 2A is reproduced with permission from Alexander IJ: *The Foot: Examination and Diagnosis,* ed 1. New York, NY, Churchill Livingstone, 1990, p 63.

Hammer Toe

Synonyms

None

Definition

Hammer toe is a deformity of the lesser toes in which there is flexion of the proximal interphalangeal (PIP) joint. A passive extension of the metatarsophalangeal (MP) joint occurs when the toe is flat on the ground (Figure 1). The distal interphalangeal (DIP) joint is usually not affected. A hammer toe is usually caused by chronic tight shoe wear that crowds the toes, but may also be seen after trauma. A hammer toe is considered fixed if the deformity cannot be passively corrected and flexible if correctable.

Clinical symptoms

Patients have an obvious deformity and report pain in the toe and difficulty with shoe wear.

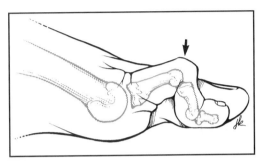

Figure 1A
Hammer toe

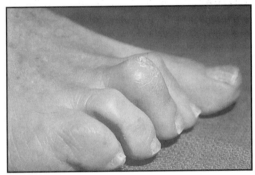

Figure 1B
Clinical appearance of a hammer toe

Tests

Exam
View the patient barefoot, both standing and nonweightbearing. The toe will appear cocked up, with extension at the MP joint and flexion at the PIP joint when the patient is standing. Extension of the MP joint will largely resolve in the nonweightbearing position. There may be associated corns or calluses on the dorsum of the PIP joint or the tip of the toe. The patient may also have pain due to a secondary metatarsalgia, which is often associated with an overlying plantar callus.

Diagnostic
Standing AP and lateral radiographs may be helpful in ruling out chronic MP subluxation, dislocation or an acute fracture, but are otherwise of little help in planning conservative care.

Differential diagnosis

Claw toe (contracture of the MP and PIP joints)

Mallet toe (flexion contracture of DIP joint)

Adverse outcomes of the disease

Without treatment, patients will have chronic pain, difficulty finding comfortable shoes, and a continual problem with hard corns on the dorsal aspect of the toe. Ulceration, infection, and osteomyelitis can develop.

Treatment

Shoes with soft, roomy toe boxes can help to accommodate the deformity. Shoes should be $1/2''$ longer than the longest toe. Patients should be advised to avoid constricting, high-heeled shoes. Early in the process, a home exercise program of passive manual stretching of the toe and foot intrinsic exercises, such as picking up marbles with the toes, or laying a towel flat on the floor and crumpling it under the foot using the toes, may be beneficial. Commercially available straps, cushions, and nonmedicated corn pads can also provide symptomatic relief.

"In depth" shoes, which have an extra $3/8''$ depth in the toe box, accommodate a fixed hammer toe and are available in specialty shoe stores. Shoe repair shops can also help by stretching a small pocket in the toe box of a shoe using a cobbler's swan, a ball-and-ring device that will crimp an outward bulge in the shoe.

Adverse outcomes of treatment

Persistent pain and calluses are possible even with treatment. Taping, strapping, and appliances should be avoided in patients with insensate feet, as these methods may cause pressure ulcers in the affected toe. Likewise, patients with vascular disease are susceptible to further circulatory compromise from taping and appliances.

Referral decisions/Red flags

Failure of conservative treatment or the presence of ulceration with suspected septic arthritis or osteomyelitis all indicate the need for further evaluation.

Acknowledgements

Figure 1A is reproduced with permission from Alexander IJ: *The Foot: Examination and Diagnosis,* ed 1. New York, NY, Churchill Livingstone, 1990, p 70.

Figure 1B is reproduced with permission of California Pacific Medical Center, San Francisco, CA.

Ingrown Toenail

Synonyms

Paronychia

Infected toenail

Onychocryptosis

Definition

An ingrown toenail occurs secondary to one or a combination of improper nail trimming, tight shoes, hereditary predisposition, subungual pathology, congenital incurved nail, thickened nail, or direct trauma (Figure 1). There are three types of ingrown nails: 1) normal nail plate with improper trimming develops a spur growing into the nail fold; 2) lateral nail margin grows inward (incurved nail); and 3) soft-tissue hypertrophy over a normal nail plate. Skin breakthrough creates a portal of entry for normal foot bacteria or a secondary fungal infection.

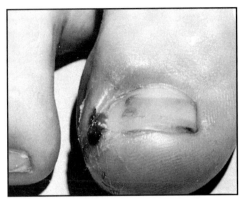

Figure 1
Ingrown toenail

Clinical symptoms

Stage I (inflammatory stage) is characterized by induration, swelling, and tenderness along the nail fold. In stage II (abscess stage), the patient has purulent or serous drainage, increased tenderness, and increased erythema. In stage III (granulation stage), granulation tissue grows onto the nail plate, inhibiting drainage. This stage is less painful than stage II.

Tests

Exam
The diagnosis is clinical; visual inspection is the basis for staging the condition.

Diagnostic
Radiographs may be obtained of stage II and stage III ingrown toenails to rule out a subungual exostosis and osteomyelitis.

Differential diagnosis

Felon (deep abscess on the plantar aspect of the toe)

Onychomycosis (fungal infection of the nail)

Paronychia (superficial abscess)

Subungual exostosis (osteochondroma beneath the nail)

Adverse outcomes of the disease

Progressive pain, osteomyelitis, nail plate deformity, source of systemic infection, and seeding prosthetic joints or heart valves are all adverse outcomes of the disease.

Treatment

Conservative

Stage I: Warm soaks, proper nail trimming, accommodative shoe wear, and clean socks are all necessary. With a blunt nail instrument, insert cotton or waxed dental floss beneath the nail fold to lift the nail edge from its embedded position. Exchange packing daily. The patient should wear nonconstrictive shoes or sandals.

Stage II: Initial treatment should include foot soaks along with broad-spectrum oral antibiotics. Partial or complete excision of the nail under digital block should be performed if the patient has severe pain, if there is a risk of secondary infection to a prosthetic joint, or if the course of oral antibiotics fails (see Digital Anesthetic Block). Partial nail excision is preferred (see Complete Nail Plate Avulsion). Complete nail excision increases the risk of upward deformation of the nail bed (clubbing). An avulsed nail requires 3 to 4 months to regrow.

Stage III: Partial or complete nail plate excision with or without germinal matrix removal is indicated.

Surgical

For hypertrophied lateral and/or medial nail folds, wedge resection and suturing of the nail fold are done without disturbing the nail plate.

Nail plate ablation

Once the nail plate is partially or completely excised, and infection is controlled with oral antibiotics, the germinal matrix can be permanently ablated with either a combination sodium hydroxide/acetic acid or phenol solution or with surgery.

Adverse outcomes of treatment

Adverse outcomes include recurrence (50% to 70% with excision only), nail plate deformity, upturned nail or clubbed nail after complete nail plate excision, and poor cosmesis.

Referral decisions/Red flags

Failure of conservative treatment or the presence of stage III disease are indications for further evaluation.

Acknowledgements

Figure 1 is reproduced with permission from Lutter LD, Mizel MS, Pfeffer GB (eds): *Orthopaedic Knowledge Update: Foot and Ankle.* Rosemont, IL, American Academy of Orthopaedic Surgeons, 1994, pp 41–59.

foot and ankle

procedure

Complete Nail Plate Avulsion

Anatomy

A recurrent ingrown toenail or paronychial infection makes it necessary to remove a portion of the nail plate, usually in the great toe. The lateral or medial margins, or both, of the nail may be involved. Removal of the entire nail plate is described below. Partial removal involves undermining the lateral or medial third, vertically cutting the nail at the junction of the lateral or medial third, and avulsing the small segment adjacent to the nail fold.

Materials

1/4-inch Penrose drain or a strip cut from a rubber glove (optional)

Materials to administer a digital block (see procedure titled Digital Anesthetic Block)

Skin preparation solution (iodine or other similar antiseptic solution)

Strong small scissors or nail cutter

Small hemostat

Nonadherent sterile gauze

Sterile dressing material

Step 1

Wear protective gloves at all times during the procedure and use sterile technique.

Step 2

Wrap a 1/4-inch strip of rubber around the base of the toe to act as a tourniquet and control bleeding (optional).

Step 3

Follow the steps in the procedure titled Digital Anesthetic Block to administer a digital block.

Step 4

Cleanse the toe with an antiseptic solution.

Step 5

Using a small scissors or hemostat, elevate the nail plate from the underlying nail bed (Figure 1).

Step 6

Separate the proximal cuticle (nail fold) from the nail plate.

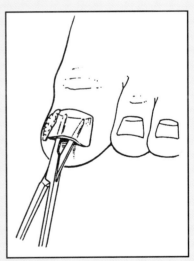

Figure 1
Elevate the nail plate from the nail bed

foot and ankle

Complete Nail Plate Avulsion (continued)

Step 7

Grasp the free portion of the nail plate with a hemostat and avulse it (Figure 2), then palpate the nail bed to ensure that no spikes of nail tissue remain.

Step 8

If a tourniquet was used, remove it and apply compression to stop local bleeding.

Step 9

Apply nonadherent sterile gauze over the exposed nail bed and wrap the entire toe with a sterile dressing.

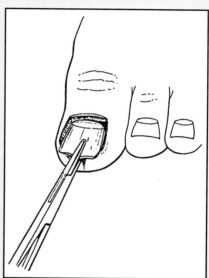

Figure 2
Avulse the nail plate

Adverse outcomes

Because of a high recurrence rate of ingrown toenails after partial nail plate avulsions, permanent ablation of the nail matrix may be required.

Aftercare/patient instructions

Instruct the patient to remove the dressing and gauze in 48 hours and replace them with an adhesive strip. Also instruct the patient to pack a wisp of cotton from a cotton ball under the advancing edge of the newly growing nail over the next few months to prevent the advancing edge of the nail from digging into the exposed soft tissue.

Mallet Toe

Synonyms

None

Definition

Mallet toe is a flexion deformity at the distal interphalangeal (DIP) joint, with relatively normal alignment at the proximal interphalangeal (PIP) and metatarsophalangeal (MP) joints (Figure 1). Usually the result of a "jamming" type injury, this condition may also occur over time with tight shoe wear. The second toe is most commonly affected, as it is the longest.

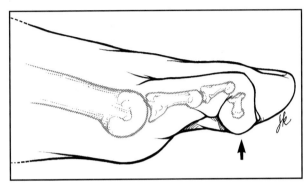

Figure 1
Mallet toe (Arrow indicates usual area of callus formation)

Clinical symptoms

Patients have obvious deformity, pain, and a callus at the tip of the toe. Ulceration may be present in patients with diabetes, as they may also have decreased sensibility in the foot.

Tests

Exam
Examination reveals flexion of the DIP joint, usually of the second toe. A periungual corn or callus may also develop over the tip of the toe, which sits plantar flexed.

Diagnostic
AP and lateral radiographs of the affected toes may be indicated to rule out fractures in acute situations or osteomyelitis in patients who have ulcerations.

Differential diagnosis

Claw toe (extension contracture of the MP joint; flexion contracture of the PIP joint)

Fracture of the distal phalanx

Hammer toe (flexion contracture of the PIP joint)

Psoriatic arthritis

Rheumatoid arthritis

Adverse outcomes of the disease

Patients may have persistent pain, ulceration of the callus at the tip of the toe, and/or infection.

Treatment

File the callus at the tip of the toe and then advise the patient to use a pumice stone on a daily basis. Use of extra-depth shoes and/or a cushioned pad over the tip of the toe is also recommended (Figure 2).

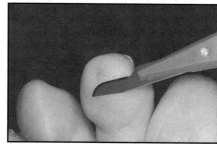

Figure 2A
Treatment of mallet toe

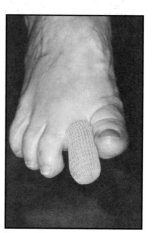

Figure 2B
Cushioned toe cap

Adverse outcomes of treatment

Persistent or recurrent deformity is possible. Skin ulceration or circulatory compromise can result from excessively tight pads or splints.

Referral decisions/Red flags

Failure of conservative management, ulceration, or acute traumatic mallet toe all indicate the need for further evaluation.

Acknowledgements

Figure 1 is reproduced with permission from Alexander IJ: *The Foot: Examination and Diagnosis,* ed 1. New York, NY, Churchill Livingstone, 1990, p 71.

Morton's Neuroma

Synonyms

Plantar neuroma

Interdigital neuroma

Definition

A Morton's neuroma is not a true neuroma, but rather a perineural fibrosis of the common digital nerve as it passes between the metatarsal heads. The fibrosis is secondary to repetitive irritation of the nerve. The condition is most common between the third and fourth toes (third web space) (Figure 1), but may also occur in the second web space (between second and third toes). Although rare, a neuroma in either the first or fourth intermetatarsal space may also occur. This condition has a female to male ratio of 5:1, probably related to shoe wear. The simultaneous occurrence of two neuromas is extremely rare.

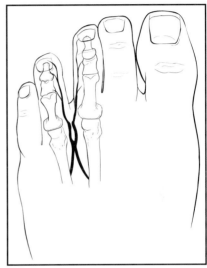

Figure 1
Interdigital neuroma between the metatarsal heads

Clinical symptoms

Plantar pain in the forefoot is the most common presenting symptom. Dysesthesia into the affected two toes or burning plantar pain that is aggravated by activity is also common. Occasionally, there will be numbness in the adjacent toes of the involved web spaces. Night pain is rare. Many patients state that they feel as though they are "walking on a marble." Relief is often obtained by removing the shoe and rubbing the ball of the foot. Symptoms are aggravated by wearing high-heeled and tight, restrictive shoes.

Tests

Exam

Firmly squeeze the metatarsal heads together with one hand while applying direct pressure to the interspace with the other hand. Isolated pain on the plantar aspect of the web space is consistent with an intermetatarsal neuroma.

Inspect the plantar surface for calluses, and then palpate dorsally along metatarsal shafts and heads to evaluate for stress fractures or metatarsalgia. Individually stress the tarsometatarsal joints by grasping the midfoot with one hand and moving each metatarsal in a dorsoplantar direction to rule out midfoot arthritis. Similarly, grasp metatarsal shafts and, while keeping the toes parallel to the metatarsals, try to displace the digits dorsally, then plantarly. Pain or excess motion with this maneuver indicates synovitis or inflammation of the metatarsophalangeal (MP) joints.

foot and ankle

Diagnostic
Radiographs are normal in Morton's neuroma; an MRI scan and ultrasound may detect a neuroma, but are unreliable and rarely, if ever, indicated.

Differential diagnosis

Hammer toe

Metatarsalgia (plantar tenderness over the metatarsal head)

Metatarsophalangeal synovitis (tenderness and swelling directly over the MP joint)

Stress fracture (dorsal metatarsal tenderness)

Adverse outcomes of the disease

Adverse outcomes include chronic, intermittent pain and the need for activity modification.

Treatment

Patients should be advised to wear a low-heeled, soft-soled shoe with a wide toe box. Pain relief may also be possible using metatarsal pads or bars to elevate and spread the metatarsal heads (see Application of a Plantar Pad). A mixture of 1 to 2 mL of lidocaine without epinephrine and 1 mL (10 mg/mL) of corticosteroid injected just proximal to the metatarsal heads can be both diagnostic and therapeutic (see Morton's Neuroma Injection). An injection can diminish symptoms sufficiently to avoid surgery in up to 50% of patients. If symptoms persist or recur, surgical excision of the neuroma or division of the transverse metatarsal ligament is indicated.

Adverse outcomes of treatment

Persistent symptoms following conservative treatment are obviously undesirable. Symptoms may become worse if a painful stump of nerve develops following surgical excision of the neuroma.

Referral decisions/Red flags

Persistent pain despite shoe or insert modifications or injection indicates the need for further evaluation.

Acknowledgements

Figure 1 is adapted with permission from McElvenny RT: The etiology and surgical treatment of intractable pain about the fourth metatarsophalangeal joint (Morton's toe). *J Bone Joint Surg* 1943;25A:675–679.

procedure

Application of a Plantar Pad

Materials

Felt or gel pad

Temporary marker
(lipstick, eyeliner)

Step 1

Locate the neuroma on the plantar aspect of the foot in the soft tissue between the involved metatarsal heads.

Step 2

Ask the patient to mark the painful spot on the bottom of the foot with a material that transfers easily, such as lipstick or eyeliner.

Step 3

Instruct the patient to stand, without socks, in a shoe to transfer the mark to the inside of the shoe.

Step 4

Place the pad into the shoe directly over the mark. This ensures that the metatarsal heads are kept apart and away from the neuroma when the patient is bearing weight. Note that this placement differs from placement of a pad for metatarsalgia or a discrete plantar keratosis. Felt or gel pads are inexpensive and effective, and come in different sizes to accommodate different sized shoes and feet.

Aftercare/patient instructions

Advise the patient that if the pad is effective in treating neuroma symptoms, but the patient needs a more permanent device that can be transferred from shoe to shoe, a custom orthosis can be fabricated by a certified pedorthist.

foot and ankle

procedure

Morton's Neuroma Injection

Materials

Sterile gloves

Alcohol

Ethyl chloride spray

3-mL syringe

18-gauge needle

21-gauge, 1- to
    1½ -inch needle

1 mL of 10 mg/mL
    corticosteroid
    preparation

2 mL of 1% lidocaine
    or 0.5% bupiva-
    caine, both *without*
    epinephrine

Adhesive dressing

Step 1

Wear protective gloves at all times during the procedure and use sterile technique.

Step 2

Use the 18-gauge needle to draw 1 mL of the 10 mg/mL cortico-steroid preparation and 2 mL of the local anesthetic into the syringe, then switch to the 21-gauge needle to preserve sterility.

Step 3

Cleanse the dorsal skin between the metatarsal heads with alcohol.

Step 4

Freeze the dorsal skin with an ethyl chloride spray.

Step 5

Place the needle in line with the metatarsophalangeal joint, which is approximately 1 to 2 cm proximal to the web of the toe (Figure 1).

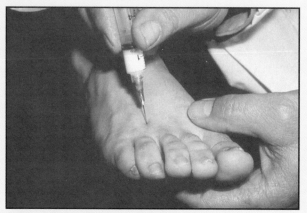

Figure 1
Proper location for injection of Morton's neuroma

foot and ankle

Morton's Neuroma Injection (continued)

Step 6

Insert the needle into the plantar aspect of the foot so that the tip gently tents the skin. Withdraw the needle approximately 1 cm so that the tip is where the neuroma is found, at the level of the plantar metatarsophalangeal joint.

Step 7

Inject the corticosteroid/anesthetic mixture around the neuroma, taking care not to inject into the plantar pad.

Step 8

Dress the puncture wound with a sterile adhesive bandage.

Adverse outcomes

Atrophy of the plantar fat pad from the corticosteroid is possible. Avoid injecting the patient with a corticosteroid more than two times. If the first injection did not help, a second injection is unlikely to be beneficial. If two injections are necessary, separate them by a span of several months.

Aftercare/patient instructions

Advise the patient that there may be a feeling of fullness on the bottom of the foot that lasts for approximately 24 hours. Transient numbness in one of the toes distal to the injection site may occur. Explain to the patient that symptoms may return when the anesthetic agent wears off in a few hours and that it may take a few weeks for the corticosteroid to have an effect on the symptoms.

foot and ankle

The Essentials of Musculoskeletal Care   459

Orthotic Devices

Synonyms

None

Definition

Orthoses, or orthotic devices, are appliances that help improve foot function. Proper shoe fit, shoe modifications, inserts, pads, and orthoses can significantly relieve common foot complaints. Proper use of orthoses depends on an adequate physical examination and an understanding of foot types and functions. An orthosis can support an area of collapse, cushion an area of pressure, accommodate fixed deformity, limit motion, equalize leg length, and reduce shear.

Foot types and functions

During gait, the normal foot changes from a supple, shock-absorbing structure to a rigid lever for push-off. At heel strike, the foot is supple, allowing shock absorption and accommodations to uneven ground. At midstance, the foot begins conversion to a rigid lever for push-off. Pronation is the normal process of the foot becoming flat as it bears weight during gait.

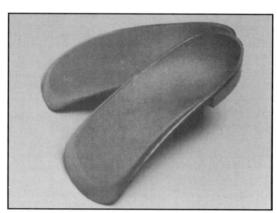

Figure 1
Custom orthoses

A cavus (highly arched) foot is rigid, cannot unlock during early stance, and lacks shock absorption. A pes planus (flatfoot) foot is extremely supple and often does not effectively supinate to form a rigid lever for push-off.

Many types of simple orthotic devices can be dispensed directly from the physician's office. Many orthotic devices are available at drug stores, sporting goods stores, and shoe stores. Custom orthoses are used for more complex problems or after "off-the-shelf" devices fail (Figure 1). These can be fabricated by a certified pedorthist.

Types of orthotic devices

Pads/Inserts

1. Full-contact insert (full length): A prefabricated rubber insert that reduces shock by absorbing normal and shear forces.

2. Full-contact orthosis: An orthosis that is molded over a positive model of the foot or molded to the foot. These are usually made, in part, of prefoam materials to accommodate a deformity, alter foot biomechanics, or prevent recurrent neuropathic ulcers plantar to bony prominences. They are often "posted" to correct the abnormality (ie, bring the ground up to the foot).

3. Soft orthosis: An orthosis that is used primarily for cushioning, but offers little control.

4. Semirigid orthosis: Most common type of orthosis; it provides strength and durability and helps to alter the biomechanics of the foot.

5. Rigid orthosis: An orthosis that offers maximum durability and support, but requires a precise fit as it provides little flexibility.

6. Heel insert (felt, foam, gel, or silicone): Prefabricated, shock-absorbing devices that are available over the counter (Figure 2).

7. Heel wedge: A device that is made up of tapered material to support varus or valgus hindfoot.

8. Metatarsal pad: A pad that is fixed to an insert or the bottom of the shoe (Figure 3).

9. Scaphoid pad (arch cookie): A medial longitudinal arch pad that provides support for a flatfoot.

10. Toe crest: An insert that elevates toes to relieve pressure at toe tufts (for mallet toe).

11. Toe separator: A foam pad that is placed between toes to decrease friction. It is used for calluses and corns.

12. University of California Biomechanics Laboratory (UCBL) orthosis: An orthosis that stabilizes the hindfoot by using medial and lateral flares. Often used for midfoot plantar fasciitis or a collapsed foot (Figure 4).

Figure 2
Heel insert

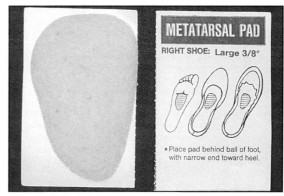

Figure 3
Metatarsal pad

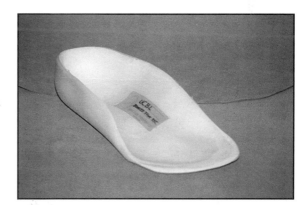

Figure 4
UCBL orthosis

foot and ankle

Shoes/Modifications

1. Extra-depth shoes: A shoe that accommodates forefoot problems, such as a hammer toe, and allows room for inserts.

2. Heel elevation: A device that elevates the heel inside the shoe a maximum of $1/2''$. Any elevation greater than $1/2''$ needs to be added to the external heel.

3. Metatarsal bar: An internal or external transverse bar that unloads the forefoot.

4. Reverse heel: A device used to decrease forefoot pressure and treat metatarsalgia.

5. Rocker bar: A longer convex metatarsal bar that reduces forces in segments of the foot.

6. Running shoes: Shoes that are designed to lose 50% of shock absorbency at 300 to 500 miles. They should be replaced at least every 6 months.

7. Solid Ankle Cushion Heel (SACH): Soft material that replaces the posterior portion of the shoe's heel to reduce shock at heel strike.

8. Shoe fit: Measurement of the bare foot standing width of the forefoot that is traced on a chart. Forefoot width of the shoe should be within $1/2''$ of this measurement.

9. Thomas heel: A device that supports the pronated foot with medial extension.

Application of orthotic devices

Table 1 indicates what type of orthotic therapy is needed for specific foot conditions.

Table 1
Recommended orthotic treatment for specific diagnoses

Diagnosis	Orthotic therapy
Bunions and/or bunionettes	Wide toe box, stretch shoes, soft seamless uppers, "bunion shield" type pad
Cavus foot (rigid)	Soft orthotic cushions to evenly distribute pressures
Flatfoot (adult)	More rigid, prefabricated arch support, with or without medial heel wedge, if symptomatic
Flatfoot (child)	No special orthotic or shoe treatment indicated. Normal in infancy, with more than 97% correcting spontaneously
Hallux rigidus	1/2" length prefabricated stiff insert or Morton's extension inlay, rocker-bottom sole
Hammer toe or claw toe	Accommodative shoe wear, toe crest
Interdigital neuroma	Wide toe box, metatarsal pad
Metatarsalgia	Wide shoes, metatarsal pads, bars
Neuropathic ulceration (diabetic foot)	Full-contact cushioned orthosis, extra-depth or custom shoes, rocker-bottom sole
Pes planus or pronator	Asymptomatic: no special orthotic or shoe treatment indicated. Symptomatic: semirigid insert or longitudinal arch pad, medial heel wedge, extended heel counter
Proximal plantar fasciitis (heel pain syndrome)	Prefabricated heel insert (silicone, rubber, gel, or felt), stretching exercises
Recurrent ankle sprains	Lateral heel flare or wedge
Runner's painful knee	Sport orthotic inlay: full-length, soft, prefabricated
Posterior tibial dysfunction	Semirigid orthosis, extended medial counter, possible UCBL or ankle-foot orthosis (AFO)

Acknowledgements

Figure 1 is reproduced with permission from Prolab, San Fransisco, CA.
Figure 2 is reproduced with permission from Hapad, Inc, Bethel Park, PA.

Plantar Fasciitis

Synonyms

Heel spur

Heel pain syndrome

Plantar heel pain

Subcalcaneal pain

Proximal plantar fasciitis

Definition

Plantar fasciitis refers to the plantar heel pain that occurs where the plantar fascia arises from the medial calcaneal tuberosity (Figure 1). There is inflammation in both the bone and plantar fascia, with histologic evidence of chronic degeneration in the fascia fibers that arise from the bone.

This condition affects women twice as often as men. It is not specifically associated with any one foot type, but it does affect overweight individuals more often than those of normal weight. For patients with pain in both heels, there is a positive association with the seronegative spondyloarthropathies.

Clinical symptoms

Patients report focal pain and tenderness directly over the medial calcaneal tuberosity and 1 to 2 cm distally along the plantar fascia (Figure 2). The pain is often most severe upon awakening or when rising from a resting position, as the first few steps stretch the plantar fascia. Nonweightbearing typically relieves symptoms.

Figure 1
Location of plantar fasciitis

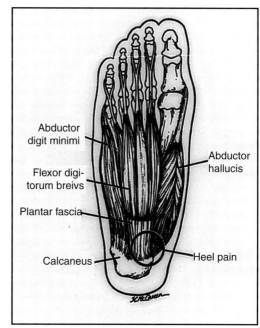

Figure 2
Area of maximum heel pain

Abductor digit minimi

Flexor digitorum breivs

Plantar fascia

Calcaneus

Abductor hallucis

Heel pain

Tests

Exam
Examination reveals tenderness directly over the plantar medial calcaneal tuberosity and 1 to 2 cm distally along the plantar fascia (Figure 3). Often, considerable pressure will need to be applied to this area during the examination to reproduce weightbearing stress and the patient's symptoms. Patients will also likely have tightness in the Achilles tendon, which contributes to increased tension on the plantar fascia during gait. Passive dorsiflexion of the toes (Windlass mechanism) does not usually aggravate the symptoms.

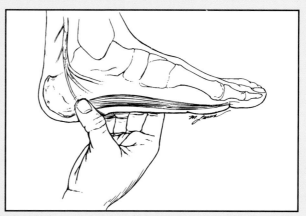

Figure 3
Point of maximum tenderness in a patient with entrapment of the first branch of the lateral plantar nerve

Diagnostic
Radiographs are not necessary as part of the initial evaluation if the patient's history and examination are consistent with a diagnosis of plantar fasciitis (pain upon rising and focal plantar pain). Standing lateral radiographs should be obtained prior to an injection of corticosteroid or for patients who continue to have symptoms following 6 to 8 weeks of conservative treatment. Standing lateral radiographs may also be indicated for patients who have systemic symptoms or pain at rest. A bone scan will often show increased uptake over the medial calcaneal tuberosity, consistent with the local inflammation. An MRI scan or CT scan is rarely indicated.

Differential diagnosis

Acute traumatic rupture of the plantar fascia

Calcaneal stress fracture (rare)

Calcaneal tumor (rare)

Distal plantar fasciitis

Entrapment of the first branch of the lateral plantar nerve (maximal tenderness over medial plantar hindfoot)

Fat pad atrophy/contusion

Sciatica

Tarsal tunnel syndrome

Adverse outcomes of the disease

Without treatment, patients may experience chronic heel pain, significant changes in activity level, and forefoot, knee, hip, and back symptoms due to altered gait.

foot and ankle

Treatment

Ninety-five percent of patients with plantar fasciitis can be cured with conservative treatment. Patients should not undergo surgery before a minimum course of conservative treatment, typically 6 months, has been completed.

Initial treatment should consist of use of an "off-the-shelf" orthotic device, such as a silicone, rubber, or felt heel pad (Figure 4), along with a home program of Achilles and plantar fascia stretching.

With one type of stretching exercise, the patient should be instructed to lean forward against a wall, keeping one knee straight with the heel on the ground while bending the other knee (Figure 5). This maneuver can be done with either one or both legs, depending upon whether one or both heels are painful. As the patient bends forward, he or she should report feeling the heel cord and the arch of the foot of the straight leg stretch. The patient should stretch and hold for 10 seconds, then relax and straighten up. Patients should repeat this exercise 20 times for each leg if both heels hurt.

With the second exercise, the patient should be instructed to lean forward onto a table, chair, or countertop, spread the feet apart and place one foot in front of the other, flex the knees, and squat down slowly (Figure 6). Explain to the patient that the heels must be kept on the ground as long as possible during the squat. As the patient squats, he or she should report feeling the heel cord and arches of the foot stretch as the heels finally start to rise off the ground. The patient should stretch and hold for 10 seconds, then relax and straighten up. Patients should repeat this exercise 20 times.

Up to 95% of patients will experience significant improvement in symptoms after 8 weeks of initial treatment. Use of contrast baths, ice, NSAIDs, and/or shoes with shock-absorbing soles can also be used to decrease inflammation in the painful heel (see Contrast Baths).

If symptoms persist after 8 weeks, injection of corticosteroid into the heel may be indicated (see Heel Injection for Plantar Fasciitis). A formal physical therapy consultation should also be considered to help direct the patient's continued stretching program.

If symptoms persist, use of a nonremovable cast, night splint, or custom orthotic device should be considered. The night splint holds the ankle and foot in slight extension, which maintains the Achilles tendon and plantar fascia in a stretched position during sleep.

Figure 4
Heel pad

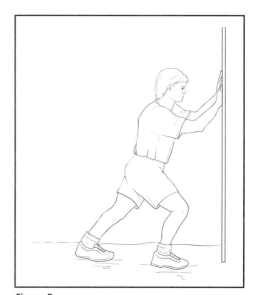

Figure 5
Achilles stretching exercise

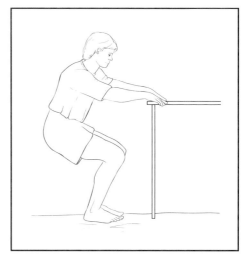

Figure 6
Plantar fascia stretching exercise

foot and ankle

Adverse outcomes of treatment

NSAIDs may cause gastric, renal, or hepatic complications. Fat pad necrosis can develop due to improper injection of corticosteroid.

Referral decisions/Red flags

Patients whose symptoms persist after 4 months of conservative treatment need further evaluation. Surgery may be considered if exhaustive conservative treatment fails.

Acknowledgements

Figure 2 is reproduced with permission from the Mayo Foundation, Rochester, MN.

Figure 3 is reproduced with permission from Schon L: Plantar fascia and Baxter's nerve release, in Myerson M (ed): *Current Therapy in Foot and Ankle Surgery.* St. Louis, MO, Mosby, 1993, pp 177–182.

Figures 5 and 6 are adapted with permission from the American Orthopaedic Foot and Ankle Society, Seattle, WA.

procedure

Heel Injection for Plantar Fasciitis

Materials

Sterile gloves

Alcohol

Ethyl chloride spray

Lateral radiograph of the heel

5-mL syringe

18-gauge needle

21-gauge 1- to 1½ -inch needle

Mixture of 1 mL of 10 to 40 mg/mL of corticosteroid preparation and 4 mL of 1% lidocaine or 0.5% bupivacaine, *without* epinephrine

Adhesive dressing

Step 1

Wear protective gloves at all times during the procedure and use sterile technique.

Step 2

Use the 18-gauge needle to draw 1 mL of 40 mg/mL corticosteroid preparation and 4 mL of the local anesthetic into the syringe, then switch to the 21-gauge needle to preserve sterility.

Step 3

Cleanse the medial aspect of the heel with alcohol.

Step 4

Spray the ethyl chloride onto the medial heel to freeze the skin.

Step 5

Measure the soft-tissue thickness beneath the calcaneus directly on the radiograph.

Step 6

Using the measurement from the radiograph as a guide, palpate the calcaneus medially where it begins to curve upward. Insert the 21-gauge needle in this area, which is approximately 2 cm from the plantar surface of the foot (Figure 1).

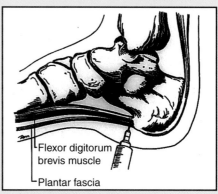

Flexor digitorum brevis muscle

Plantar fascia

Figure 1
Proper location for injection

Heel Injection for Plantar Fasciitis (continued)

Step 7

Advance the needle down to the calcaneus until it hits the bone. Walk the tip of the needle distally along the bone to the plantar surface of the calcaneus.

Step 8

Advance the needle to its hilt and inject 3 mL of the anesthetic/corticosteroid mixture. Inject the remaining 2 mL of the preparation while withdrawing the needle 2 cm, then withdraw the needle completely. Make certain that you do not inject the anesthetic/corticosteroid preparation into the medial subcutaneous tissue. The corticosteroid should not be injected superficial to the plantar aspect of the calcaneus or plantar to the plantar fascia layer.

Step 9

Dress the puncture wound with a sterile adhesive bandage.

Adverse outcomes

Injection of a corticosteroid into the superficial fat pad can cause fat necrosis with loss of cushioning of the plantar heel.

Aftercare/patient instructions

Advise the patient that transient numbness of the heel may occur. Also explain that heel pain may return in a few hours when the anesthetic agent wears off and that it may take a few weeks for the corticosteroid to have an effect on the symptoms.

foot and ankle

Plantar Warts

Synonyms

Verruca vulgaris

Definition

Warts are hyperkeratotic lesions that develop on the sole of the foot and are caused by a papilloma virus.

Clinical symptoms

Patients have painful, slightly raised lesions on the sole of the foot. These lesions may occur in clusters known as "mosaic warts."

Tests

Exam

Warts usually appear on nonweightbearing areas on the sole of the foot. Normal papillary lines ("fingerprint pattern") of the skin cease at the margin of the lesion (Figure 1). Lesions are usually very tender if pinched side to side, which does not typically occur with a corn or callus. A callus occurs beneath a bony prominence, whereas a wart can occur anywhere on the sole. Superficial paring of a wart usually reveals punctate hemorrhage and a fibrillated texture. A callus is avascular, has a uniform texture, and resembles yellow candle wax.

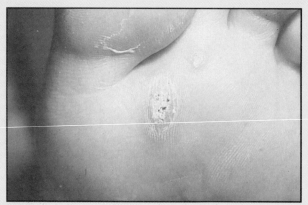

Figure 1
Clinical appearance of a plantar wart

Diagnostic

Where doubt exists, histopathologic examination of a specimen confirms the diagnosis. However, this type of testing is seldom necessary, given the characteristic gross appearance after superficial paring.

foot and ankle

Differential diagnosis

Callus (seed callus, discrete plantar keratosis)

Adverse outcomes of the disease

Plantar warts are often persistent; they spread to other areas of the foot, grow larger, and leave scars on the sole of the foot.

Treatment

Most lesions resolve spontaneously within 5 to 6 months, so aggressive treatment should be reserved for unusually large, painful, or persistent lesions. Initial treatment commonly includes superficial paring, followed by the use of a keratolytic agent, such as salicylic acid solutions in liquid or salve form. The lesion should then be covered with occlusive tape to ensure that the medication stays within the desired area and to debride the necrotic layers of tissue upon removal of the tape. Medication should be applied twice daily for about a month.

Warts that are resistant to initial treatment will sometimes respond to perilesional injection of approximately 1 mL of local anesthetic with epinephrine. Electrocautery, cryotherapy with liquid nitrogen, laser ablation, or curettage may be performed under local anesthetic. Care should be taken to avoid causing necrosis of the deep dermis, which can produce intractable, painful scarring on the sole of the foot. In curettage, for example, the subcutaneous fat should not be visible when the procedure is finished. Intralesional injection of bleomycin and radiation therapy have also been described for severe, recalcitrant lesions, but these options are probably best performed by specialists with experience in their use.

Adverse outcomes of treatment

Secondary infection may occur following treatment. Intractable scarring from excessively deep ablation is also a significant risk. Definitive treatment of plantar warts poses a significant challenge.

Referral decisions/Red flags

Persistence or recurrence of a wart warrants further evaluation.

foot and ankle

Acknowledgements
Figure 1 is reproduced with permission from California Pacific Medical Center, San Francisco, CA.

Posterior Heel Pain

Synonyms

Pump bump

Retrocalcaneal bursitis

Haglund's syndrome

Insertional Achilles tendinitis

Definition

Pain behind, rather than below, the calcaneus is characteristic of this condition. The painful structure can be one of the following: the insertion of the Achilles tendon into the calcaneus; the retrocalcaneal bursa between the distal Achilles tendon; the posterosuperior portion of the calcaneus; or a bursa between the skin and the Achilles tendon (Figure 1).

Clinical symptoms

Posterior heel pain usually affects middle-aged or elderly patients, but may also be associated with athletic overuse or be among the presenting symptoms of rheumatoid arthritis or the spondyloarthropathies. Symptoms are often worse when beginning an activity after rest, and a limp is not uncommon. Shoe wear is difficult because of direct irritation by the posterior aspect of the shoe.

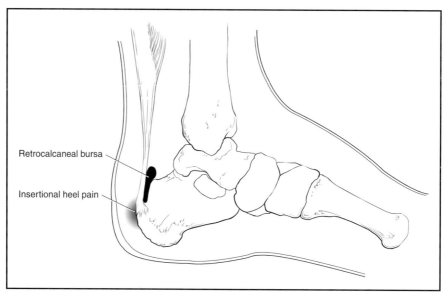

Retrocalcaneal bursa

Insertional heel pain

Figure 1
Sites of posterior heel pain

foot and ankle

Tests

Exam

Examination often reveals swelling and erythema of the posterior heel. The prominence created is called a "pump bump." A superficial bursa, if present, is often inflamed from shoe wear and locally tender and warm to the touch.

Inflammation or calcification of the Achilles insertion also occurs with local tenderness and warmth, but an inflamed superficial bursa may not be present. Swelling of the tendon often occurs and can be palpated on examination.

Retrocalcaneal bursitis is associated with pain and tenderness anterior to the Achilles tendon, along the medial and lateral aspects of the posterior calcaneus. Squeezing the bursa from side to side and/or plantar flexion of the foot will reproduce a patient's symptoms.

Diagnostic

Standing lateral radiographs may show calcification in the distal Achilles tendon or a spur projecting upward at the insertion of the tendon midway down the calcaneus. A prominent posterosuperior process of the calcaneus (Haglund's process) may also be apparent (Figure 2).

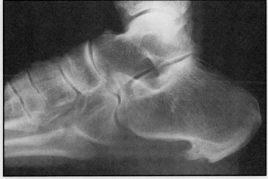

Figure 2
Radiographic appearance of a plantar heel spur

Differential diagnosis

Achilles tendon avulsion (a pop, palpable defect in tendon, positive Thompson test)

Inflammatory arthritides (often bilateral with other sites of involvement, systemic symptoms)

Plantar fasciitis (tenderness and pain beneath, rather than behind, the heel)

Stress fracture of calcaneus (midcalcaneal bony tenderness, acute overuse)

Sural neuritis (rare, history of direct trauma, positive percussion sign over the nerve lateral to the tendon)

Adverse outcomes of the disease

Achilles tendon rupture may occur with chronic insertional tendinitis. Chronic pain and limping are common.

Treatment

A program of stretching of the Achilles tendon by leaning forward against a wall with the foot flat on the floor and heel elevated with a $3/8''$ or $1/2''$ heel insert (or heeled shoe) is indicated. Contrast baths, NSAIDs, and use of an open-back shoe may also be helpful (see Contrast Baths). If the patient's symptoms are resistant to this treatment, consider cast immobilization for 4 to 6 weeks. Injection of corticosteroid should be avoided because the tendon may be damaged and subsequently rupture.

Adverse outcomes of treatment

Tendon rupture and chronic pain are both possible. NSAIDs may cause gastric, renal, or hepatic complications.

Referral decisions/Red flags

Failure of conservative treatment may require surgical excision of the Haglund's process with debridement of the retrocalcaneal bursa and Achilles insertion.

Acknowledgements

Figure 1 is adapted with permission from Pfeffer GB, Baxter DE: Surgery of the adult heel, in Jahss MH (ed): *Disorders of the Foot and Ankle: Medical and Surgical Management,* ed 2. Philadelphia, PA, WB Saunders, 1991, vol 2, pp 1382–1416.

foot and ankle

Posterior Tibial Dysfunction

Synonyms

Acquired flatfoot

Posterior tibial tendon rupture

Posterior tibial insufficiency syndrome

Definition

This condition is characterized by loss of function of the posterior tibialis tendon, whether by trauma, degenerative tear (complete or incomplete), tenosynovitis, or muscle dysfunction. The result will often be a loss of support for the arch and a subsequent acquired flatfoot deformity. The flatfoot is flexible (and correctable) initially, but a fixed deformity may develop over time.

Clinical symptoms

Patients often state that their foot rolls in or out and that they have pain and swelling at the medial ankle with weightbearing. If the flatfoot deformity is advanced, patients may have pain in the lateral hindfoot due to impingement of the calcaneus on the tip of the fibula. With the relatively sudden onset of flatfoot, wear patterns in the shoe change as the foot rolls over the medial counter.

Tests

Exam

Examination reveals swelling confined to the area around the medial malleolus and tenderness along the course of the tendons. In addition, there is often exquisite point tenderness over the tendon just distal to the medial malleolus, where the tendon most commonly tears. As the patient inverts and everts the foot, there may be a palpable click or pop associated with a facial wince; this usually indicates an incomplete tear with locking of a torn flap of tendon within the sheath. With a complete tendon rupture, the patient will be unable to plantar flex and invert the foot. In addition, the patient will have difficulty walking on tiptoes.

Patients with an acquired flatfoot will have a "too many toes" sign. When their feet are viewed from behind as they stand, there will be more toes lateral to the ankle on the affected side due to valgus of the hindfoot and abduction of the forefoot (Figure 1).

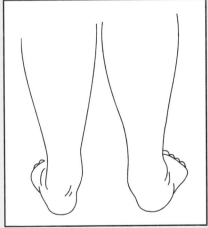

Figure 1
"Too many toes" sign

foot and ankle

When viewed from behind, the calcaneus normally inverts when the patient rises on tiptoes on a single foot. If the posterior tibial tendon is not functioning, normal inversion of the heel will not occur (Figure 2).

In mild cases of isolated posterior tibial tendinitis without tendon rupture, repeated toe rise will demonstrate weakness on the affected side. In more advanced cases, patients may not be able to rise on tiptoes on the affected side (single toe rise) because of weakness and pain.

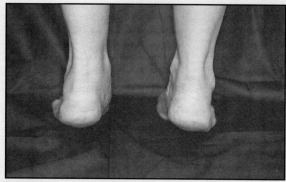

Figure 2
The left heel does not rotate inward (varus) when the patient stands on her toes

Diagnostic

Standing AP and lateral radiographs show a flatfoot with the sag in the midfoot at the talonavicular joint and/or naviculocuneiform joint. An MRI scan may be appropriate, but only as a preoperative study.

Differential diagnosis

Congenital pes planus

Lisfranc dislocation

Medial ankle laxity

Tarsal coalition

Adverse outcomes of the disease

Painful weightbearing and severe gait dysfunction are possible. With time, the deformity becomes more pronounced, and therefore, more difficult to treat; a rigid flatfoot may eventually develop.

Treatment

In the absence of a flatfoot, treatment focuses on reducing inflammation, swelling, and pain. Use of a cast or cast boot for 4 to 6 weeks, NSAIDs, and activity modification may help. A molded ankle-foot orthosis can also be used. An arch support should be used after the cast is removed. Injection of corticosteroid will weaken the already pathologic tendon and is not recommended. If these modalities fail, surgical debridement and reconstruction of the tendon may be indicated. Treatment of associated obesity may provide the greatest stress relief for this problem.

Once a flexible flatfoot starts to develop, use of a medial heel wedge, medial longitudinal arch support, or a molded ankle-foot orthosis may help. Often surgery is required to either augment the posterior tibialis tendon or partially fuse the hindfoot. Once a rigid, painful flatfoot develops, stabilization of the hindfoot by arthrodesis is the only surgical treatment option.

Adverse outcomes of treatment

NSAIDs may cause gastric, renal, or hepatic complications. Recurrence of deformity, failure of fusion, and infection are all possible.

Referral decisions/Red flags

Progression of flatfoot or failure of conservative treatment indicates the need for further evaluation.

Acknowledgements

Figure 1 is adapted with permission from the Mayo Foundation, Rochester, MN.

foot and ankle

Rheumatoid Foot and Ankle

Synonyms

None

Definition

Up to 90% of patients with rheumatoid arthritis have clinical involvement of the foot or ankle, initially presenting as painful synovitis. Of these patients, 90% will report problems with the midfoot and forefoot and 67% will have problems with the hindfoot and ankle. Capsular laxity leads to hallux valgus, subluxated or dislocated lesser toes in the forefoot, and valgus hindfoot deformities.

Clinical symptoms

Metatarsalgia commonly occurs due to distal migration of the fat pad, associated with claw toes and dorsally subluxated or dislocated metatarsophalangeal (MP) joints (Figure 1). Severe hallux valgus of the great toe and hammer toe or claw toe deformities may also occur (Figure 2).

In the hindfoot, the posterior tibial tendon is often involved, with posteromedial ankle pain and swelling of the sheath. Peroneal tenosynovitis is also common.

The ankle is usually one of the last joints to be involved in rheumatoid arthritis. When it occurs, however, effusion and pain with motion and weightbearing are common.

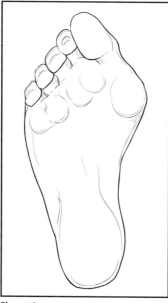

Figure 1
Plantar keratosis develops as the fat pad is pulled distally and the metatarsal heads become more prominent

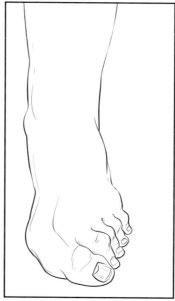

Figure 2
Hallux valgus and other common toe deformities associated with RA

Tests

Exam

The diagnosis of rheumatoid arthritis has already been made in most patients, but 17% of patients with rheumatoid arthritis initially present with foot and ankle complaints. Rheumatoid nodules are particularly painful when located plantar to the heel and hallux MP joint.

Diagnostic

Radiographs may reveal soft-tissue swelling, subchondral erosions, osteopenia, joint destruction, and often malalignment. Standing AP and lateral views of the foot may show hallux valgus, dislocated or subluxated lesser toes, joint narrowing, and loss of the arch on the lateral view (Figure 3). Lateral drift (malalignment) occurs at the MP joints, and joint erosions at the metatarsocuneiform and talonavicular joints. AP views of the talonavicular joint may demonstrate lateral subluxation. Standing views of the ankle (mortise, lateral) are helpful in assessing the status of the tibiotalar articulation.

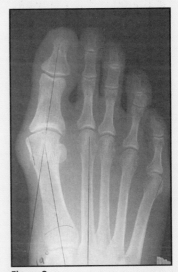

Figure 3
AP radiograph showing hallux valgus

Differential diagnosis

Osteoarthritis

Posterior tibial tenosynovitis (maximum tenderness over the posterior tibial tendon)

Psoriatic arthritis

Reiter's syndrome (other seronegative arthropathies)

Adverse outcomes of the disease

Persistent pain, plantar ulceration or severe metatarsalgia, rheumatoid vasculitis (skin sloughs) are all possible, as are loss of joint function and significant foot and ankle deformity.

Treatment

Conservative

Medical management can improve symptoms and slow the progression of disease.

An injection of corticosteroid may be considered for a painful joint or if there is significant tenosynovitis of the posterior tibial tendon.

Use of extra-depth accommodative shoes is appropriate for forefoot symptoms. A metatarsal bar or a longitudinal full-length arch support can also be used to cushion plantar prominences. Painful forefoot motion can

foot and ankle

be diminished with use of a steel-shank insert and a rocker-bottom sole. A certified pedorthist can make these modifications.

Use of a University of California Biomechanics Laboratory (UCBL) orthosis (supramalleolar orthosis) can help control a flexible flatfoot deformity due to posterior tibialis dysfunction, and a custom-molded ankle-foot orthosis provides maximum control of foot position in a severe flexible deformity. A molded ankle-foot orthosis can also be used for patients with hindfoot and/or ankle involvement.

Surgical

Forefoot. The most reliable forefoot procedure involves fusion of the great toe and resection of all metatarsal heads. In patients with significant hindfoot or ankle involvement where preservation of motion is important, a resection or implant arthroplasty of the great toe may be considered.

Hindfoot. Severe tarsal tunnel syndrome may develop due to proliferative synovitis in patients with rheumatoid arthritis. Tenosynovectomy with release of the tarsal tunnel may be indicated. Bony procedures are usually necessary. Arthrodesis is used to correct painful malalignment or arthritis. The talonavicular joint is the first joint involved, and early fusion may prevent flatfoot deformity or progression of other joint involvement. Triple arthrodesis is often indicated due to advanced deformity or diffuse involvement of the hindfoot joints.

Ankle. Though usually spared, the ankle joint may require arthrodesis. Distinguishing subtalar (hindfoot) pain from ankle pain is important. Radiographs and selective anesthetic injections may also help to differentiate between the two. Involvement of both the subtalar and ankle joints may require tibiotalocalcaneal fusion. Replacement of the ankle joint has been associated with significant rates of loosening, subsidence, and wound complications, and, therefore, should be used only very selectively.

Adverse outcomes of treatment

Foot deformity may progress despite conservative treatment, making surgical reconstruction very difficult.

Referral decisions/Red flags

Persistent pain despite conservative management and accommodative shoe wear modifications signal the need for further evaluation.

Acknowledgements

Figures 1 and 2 are adapted with permission from Netter F: *The Ciba Collection of Medical Illustrations.* Summit, NJ, Ciba-Geigy, 1987, vol 8.

Shoe Wear

Improper fit or manufacture of shoes has long been implicated as the single greatest cause of most foot deformities and problems that physicians encounter, especially in women. Shoes that do not fit properly can deform an otherwise normal foot, resulting in hammer toes, hallux valgus, bunionettes, corns, and ultimately surgery (Figure 1). A recent study indicates that most women older than age 20 years have not had their feet measured in more than 5 years. Because foot size increases with age, individuals often purchase shoes that are too small, which results in foot pain and deformity. Thus, individuals need instruction in proper shoe fit, and shoes that fit properly must be made available from manufacturers.

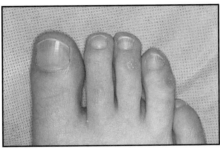

Figure 1A
Long second, third, and fourth toes become constricted by tight shoes

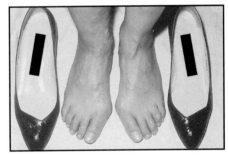

Figure 1B
Constrictive shoes

How shoes are made

The foundation of a shoe is called a last, which is a three-dimensional form (either straight or curved) on which the shoe is made. The shape of the toe box, instep, girth, and foot curvature are determined by the last. The straighter the last, the straighter the shoe, and the more medial support it provides, which can help control over-pronation.

foot and ankle

Lasting techniques

The most common methods of lasting include slip lasting, board (flat) lasting, and combination lasting. A slip lasted shoe is constructed by sewing together the upper, like a moccasin, and then gluing it to the sole. This method makes a lightweight, flexible shoe with no tortional rigidity. With flat or board lasting techniques, the upper is placed over the last and fastened to the insole with cement, tack, or staples. This construction makes a stable, but less flexible shoe. Combination lasting uses more than one lasting technique for the same shoe. Shoes made in this way are typically board lasted in the rear for stability and slip lasted in the forefoot for flexibility.

Outer sole

The outer sole, or outsole, is the plantar surface of the shoe that makes contact with the ground; it is usually attached to the midsole to form the complete sole of the shoe. Most athletic shoes have outer soles made of hard carbon rubber or blown rubber compounds. The outer sole provides pivot points and can be designed with a herring bone pattern, suction cups, radial edges, or asymmetric studs. These design patterns enhance stability and traction.

Midsoles and wedges

The midsole and heel wedge are located between the upper and the outer sole and are attached to both. The components provide cushioning, shock absorption, lift, and control.

Heel counter

The heel counter is a firm cup built into the rear of the shoe that holds the heel in position and helps control excessive foot motion.

Toe box

The toe box provides a stiff material inserted between the lining and upper in the toe area to prevent collapse and protect the toes.

Tongue

The tongue is designed primarily to protect the dorsum of the foot from dirt, moisture, and lace pressure.

Sock lining, arch support, and inserts

A sock lining covers the insole and improves comfort and appearance. This lining acts primarily as a buffer zone between the shoe and the foot. Arch supports, heel cups, and other types of padding can be added to provide support, cushioning, and motion control.

Welt

The welt is a strip of leather or other material that joins the upper with the outer sole.

How shoes are fit

While ensuring proper fit is not an exact science, there are a number of easy-to-follow guidelines that can help (Figure 2). Shoes should always be fit to the weightbearing foot at the end of the day, as this is the time and position that feet are at their largest.

Shoes do not typically stretch to conform to the shape of a foot (Figure 3). The upper should not wrinkle with flexion of the foot, and the foot should not bulge over the welt. The end of the longest toe of the largest foot should be within $1/2''$ from the end of the toe box. The forefoot should not be crowded, and the toes should be allowed to extend. There should be a relatively snug grip of the counter about the heel. Heels more than $2^1/4''$ high should be avoided, as these will exert excessive pressure on the front of the foot. Above all else, shoes should be comfortable from the moment they are tried on.

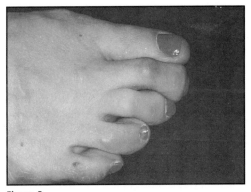

Figure 3
The foot takes the shape of the shoe, not vice versa

Ten Points of Proper Footwear*

1. Sizes vary among shoe brands and styles. Don't select shoes based on the size marked inside the shoe. Pick the shoe by how it fits your foot.

2. Select a shoe that conforms as nearly as possible to the shape of your forefoot.

3. Have your feet measured regularly. The size of your feet may change as you grow older.

4. Have both feet measured (often one foot is larger than the other). The shoe should be fitted to the larger foot.

5. Have your shoes fitted at the end of the day when your feet are the largest.

6. Stand during the fitting process because the foot lengthens as you are standing. There should be one fingerbreadth ($1/2$ in [1.3 cm]) between your longest toe and the end of the shoe.

7. The ball of your foot should fit snugly into the widest part of the shoe, but the shoe should not be too tight.

8. If the shoes don't fit, don't purchase them. Don't expect them to "stretch to fit."

9. Your heel should fit comfortably in the shoe with a minimum amount of pistoning.

10. While you are still in the shoe store, walk to make sure that the fit feels correct.

* Brochures are available from the American Orthopaedic Foot and Ankle Society, 701 16th Avenue, Seattle, Washington 98122.

Figure 2
Ten points of proper footwear

foot and ankle

Adverse outcomes from poor shoe fit

Bunions

Hammer toes

Neuromas

Calluses and corns

Ingrown toenails

Bunionettes

Avoidable surgery

Acknowledgements

Figure 1B is reproduced with permission from California Pacific Medical Center, San Francisco, CA.

Figure 2 is reproduced with permission from American Orthopaedic Foot and Ankle Society, Seattle, WA.

Stress Fracture of the Leg and Foot

Synonyms

March fracture

Insufficiency fracture

Definition

Stress fractures are skeletal breaks that occur as a result of repetitive overuse, which causes the bone to "fatigue." They typically result from a sudden increase in level of activity, particularly high-impact activity or in conjunction with an underlying metabolic bone disease such as osteoporosis. Amenorrhea associated with overtraining in female athletes also predisposes them to stress fractures. Common sites in the leg include the tibia and fibula, the navicular, and the metatarsals (especially the second). However, any bone, if subjected to repetitive unaccustomed loading, can sustain a stress fracture.

Clinical symptoms

Patients with a stress fracture typically report a relatively indolent onset of pain and occasionally mild swelling in the affected part. The pain generally decreases with rest and increases with activity.

Tests

Exam
Examination reveals well-localized tenderness, some swelling, occasional ecchymosis (particularly when involving the foot), and slight warmth.

A fever is not associated with stress fractures and indicates the possibility of infection. Pain with rest and pain at night should arouse suspicion of infectious or neoplastic processes.

Diagnostic
Plain radiographs obtained within the first 2 weeks of symptoms are usually normal. Three to 4 weeks into the course of a stress fracture, subperiosteal new bone formation or a sclerotic or lucent fracture line may appear on radiographs (Figure 1). A bone scan can be helpful in confirming the diagnosis at an early stage and will usually be positive within 5 days of the onset of pain. An MRI scan can confirm the diagnosis within 24 hours, but is not generally required.

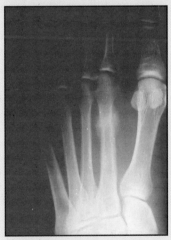

Figure 1
AP radiograph showing a sclerotic fracture line on the second metatarsal

foot and ankle

Differential diagnosis

Metabolic (multiple stress fractures)

Neoplasm (pain at rest or at night)

Osteomyelitis (fever, pain at rest or at night, elevated erythrocyte sedimentation rate)

Adverse outcomes of the disease

Displacement of the fracture fragments and delayed healing or nonunion are possible.

Treatment

If the extremity is acutely painful and moderately swollen in the initial phase, a period of rest, ice, gentle compression, and elevation can relieve symptoms.

Four or 6 weeks of decreased weightbearing is standard; crutches and casting are not always needed. A runner who experiences symptoms only when running requires a break from running. Athletes should be encouraged to cross train with nonpainful activities, such as cycling or swimming, to maintain cardio-vascular fitness and maintain compliance and morale.

Metatarsal stress fractures can usually be managed with stiff-soled shoes or postoperative shoes and the use of a cane; if symptoms persist, 2 to 4 weeks of immobilization with a cast or cast boot is indicated.

Navicular (midfoot) stress fractures are particularly difficult to diagnose and often resist treatment because of the relatively poor blood supply to that bone. A period of short leg casting, initially nonweightbearing for 4 weeks, should be recommended. If symptoms persist subsequent to the casting, then a CT scan or MRI scan is indicated to further evaluate the navicular. Open reduction, internal fixation, and bone grafting may be necessary if nonunion occurs.

Adverse outcomes of treatment

Avoid NSAIDs as they may delay fracture healing.

Referral decisions/Red flags

Failure of pain relief after 4 to 6 weeks and recurrent fractures in the same bone signal the need for further evaluation.

Tarsal Tunnel Syndrome

Posterior tibial nerve entrapment

Definition

The most common nerve entrapment in the hindfoot is a tarsal tunnel syndrome, which occurs at the medial ankle, just below the medial malleolus (Figure 1). Symptoms may be intermittent and are often vague, which differs significantly from the very specific, sensory symptoms of carpal tunnel syndrome. Diabetic neuropathy may masquerade as tarsal tunnel syndrome at the outset, especially if there is constant, burning dysesthesias.

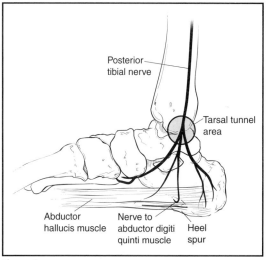

Figure 1
Anatomy of the tarsal tunnel

Clinical symptoms

Onset may be associated with proximal calf pain, especially at night. Paresthesias (tingling) or dysesthesias (burning) may occur on the bottom of the foot, along with aching and cramping in the arch; these symptoms may be worse when weightbearing. Paresthesias are usually intermittent and worsened or associated with certain activities or sports. There may be numbness, tingling, and burning on the plantar aspect of the foot in the distribution of the medial plantar nerve (medial two toes, plantar surface), or the lateral plantar nerve (lateral three toes, plantar surface), or both. The calcaneal branch supplies the plantar and medial surfaces of the heel. Chronic burning is more common in diabetic neuropathy and reflex sympathetic dystrophy.

Tests

Exam

Examination reveals tenderness over the tarsal tunnel, usually below the medial malleolus. Percussion over the posterior tibial nerve should reproduce symptoms, as may compression for 30 seconds. With long-standing entrapment, there may be clawing of the toes. There may also be diminished sensation to pinprick on the plantar aspect of the involved foot compared to the opposite side.

Diagnostic

Radiographs of the ankle and foot are necessary, but are usually normal in the absence of a history of significant trauma or fracture. Results of posterior tibial nerve conduction studies may be abnormal, but not in all instances.

Differential diagnosis

Degenerative arthritis of the subtalar (talocalcaneal) joint

Posterior tibial tendinitis (isolated swelling, tenderness just posterior to the medial malleolus)

Reflex sympathetic dystrophy

Sciatica (L5 root supplies the dorsal surface; S1 root supplies the lateral border of foot)

Adverse outcomes of the disease

Persistent pain and numbness are possible, and severe numbness may lead to plantar ulcers. Reflex sympathetic dystrophy may also develop.

Treatment

Conservative treatment is not usually helpful for this condition. A nonrigid orthotic insert for the shoe can be fabricated by a certified pedorthist to help cushion the painful foot. In patients with a flexible flatfoot and significant pronation, a medial longitudinal arch support (orthotic device) may decrease stretch on the nerve and thereby decrease symptoms. Use of NSAIDs and contrast baths may help decrease any local inflammation and compression on the nerve (see Contrast Baths). An injection of corticosteroid is usually not beneficial.

Surgical results with a tarsal tunnel release in the absence of a space-occupying lesion are often mixed. Surgery is not appropriate unless the symptoms are clear and can be reproduced by direct compression/percussion over the posterior tibial nerve in the area of entrapment.

Adverse outcomes of treatment

NSAIDs may cause gastric, renal, or hepatic complications. Symptoms may not completely resolve if there is permanent nerve damage. Surgical treatment may also fail. Injected corticosteroids can sometimes cause subcutaneous atrophy that can be unsightly. Infection is also a risk, but can be largely avoided by use of careful sterile technique.

Referral decisions/Red flags

Persistent pain, diminished sensibility, or presence of a mass below within the tarsal tunnel all indicate the need for further evaluation.

Acknowledgements

Figure 1 is adapted with permission from Baxter DE: Functional nerve disorders in the athlete's foot, ankle, and leg, in Heckman JD (ed): *Instructional Course Lectures 42*. Rosemont, IL, American Academy of Orthopaedic Surgeons, 1993, pp 185–194.

foot and ankle

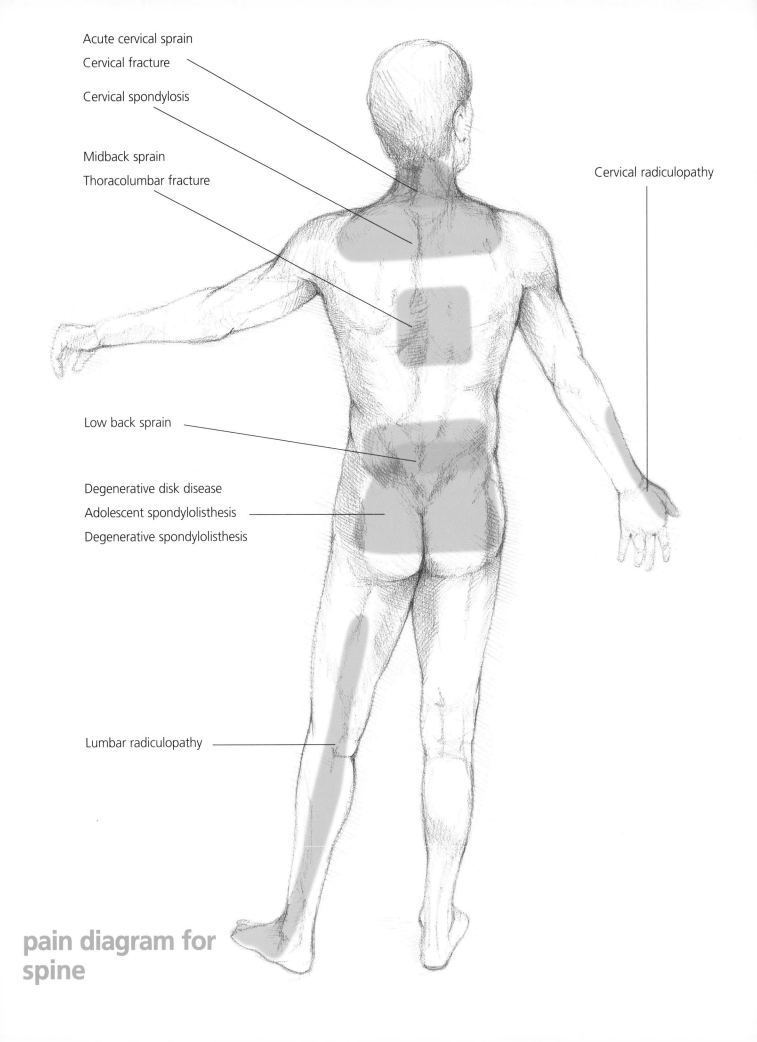

Acute cervical sprain

Cervical fracture

Cervical spondylosis

Midback sprain

Thoracolumbar fracture

Cervical radiculopathy

Low back sprain

Degenerative disk disease

Adolescent spondylolisthesis

Degenerative spondylolisthesis

Lumbar radiculopathy

pain diagram for
spine

s p i n e

Section Editor

**Gregory S. McDowell, MD
Orthopedic Surgeons, P.S.C.
Billings, Montana**

spine— an overview

Common degenerative disorders of the spine do not cause loss of life or limb, but they do alter function and comfort and often threaten employment and recreational activities. In this section of *Essentials of Musculoskeletal Care,* common spinal disorders are discussed, in which overall care of the patient is emphasized.

Low back pain affects 60% to 80% of the adult population at some time; therefore, it may be less a disease than a natural consequence of aging (Table 1). Patients between the ages of 30 and 50 years find chronic back pain particularly frustrating as these years coincide with peak economic, social, and family demands. Most episodes of back and neck pain resolve within a few weeks with little residual effect, but back pain is often a patient's first persistent illness. Therefore, patients with persistent pain usually seek treatment and are often dissatisfied with that treatment because of the recurring nature of their problem.

For most back problems, it is important to perform a thorough medical history and establish a diagnosis (Table 2). Nonsurgical treatment should be provided before further intervention is considered. During this time, patients should continue working, although job modifications may be needed. Ultimately, some patients may need to consider changing jobs, a move that can have significant economic consequences. Educating patients about their spinal problems and the impact these problems have on their lives is a challenge that demands patience and concern.

With back pain comes other associated problems that are not of a physical nature. Litigation is all too common following vehicular injury, and workers' compensation issues cloud the treatment process with concerns of causation and aggravation.

The most significant adverse outcomes of spinal disease are myelopathy or paralysis. In addition to trauma and tumor, these problems may develop in association with a cauda equina syndrome or with a sudden progression of an adolescent spondylolisthesis. Extra-spinal causes of back pain include pancreatitis, kidney stones, pelvic infections, and tumors or cysts of the reproductive tract.

The most serious etiology of back pain is a malignant tumor, most of which are metastatic, and some of which cause bony collapse and paralysis. The tumors that most commonly metastasize to bone are prostate, breast, lung, renal, adrenal, and thyroid.

Type of pain

Night pain that interrupts sleep, along with fever and weight loss, may indicate a malignancy or infection. Acute, posttraumatic pain may indicate an associated fracture. Low back pain is distinctly uncommon in children and always warrants evaluation.

Table 1
Common spinal conditions by age group

Age group	Common conditions
Younger than 10 years	Intervertebral diskitis, myelomeningocele, osteoblastoma, leukemia, congenital kyphosis, and scoliosis
Teens	Spondylolisthesis, kyphosis (Scheuermann's disease)
Twenties	Disk injuries (central disk protrusion, disk sprain), spondylolisthesis, spinal fracture
Thirties	Cervical and lumbar disk herniation or degeneration
Forties	Cervical and lumbar disk herniation or degeneration, spondylolisthesis with radicular pain
Fifties	Disk degeneration, herniated disk, metastatic tumors
Sixties and older	Spinal stenosis, disk degeneration, herniated disk, spinal instability, metastatic tumors

Location of pain

Neck pain
Pain located in the neck, trapezial, and interscapular areas most commonly occurs in association with degeneration of the intervertebral disk. Pain from an acute injury (whiplash) is usually self-limiting and may be associated with a transient numbness in the arm. Few patients have symptoms that persist for months.

Neck and radicular arm pain
When accompanied by referred pain into the arm, neck pain may be the result of an entrapment of a cervical nerve root by a herniated disk or bony spur. Nerve root entrapment at C4-5 results in pain that radiates to the lateral shoulder, but this occurs relatively infrequently. Nerve root entrapment at C5-6 causes pain and/or numbness of the thumb and radial aspect of the forearm. In addition, there may be loss of strength in the biceps, wrist extensors, and pronators, and loss of the biceps and brachioradialis reflexes. Nerve root entrapment at C6-7 commonly causes pain and numbness in the triceps area, index finger, and long finger. There may be loss of strength and loss of deep tendon reflexes in the triceps (Figure 1).

Many patients with a herniated cervical disk are more comfortable when they place the hand of the symptomatic arm on their head, since this position reduces the tension on the nerve. In contrast, patients with intrinsic shoulder problems feel more comfortable with their arms at their sides.

Table 2
Common presentations of spinal problems

Problem	Associated signs and symptoms	Possible diagnosis
Neck pain	Paravertebral discomfort relieved with rest and aggravated with activity	Acute neck sprain
	Limited motion or morning stiffness	Cervical spondylosis
Neck and arm pain	A younger patient with an abnormal upper extremity neurologic examination	Cervical radiculopathy due to herniated nucleus pulposus
	An older patient with limited motion and pain on extension	Cervical radiculopathy due to cervical spondylosis
	Urinary dysfunction with global sensory changes, weakness, and an abnormal gait	Cervical myelopathy secondary to cervical spondylosis or trauma
	Shoulder pain and a positive impingement sign	Shoulder pathology
	Tinel's sign and nondermatomal distribution of symptoms	Peripheral nerve entrapment
Back pain	Paravertebral discomfort relieved with rest and aggravated with activity	Acute low back sprain
	Limited motion and stiffness	Degenerative disk disease, ankylosing spondylitis
	Unrelenting night pain and weight loss	Tumor
	Fevers, chills, and sweats	Infection or intervertebral disk infection
Back and leg pain	A younger patient with an abnormal lower extremity neurologic examination	Lumbar radiculopathy due to herniated nucleus pulposus
	An older patient with poor walking tolerance and a stooped gait	Spinal stenosis
	Tenderness over the lateral hip and discomfort at night	Trochanteric bursitis

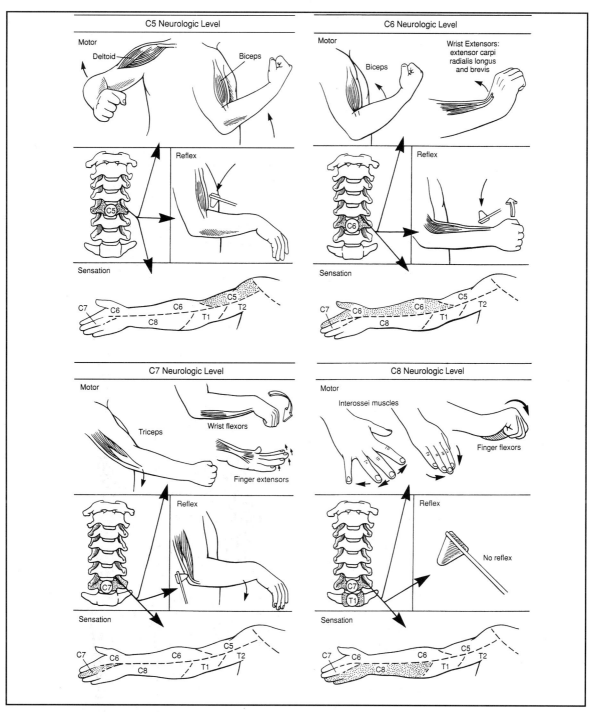

Figure 1
Neurologic evaluation of the upper extremity (C5-6, C6-7)

spine

With peripheral nerve entrapment syndromes, such as acute carpal tunnel syndrome (positive Phalen's sign), patients may report arm pain and a lesser degree of neck pain.

Low back pain

Low back pain typically occurs in the midline at about the L4 or L5 level. Many patients indicate that this pain radiates to the sacroiliac joint area, and at times even into the buttocks.

Back problems that affect patients between the ages of 20 and 30 years are often related to sprains of the soft-tissue structures of the back, including the hard portion of the intervertebral disk called the annulus fibrosus. The annulus becomes weakened by repeated small tears. When these problems occur in patients in this age group, the likelihood of disk herniations occurring later, when these patients reach age 30 to 50 years, is much greater.

Some patients with spondylolisthesis have back pain only and may have to consider changing jobs if their occupation aggravates their symptoms. With seronegative spondyloarthropathies (such as ankylosing spondylitis), patients often have back pain and morning stiffness that lasts longer than 30 minutes. Radiographs are usually normal early in the disease.

With aging and degeneration of the intervertebral disk, associated arthritis may develop in the facet joints and contribute to the chronic back pain often seen after age 40 years. Back pain associated with spinal stenosis may be aggravated by spinal extension, but there are few or no leg complaints. DISH often becomes symptomatic after age 50 years and primarily affects men.

Back and radicular leg pain

Unilateral leg pain is common with a herniation of an intervertebral disk that is most likely to occur at the L4-5 or L5-S1 vertebral levels. This pain is typically worse with sitting and is associated with sciatic or femoral tension signs (flip sign or femoral nerve stretch). Problems in this area can also affect reflexes and sensation in the lower extremity (Figure 2). Trochanteric bursitis in women classically mimics sciatica, and the two problems often coexist.

Bilateral leg pain may indicate spinal stenosis, a large central disk rupture (and possible cauda equina syndrome), or spondylolisthesis, especially after the age of 40 years. Exercise-induced leg pain, unilateral or bilateral, may indicate spinal stenosis. These patients may also have muscle spasms in the calf and fatigue in their legs. Spinal stenosis is typically more severe at the L3-4 and L4-5 interspaces.

Extra-spinal causes of nerve root entrapment or irritation include hip disease, ovarian cysts, and retroperitoneal lesions.

Deformity

Most spinal deformities are evident on forward bending. Kyphosis is best seen from the side, and scoliosis is best seen along the spine from the top or bottom of the spine. Look for café-au-lait spots (five or more) in neurofibromatosis, for neurologic changes, and for tight hamstrings with an associated spondylolisthesis.

Scoliosis deformities are associated with rotation of the spinal column that is more obvious when the patient bends forward. Patients whose left rib cage becomes more prominent on forward flexion require an extensive work-up for intraspinal pathology. Scoliosis can progress rapidly during adolescence, and treatment options quickly progress from brace to surgery.

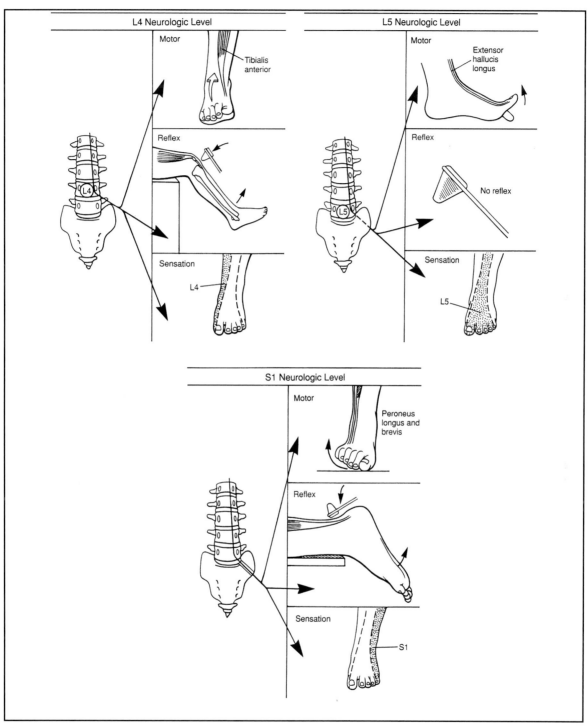

Figure 2
Neurologic evaluation of the lower extremity (L4-5)

Onset of a spinal deformity in adulthood suggests tumor or spinal instability and is often accompanied by compromise of the spinal nerve roots.

Trauma

All spinal trauma must be evaluated radiographically. Some bony injuries are subtle, especially in the cervical spine, and the potential consequences can be devastating. Spinal fractures are often accompanied by other life-threatening visceral, head, or skeletal injuries. Quadriplegia can occur even in the absence of cervical spine fractures.

Special imaging of the spine should be left to the discrimination of the consultant to avoid unnecessary reduplication of these expensive tests.

Gender

Women have an increased incidence of the following spinal conditions: scoliosis in adolescence, metastatic breast cancer and trochanteric bursitis in later adulthood, and osteoporosis (with vertebral body fractures) after menopause.

Men have an increased incidence of the following spinal conditions: kyphosis in adolescence, ankylosing spondylitis in adulthood, multiple myeloma and DISH in later adulthood, and metastatic carcinomas from the prostate and lung.

In conclusion, spinal problems may be associated with radicular pain into a limb and may have serious etiologies. Most, however, are related to degenerative changes associated with aging or growth. For various reasons some spinal injuries are missed in the multiply injured patient even after an appropriate work-up.

spine

Acknowledgements

Figures 1 and 2 are reproduced with permission from Klein JD, Garfin SR: History and physical examination, in Weinstein JN, Rydevik BL, Somtag VKH (eds): *Essentals of the Spine*. New York, NY, Raven Press, 1995, pp 71–95.

spine— physical exam

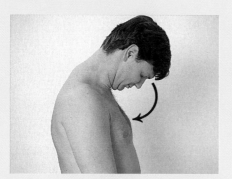

IA

IB

Figure I
Flexion-extension range of motion

With the patient seated, ask the patient to maximally flex (A) and extend (B) the neck, and measure or estimate each motion in degrees.

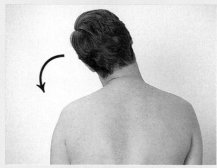

2

Figure 2
Lateral bending

With the patient seated, ask the patient to bend the neck as far to the right and left as possible, and measure in degrees.

spine

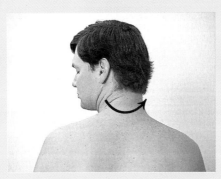

3

Figure 3
Cervical rotation

With the patient seated, ask the patient to rotate the neck to the left and right sides, and measure in degrees.

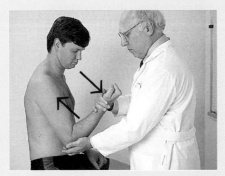

4

Figure 4
Elbow flexor strength

With the patient seated, ask the patient to flex the elbow as you apply resistance, and compare the two sides. A ratchety, giving way motion is a nonorganic sign. True weakness is uniform. This tests the C5 innervated muscles.

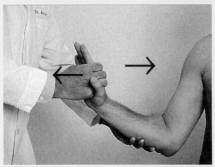

5

Figure 5
Wrist extensor strength

With the patient seated, ask the patient to extend the wrist against resistance, and compare with the opposite side. This tests the C6 innervated muscles.

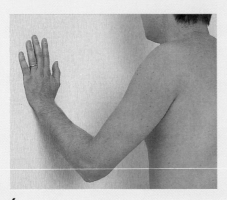

6

Figure 6
Elbow extensor strength

With the patient standing, ask the patient to do a push-up against the wall. This tests the C7 innervated muscles.

spine

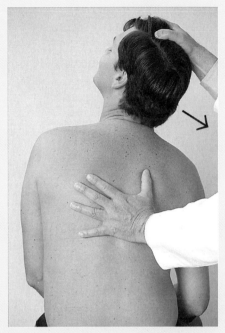

7

Figure 7
Spurling's test

With the patient seated, ask the patient to extend the neck while tilting the head to the side. This narrows the neural foramen and will increase or reproduce radicular arm pain associated with cervical disk herniations or cervical spondylosis.

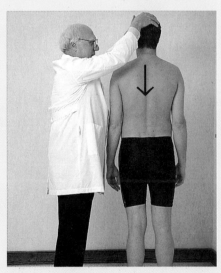

8

Figure 8
Axial loading

With the patient standing, push down on the patient's head; this may provoke neck pain in some patients with disk pathology, but increased low back pain from this maneuver is usually a nonorganic finding.

spine

9

Figure 9
Hoffman's reflex

With the patient seated, and the patient's relaxed hand cradled in yours, flick the long finger nail and look for index finger and thumb flexion as a sign of long-tract spinal cord involvement in the neck.

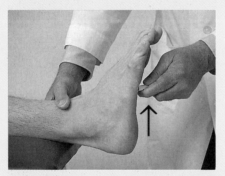

10

Figure 10
Babinski's sign (flexor plantar response)

With the patient supine, stroke lightly upward on the plantar surface of the foot and look for great toe extension (withdrawal response) and fanning of the lesser toes as a sign of long-tract spinal cord involvement.

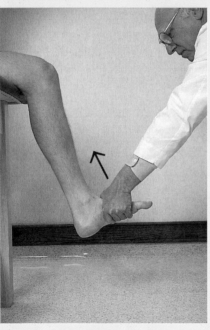

11

Figure 11
Ankle clonus

With the patient seated, dorsiflex the ankle suddenly and observe for rhythmic beating (clonus), noting the duration of and the number of "beats." This is another sign of long-tract spinal cord involvement.

spine

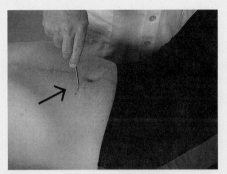

12

Figure 12
Superficial abdominal reflex

With the patient supine, stroke lightly toward the umbilicus. The normal muscular response is to pull the umbilicus toward the stimulated side. Absence of this motion may be a sign of spinal cord involvement.

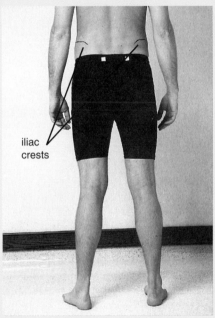

iliac
crests

13

Figure 13
Pelvic tilt

With the patient standing and facing away, look for asymmetries along the top of the iliac crest bilaterally. An unlevel pelvis usually indicates leg length inequality. The white squares indicate the approximate location of the posterior superior iliac spines.

spine

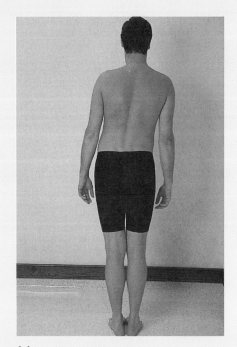

14

Figure 14
Lumbar list

With the patient standing, position yourself behind the patient and look for a shift of the trunk to the right or left. Record the direction of the list.

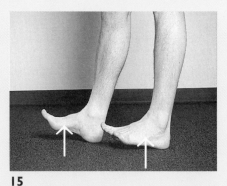

15

Figure 15
Heel walking

Ask the patient to walk across the room on the heels. Patients with tight heel cords (often in association with high arched feet) will have difficulty with this test. Indicate whether the test result is positive due to heel cord tightness (patient can't lean far into the wall), exaggerated responses, lack of coordination (positive Romberg's test), or weakness of the anterior tibial muscles (confirm by having the patient dorsiflex the ankle against your resistance). This tests the L4 innervated muscles.

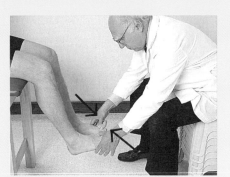

16

Figure 16
EHL (great toe extensor) weakness

With the patient seated, ask the patient to dorsiflex the great toe and compare strength on both sides. Indicate whether the test result is positive due to exaggerated responses (sudden giving way, grabbing the examiner's hand) or weakness of the muscle (a smooth giving way). This tests the L5 innervated muscles.

spine

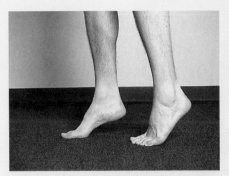

17

Figure 17
Toe walking

Ask the patient to walk across the room on tiptoes. Indicate whether the test result is positive due to exaggerated responses, lack of coordination (positive Romberg's test), or weakness of the gastrocnemius-soleus complex muscle group. This tests the S1 innervated muscles.

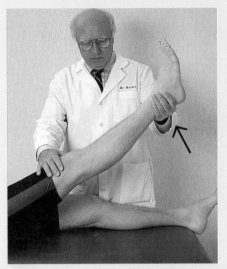

18A

Figure 18
Straight leg raise

With the patient supine, ask the patient to relax the leg as you cradle the foot in the palm of your hand, and elevate the leg until either the knee starts to bend or the patient indicates severe pain in the buttock or back (A). At that point, dorsiflex (B) the ankle to see if this motion increases pain. Dorsiflexion increases sciatic tension and increases sciatic pain. Plantar flexion relieves sciatic tension, and increased pain in the back with this test is probably nonorganic. Record the degrees of elevation at which pain occurs for each maneuver.

18B

spine

The Essentials of Musculoskeletal Care   505

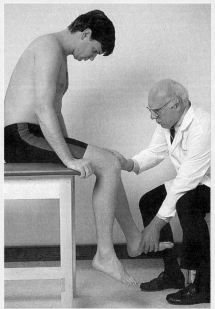

19A

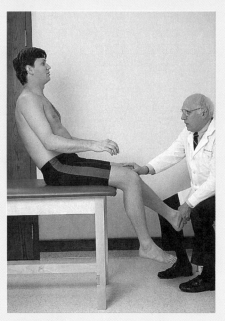

19B

Figure 19
Flip sign

With the patient seated with hands resting on the edge of the table, and the table against the wall, examine the patient's reflexes, test toe extensor strength, and then knee extensor strength. Finally, distract the patient (A) by asking whether the patient has knee troubles (focusing the patient's attention on the knee and away from the back), then lift the foot and extend the knee. Measure the degree of knee flexion reached when back pain occurs. Patients with sciatic tension will immediately flip backwards in acute pain (B), and may hit the wall. The flip sign correlates with straight leg raising pain under 45° of elevation.

spine

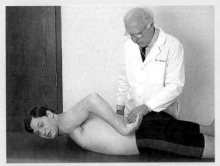

20

Figure 20
Exaggerated responses

During the examination, note whether the patient grabs your hand, exhibits histrionic behavior or generalized hypersensitivity, or perhaps limps in an exaggerated manner, half collapsing while walking or moving.

spine

Acute Cervical Sprain

Definition

An acute cervical sprain is characterized by nonradicular neck and shoulder pain (commonly localizing to the trapezius) that occurs either suddenly or following trauma. Cervical sprain is a common condition and is often self-limiting. While the symptoms are usually due to myofascial injury, they may indicate an injury to the intervertebral disk or facet joint.

Clinical symptoms

Neck pain, which can occur anywhere from the base of the skull to the cervicothoracic junction, is the most common symptom. Pain is often worse with motion and is accompanied by paraspinous muscle spasm and trapezius pain (Figure 1). Headaches are also common in the early phase and may persist for months. They may begin in the cervical or occipital area; those that develop in the latter area may be due to an occipital nerve irritation. Increased irritability, fatigue, sleep disturbances, and difficulty concentrating are sometimes seen. Work tolerance may be temporarily impaired.

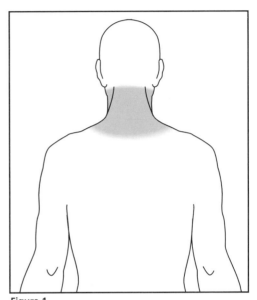

Figure 1
Typical pain diagram for a patient with an acute cervical sprain

spine

Tests

Exam

Examination reveals areas of tenderness in the paraspinous muscles, spinous processes, or inter-spinous ligaments. Limited motion is common, involving either rotation, right or left lateral bend-ing, or flexion and extension. Pain is often produced at the extremes of these motions or when applying a compressive load to the top of the head (axial compression test). The neurologic exami-nation is typically normal, and usually there is no visible deformity in the neck.

Diagnostic

AP and lateral radiographs are appropriate if the patient has a history of trauma, associated neuro-logic deficit, or if the patient is elderly. Odontoid views are appropriate with a history of acute trauma. Anterior widening of the retropharyngeal air shadow indicates soft-tissue swelling and possible disruption of the intervertebral disk or anterior longitudinal ligament and requires further evaluation. The width of the prevertebral soft tissue at the level of C3 does not exceed 7 mm in nor-mal adults. Often the normal lordotic curve is lost. There may be preexisting degenerative changes with bony spurs and narrowing of the intervertebral disk at one or more levels. Degenerative changes are most commonly seen at C5–C6 or C6–C7.

Differential diagnosis

Cervical disk herniation (commonly with neurologic abnormality)

Cervical spine tumor or infection (night pain, weight loss, history, fever, chills, sweats)

Dislocation or subluxation of the spine (abnormal radiograph)

Inflammatory conditions of the cervical spine such as rheumatoid arthritis (abnormal radiograph, laboratory findings, history)

Sociogenic/malingering/secondary gain (inconsistent or exaggerated findings)

Spinal fracture (abnormal radiograph)

Adverse outcomes of the disease

Symptoms are likely to completely resolve in most patients within the first 4 to 6 weeks. With whiplash, resolution is typically delayed, but most symptoms resolve within 6 to 12 months with few residual com-plaints.

Patients with subtle disk disruptions and injuries coexisting with degenerative conditions of the joints and disks of the cervical spine may have intractable pain. In some instances, radiculopathy due to lateral nerve root entrapment or myelopathy due to central spinal stenosis can develop.

spine

Treatment

Providing reassurance is an important first step in treating these conditions. Acute care involves 1 or 2 weeks with a soft cervical collar, appropriate pain medications, and/or short-term NSAIDs. Muscle relaxants may help if the patient has an acute spasm. Commercially available cervical pillows and cervical rolls help with sleep, and amitriptyline is often of value if symptoms persist. Avoid narcotic medications after the first week or two.

Massage, cervical traction, and manipulation of the spine may also be beneficial, as may modalities such as heat, ice, and ultrasound; however, the most benefit from these techniques will be derived within 4 weeks.

Advise patients to begin aerobic activities, such as walking, as soon as possible, and add isometric exercises as their comfort improves, preferably in the first 2 weeks. Encourage an early return to normal life activities and work.

Adverse outcomes of treatment

NSAIDs may cause gastric, renal, or hepatic complications. If the patient's condition fails to improve, reactive depression can develop. Chronic pain syndrome and drug dependence are also possible if treatment fails.

Referral decisions/Red flags

Unresponsive pain syndromes, nerve root deficit or myelopathy, and diagnostic dilemma all indicate the need for further evaluation.

spine

Adolescent Spondylolisthesis

Synonyms

None

Definition

Spondylolisthesis occurs when one vertebral body slips in relation to the one below. In the adolescent form of the condition, this usually occurs between L5 and S1, and there is usually a defect in the region of the junction of the lamina with the pedicle (pars intra-articularis), leaving the posterior element without a bony connection to the anterior element. Most likely this condition is a fatigue fracture that occurred in the preadolescent years and failed to heal. As a consequence, the vertebral body slides forward, producing the "slip" or "listhesis." Patients who participate in activities that place severe stress on this area, such as gymnastics and football, may have a higher incidence of this condition.

Clinical symptoms

This condition may be asymptomatic or minimally symptomatic; however, patients may have back pain that radiates posteriorly to or below the knees and is worse with standing (Figure 1). Frequently, they may experience spasms in the hamstring muscles manifested by the inability to bend forward and have very limited straight leg raise. True nerve compression symptoms are rare.

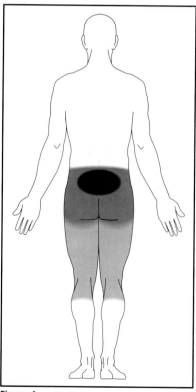

Figure 1

Typical pain diagram for a patient with adolescent spondylolisthesis

The Essentials of Musculoskeletal Care

Tests

Exam

Examination may reveal diminished lumbar lordosis and flattening of the buttocks. Palpating the spinous processes reveals a "step-off" with the spinous process of the slipped vertebra, which is "left behind," more prominent than the one above.

Hamstring spasm is manifested by marked limitation in the ability to bend forward or passive straight leg raise with the patient supine. Verifiable neurologic deficits are rare.

Diagnostic

Lateral radiographs may demonstrate forward translation of L5 relative to S1 (expressed as a percentage of the AP width of the vertebral body) (Figure 2). A defect in the pars interarticularis is evident on the oblique views (an absent neck in the "Scotty dog") (Figure 3).

Evaluate increasing slippage with radiographs taken at 6-month intervals (or sooner if symptoms increase) until growth is complete.

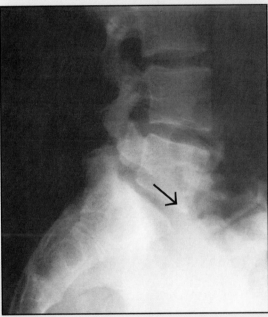

Figure 2
Lateral radiograph of a grade 2 to 3 spondylolisthesis at the lumbosacral junction

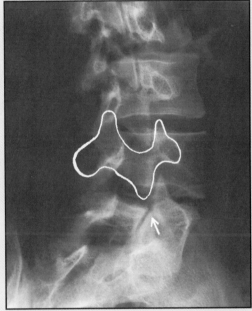

Figure 3
Oblique radiograph shows a normal "Scotty dog" appearance to the lamina (outlined area). In the vertebra below, the neck of the Scotty dog is broken (see arrow) and is the site of the pars interarticularis defect of spondylolisthesis

Differential diagnosis

Intervertebral disk injury (no "step-off" or slip, and no defect seen on plain radiograph)

Intervertebral diskitis (elevated erythrocyte sedimentation rate and fever)

Osteoid osteoma (night pain, abnormal bone scan, pain relieved with aspirin)

Spinal cord tumor (sensory findings, upper motor neuron signs)

Adverse outcomes of the disease

Progressive or complete slip of the vertebral body, chronic back pain and disability, neurologic paralysis of the lower lumbar nerve roots, or bowel and bladder involvement (cauda equina syndrome) are all possible.

Treatment

Flexion exercises and observation with periodic radiographs (a standing spot lateral view is usually adequate) are indicated until growth is nearly complete. Activities that may aggravate the condition, such as gymnastics or football line play, should be discontinued. A custom-fitted thoracolumbosacral orthosis may control pain but probably will not decrease slippage.

Adverse outcomes of treatment

Despite treatment, unrecognized progression of the forward slip can develop.

Referral decisions/Red flags

Patients with significant pain and/or obvious slippage need further evaluation. Slips with greater than 50% translation or those in which symptoms cannot be relieved through conservative treatment require spinal fusion.

Cervical Fracture

Synonyms

None

Definition

Cervical fractures typically occur as a result of high-energy trauma, such as motor vehicle accidents, and are often seen in patients with polytrauma. Concomitant head, chest, visceral, and extremity injuries are common. Cervical radiographs are part of the trauma evaluation and are mandatory in the unconscious patient. Soft-tissue injuries may result in malalignment or subluxation without obvious fracture.

Clinical symptoms

Point tenderness, pain with motion, and guarding are the most common symptoms. Patients may have associated radicular pain if there is encroachment on the neural elements. Certain positions may aggravate their symptoms.

Signs of spinal cord involvement may include gait disturbance, extremity and trunk pain or sensory changes, weakness, or bowel and bladder disturbance. While uncommon, these may be present if there is subluxation of one vertebra onto another or if there is posterior displacement of a disk or fracture fragment into the spinal canal.

Tests

Exam
Traumatic injuries to the cervical spine must be immobilized until cleared radiographically and clinically. Most patients with an acute neck injury will have well-localized severe pain. Palpation of the posterior spinous processes may reveal a "step-off," with one spinous process forward onto another. The last two cervical spinous processes are prominent and should not be confused with forward subluxation of a vertebra. There may be ecchymosis posteriorly, or swelling, which may accompany a tear of the interspinous ligament from a flexion injury. A complete neurologic examination is mandatory to assess nerve root function and spinal cord function (Figure 1). This should include rectal and genital examinations to evaluate sphincter function and sensation in this area.

Diagnostic
AP, lateral, and odontoid radiographs must show the vertebral bodies from the occiput down to and including the top of T1 (Figure 2). Some injury patterns are dramatic (Figure 3), while others are more subtle. If the area from C7 to T1 is not visible, obtain a swimmer's view to see the cervicothoracic junction. On the open mouth odontoid view, look for widening of the lateral mass(es) or fracture of the odontoid process.

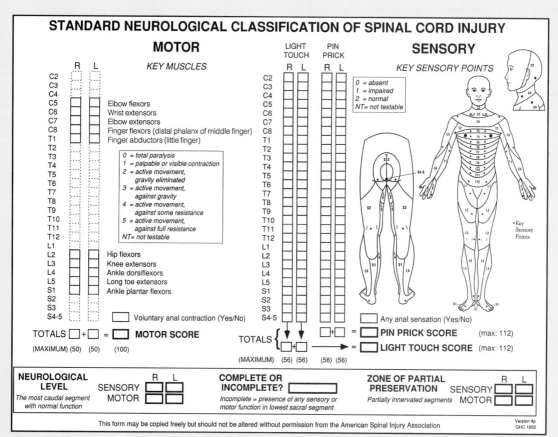

Figure 1

ASIA assessment form for the spinal cord injured

On the lateral view, check the distance between the posterior edge of the retropharyngeal air shadow and the anterior lip of the adjacent vertebral body; if there is trauma in this area, there will be swelling and the distance between these two structures will be increased. At C3, the distance should not exceed 7 mm.

Oblique views and CT scans help to better identify and understand subluxations, dislocations, and fractures of the posterior elements, including the facet joints. All these studies can be performed with the collar in place.

Flexion-extension views should be obtained for patients who are alert following significant trauma, have symptoms without neurologic deficit, and have normal radiographs; they should also be ordered for patients seen in follow up who have had symptoms for more than 1 to 2 weeks. These views rule out instability associated with a ligament injury or an occult fracture. Cervical collars can be carefully removed for these films if the static radiographs in a cooperative and neurologically intact patient are normal.

Lateral views should be obtained with the patient flexing and extending the neck unassisted. Translation of one vertebral body onto another that exceeds 3.5 mm may indicate an instability. In a lateral flexion view, the distance between the back of the atlas (C1) and the front of the odontoid (C2) should not exceed 3 mm in the adult and 4 mm in the child.

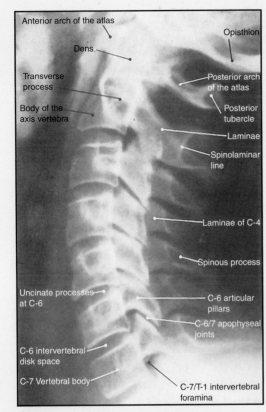

Figure 2
Lateral radiograph of the cervical spine with normal anatomic landmarks

In Figure 2, the following labels appear:
- Anterior arch of the atlas
- Dens
- Transverse process
- Body of the axis vertebra
- Uncinate processes at C-6
- C-6 intervertebral disk space
- C-7 Vertebral body
- Opisthion
- Posterior arch of the atlas
- Posterior tubercle
- Laminae
- Spinolaminar line
- Laminae of C-4
- Spinous process
- C-6 articular pillars
- C-6/7 apophyseal joints
- C-7/T-1 intervertebral foramina

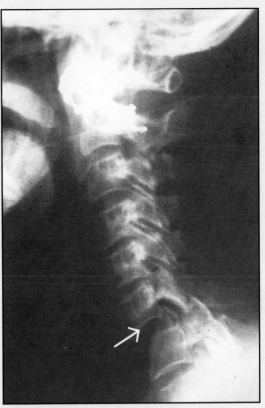

Figure 3
Facet dislocation C6 or C7

Differential diagnosis

Acute disk herniation (normal radiographs with neurologic deficit)

Burst fractures C3-C7 (involvement of the posterior cortex with bony retropulsion into the canal)

"Clay-shoveler's fracture" (C7 spinous process, need flexion-extension views)

Facet subluxation or dislocation, C3-C7 (unilateral or bilateral, usually root impingement)

Hangman's fracture (traumatic spondylolysis of the axis–C2)

Jefferson's fracture (C1 burst fracture)

Minor compression fractures, C3-C7 (less than 25% compression of the anterior cortex)

Odontoid fracture (C2, 15% of all cervical fractures, often missed)

Whiplash (normal radiograph or hypolordotic radiograph without deficit)

spine

Adverse outcomes of the disease

The most serious outcomes are quadriplegia and quadriparesis. Nerve injury may also occur, especially with injuries to the facet joints or with subluxation and dislocation of vertebrae. Chronic pain and headache accompany some of these injuries, but these are usually tolerated as time passes.

Treatment

Minor compression fractures are properly treated with a cervical brace for 6 to 8 weeks until solid, but many other fractures, subluxations, and dislocations may require manipulation, traction, surgery, or a combination thereof.

Adverse outcomes of treatment

Quadriparesis and quadriplegia, monoparesis or monoplegia, and persistent pain or deformity may result from nearly any type of treatment.

Referral decisions/Red flags

Patients who have neurologic deficit need further evaluation. Cervical fractures should be evaluated or reviewed to determine stability, even without neurologic deficit. A high index of suspicion for occult fracture should remain with uncooperative or unconscious patients.

Acknowledgements

Figure 1 is reproduced with permission from the American Spinal Injury Association, Chicago, IL.

Figure 3 is reproduced with permission from Slucky AV, Eismont FJ: Treatment of acute spine injury of the cervical spine, in Jackson DW (ed): *Instructional Course Lectures 44.* Rosemont, IL, American Academy of Orthopaedic Surgeons, 1995, pp 67–80.

Cervical Radiculopathy

Synonyms

Synonyms

Herniated cervical disk

Herniated nucleus pulposus

Lateral recess entrapment of cervical nerve root

Radiculopathy due to spinal stenosis

Definition

Cervical radiculopathy creates referred neurogenic pain in the distribution of a cervical nerve root or roots, with or without associated numbness, weakness, or loss of reflexes. The usual cause in younger patients is herniation of a cervical disk, which entraps the root as it enters the foramen. In older patients, a combination of foraminal narrowing due to vertical settling of the disk space and arthritic involvement of the facet and uncovertebral joint causes lateral entrapment of the root.

Clinical symptoms

Pain, numbness, tingling, and paresthesias (pins and needles) in the upper extremity in the distribution of the involved root are common (Figure 1). Patients may state that they can relieve pain by placing their hands on top of their head, as this removes tension from the involved nerve.

Other common symptoms include weakness, lack of coordination, changes in handwriting, diminished grip strength, and difficulty with fine manipulative tasks. Muscle spasms or fasciculations in the involved myotomes sometimes occur, as do neck pain, headache, and pain that radiates into the trapezius and paraspinous muscles and the interscapular area. Referred pain over the pectoral region (cervical angina), jaw pain, retroauricular pain, and facial pain are rare. Trunk or leg dysfunction, gait disturbances, bowel or bladder changes, and signs of upper motor neuron involvement indicate myelopathy and more commonly occur in association with central cervical spinal canal stenosis.

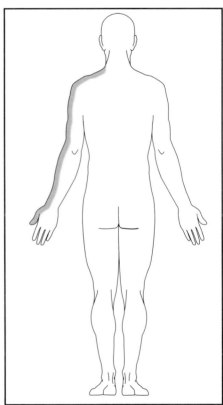

Figure 1

Typical pain diagram for a patient with cervical radiculopathy

Table 1
Clinical features of common cervical syndromes

Disk	Pain	Sensory change	Motor weakness, atrophy	Reflex change
C4-5 (C5 root)	Base of neck, shoulder, anterolateral aspect of arm	Numbness in deltoid region	Deltoid, biceps	Biceps
C5-6 (C6 root)	Neck, shoulder, medial border of scapula lateral aspect of arm, radial aspect of forearm	Dorsolateral aspect of thumb and index finger	Biceps, wrist extensors pollicis longus	Biceps, brachioradialis
C6-7 (C7 root)	Neck, shoulder, medial border of scapula lateral aspect of arm, dorsum of forearm	Index, middle fingers, dorsum of hand	Triceps	Triceps

Tests

Exam

Extension and axial rotation (Spurling's maneuver) will often cause pain in the arm or shoulder. A complete neurologic examination should be performed (Table 1). Classic findings include the following:

- C4-5 interspace; C5 nerve root (pain radiating to lateral arm, absent biceps reflex, weakness of the deltoid and/or biceps).
- C5-6 interspace; C6 nerve root (pain radiating to thumb, absent brachioradialis and/or biceps reflex, weakness of the wrist extensors).
- C6-7 interspace; C7 nerve root (pain radiating to the extensor surface of the middle finger, absent triceps reflex, weakness of the triceps and/or finger extensors).

Signs of upper motor neuron involvement may suggest spinal cord compression. Careful examination for signs of shoulder pathology, vascular disturbances, and peripheral nerve entrapment is also necessary.

Diagnostic

Plain radiographs may identify regions of spondylosis or degenerative involvement of the disk and the facet. An MRI or CT scan with intrathecal contrast confirms the diagnosis and can help in resolving a diagnostic dilemma, but these are not necessary in routine care (Figures 2 and 3). Electromyography and nerve conduction studies help in some instances to identify the location of neurologic dysfunction.

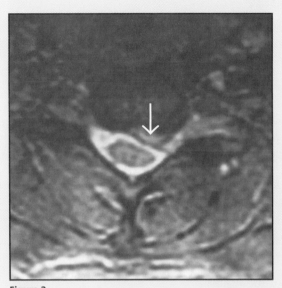

Figure 2
Axial MRI image of a cervical disk herniation in a younger patient

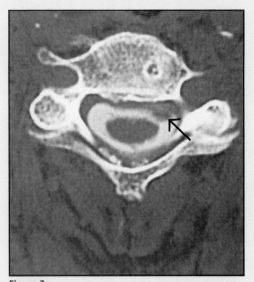

Figure 3
Axial CT scan with intrathecal contrast of a cervical disk herniation in a younger patient

Differential diagnosis

Adhesive capsulitis (restricted passive and active motion of the shoulder)

Demyelinating conditions (varying symptoms, intensity, and location)

Myocardial ischemia (abnormal ECG or stress tests)

Peripheral nerve entrapment (Phalen's test, Tinel's test at elbow or wrist)

Rotator cuff disease (painful wince with circumduction movements)

Thoracic outlet syndrome (Adson's test or military brace maneuver)

Adverse outcomes of the disease

Muscle paralysis, weakness, or chronic pain syndromes may develop, or the condition may progress to a myelopathy with cord involvement, although the latter is rare.

Treatment

Spontaneous resolution of all or most symptoms occurs within 6 to 12 weeks in most patients. With radicular pain, use of a short course of oral steroids and adequate non-narcotic pain medication is indicated. Avoid the use of oral narcotics to prevent narcotic addiction in patients in whom chronic pain syndromes develop. Cervical traction and physical therapy are helpful in the first 2 to 4 weeks. Spinal manipulation in the presence of a herniated disk with neurologic deficit may worsen the condition.

Adverse outcomes of treatment

In rare cases, quadriplegia results following manipulation of the neck in patients who have a herniated cervical disk.

Referral decisions/Red flags

Failure of conservative treatment, atrophy or motor weakness, or signs of myelopathy may require surgical evaluation. Patients with any signs that suggest a demyelinating condition, infection, or tumor require further evaluation.

Cervical Spondylosis

Synonyms

Degenerative disk disease of the cervical spine

Degenerative joint disease of the cervical spine

Cervical arthritis

Definition

In patients with cervical spondylosis, narrowing (stenosis) of the cervical spinal canal or neural foramen is produced by ingrowth of bony spurs, buckling or protrusion of the interlaminar ligaments, or by herniation of disk material. Symptoms include chronic neck pain, stiffness, and occasionally radicular pain or myelopathy.

Clinical symptoms

Neck pain is usually worse with motion, and what motion is present is limited (stiffness). Some patients report grinding or popping in the cervical region with motion. Paraspinous muscle spasm may occur, as may headaches that seem to originate somewhere in the neck. Increased irritability, fatigue, sleep disturbances, and impaired work tolerance may also develop. Pain may occur in the arms with lateral recess stenosis and nerve root entrapment (Figure 1).

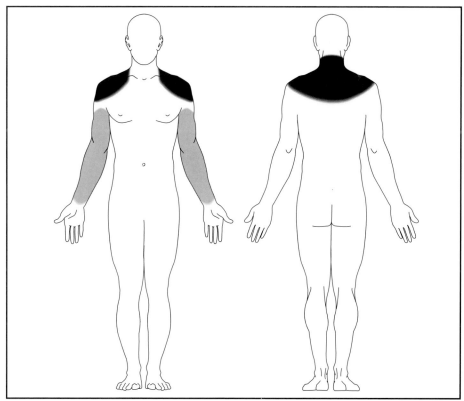

Figure 1
Typical pain diagram for a patient with cervical spondylosis

spine

Tests

Exam

Patients may have tender spots along the lateral neck or along the spinous processes posteriorly. Neck motion may be limited, especially in lateral bending and flexion/extension with myelopathy. Flexion of the neck may produce electric shocks that travel down the arms and/or legs.

With radiculopathy, findings that mimic a herniated cervical disk may be seen (Hoffmann's sign). Other upper motor neuron signs, including clonus, hyperreflexia, and Babinski's reflex (up turned toes), may be positive, and the patient may have a gait disturbance and weakness in all four extremities. Symptoms more commonly affect the upper extremities than the lower extremities.

Diagnostic

AP and lateral radiographs are appropriate (Figure 2). In the lateral view, findings include sclerosis in the intervertebral disk area with osteophytes (bone spurs) projecting anteriorly. At times, the vertebrae may fuse through these spurs. Osteophytes may also project from the posterior portion of the vertebral body into the spinal canal, producing stenosis of the cervical canal (Figure 3). Anterior subluxation of one vertebra onto the vertebra below increases the chance of cervical canal stenosis and associated neurologic findings (Figure 4). Degenerative findings are most common at the C5-C6 and C6-C7 disk spaces.

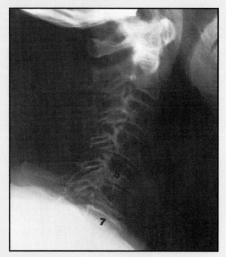

Figure 2
Lateral radiograph of an older patient with advanced cervical spondylosis

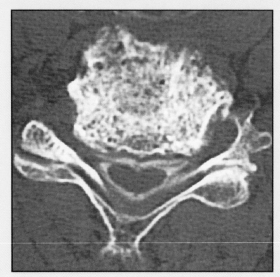

Figure 3
Axial CT scan with intrathecal contrast of cervical spondylosis in an older patient

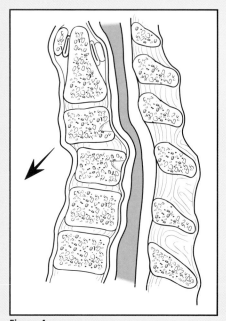

Figure 4
Degenerative disk changes and degenerative subluxations causing canal stenosis

Differential diagnosis

Metastatic tumor (night pain that prevents sleep)

Spinal cord tumor (myelopathy)

Syringomyelia (loss of superficial abdominal reflexes, insensitivity to pain)

Vertebral subluxation (in rheumatoid arthritis or following acute trauma)

Adverse outcomes of the disease

Chronic pain, myelopathy or mixed myeloradiculopathy are possible. Sequelae of spondylosis may occur when the condition is more advanced.

Treatment

Supportive treatment and reassurance may be adequate, but the symptoms will probably last several months, perhaps longer. Amitriptyline and NSAIDs or other non-narcotic pain medication will often be adequate. Avoid prescribing narcotic analgesics. Patients may also benefit from a cervical pillow or cervical roll.

Surgery may be the only option for some patients, especially those with intractable pain or progressive neurologic findings or symptoms.

Adverse outcomes of treatment

NSAIDs may cause gastric, renal, or hepatic complications. Sedation from tricyclics and adverse reactions with MAO inhibitors are possible. Narcotic addiction is also possible. Monoparesis or loss of specific nerve root function may occur, and although uncommon, quadriparesis or quadriplegia is also possible.

Referral decisions/Red flags

Intractable neck pain that is not responsive to treatment, neurologic symptoms that affect either the upper or lower extremities, lack of coordination, and electric shocks related to neck motion all indicate the need for further evaluation.

spine

Acknowledgements

Figure 4 is adapted with permission from Bohlmann HH: Cervical spondylosis with moderate to severe myelopathy. *Spine* 1977;2:151–162.

Degenerative Spondylolisthesis

Synonyms

None

Definition

Degenerative spondylolisthesis is more common in women older than age 40 years and is characterized by slippage of a vertebral body onto the one below, and associated degeneration and narrowing of the involved disk. The lamina and pars interarticularis are intact, but the slippage is allowed by alterations in the facet joints in conjunction with changes in the intervertebral disk. Usually, the upper vertebral body slides anteriorly onto the one below, creating a narrowing of the central spinal canal and lateral recesses. If it slides posteriorly, the neural foramen may become occluded.

Clinical symptoms

Back pain may be the principal complaint in association with symptoms of spinal stenosis. Mechanical symptoms of back pain caused by bending, lifting, or twisting are common because of the instability and slippage of the vertebrae. When present, leg pain simulates the pain associated with a herniated disk or spinal stenosis.

Tests

Exam

Examination reveals diminished knee and/or ankle reflexes. Strength testing after walking on a treadmill may reveal weakness in toe or heel walking, or in great toe dorsiflexion strength. Pinprick and temperature sensation are often abnormal, while light touch sensitivity is frequently normal if the patient has concomitant spinal stenosis. Often, more than one nerve root is involved, presenting a confusing or inconsistent picture. There may also be a palpable step-off when palpating the spinous processes.

Diagnostic

AP and lateral radiographs are adequate. The lateral view will show slippage of one vertebra onto another, usually with the superior vertebra displaced forward onto the one below by 3 to 6 mm (Figure 1).

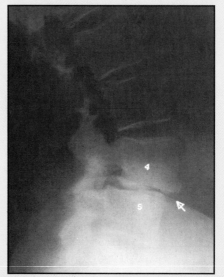

Figure 1
Lateral radiograph of a patient with degenerative spondylolisthesis at L4 on L5

Differential diagnosis

Iatrogenic instability (caused by diskectomy or decompression)

Pathologic fracture from tumor

Posttraumatic instability (fracture or disk disruption)

Spondylolytic spondylolisthesis (pars defect)

Adverse outcomes of the disease

Disabling back pain or functional neurologic impairment associated with spinal canal stenosis (forward slip) or foraminal stenosis (posterior slip) are possible. Without treatment, occupational disability is also a possibility.

Treatment

Flexion exercises, stretching exercises, occasional corset wear, and intermittent use of NSAIDs may be of benefit to these patients. Most patients require lifestyle changes, including avoiding repetitive bending, heavy lifting, and trunk twisting. Patients in heavy manual labor or sedentary occupations may require vocational change.

Adverse outcomes of treatment

NSAIDs may cause gastric, renal, or hepatic complications and may cross-react with other medications.

Referral decisions/Red flags

Patients with cauda equina syndrome, spinal stenosis symptoms with walking less than two blocks, perianal numbness, and/or bowel or bladder impairment require further evaluation.

spine

Low Back Sprain

Synonyms

Myofascial low back pain

Pulled low back

Low back strain

Lumbar sprain

Definition

Back sprain is an episode of low back pain that significantly impairs function from a few days to as long as 4 weeks. Symptoms are most often precipitated by repeated twisting or lifting heavy objects. There are multiple structures in the spine that have been implicated as the cause of pain. An episode may occur as a result of an incomplete tear in the annulus fibrosus of the intervertebral disk. There is no rupture of disk material into the spinal canal, but the disk may leak substances that induce inflammation or bulge posteriorly and cause some irritation of the lower lumbar roots. Tendons, ligaments, and muscles, or joint capsules may also be injured. Localizing an injury to a specific structure is often difficult, if not impossible.

Clinical symptoms

Patients will have low back pain that may radiate into the buttocks (Figure 1). Often, patients will have difficulty standing erect and may need to frequently change position for comfort. Since this condition most often affects young adults and is their first major episode of pain, they may show signs of nonorganic behavior, such as exaggerated responses, generalized hypersensitivity to light touch, or facial grimacing.

Risk factors include lifting and twisting, operating vibrating equipment, and sitting for prolonged periods. Other risk factors include poor fitness, poor work satisfaction, and low pay, smoking, and the personality disorders of hysteria and hypochondriasis.

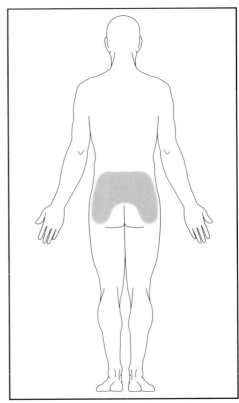

Figure 1
Typical pain diagram for a patient with low back sprain

Tests

Exam

Examination reveals diffuse tenderness in the low back or sacroiliac region, normal reflexes, and normal motor strength. Range of motion testing may elicit complaints of pain. If there is a significant midline bulge of the intervertebral disk, the patient may have mild bilateral discomfort with straight leg raising and difficulty standing erect (remaining stooped).

Diagnostic

Plain radiographs are usually not helpful for patients with acute low back strain, as they typically show changes appropriate for the patient's age: little or no disk space narrowing in adolescents and young adults and variable disk space narrowing and/or spurs in adults older than age 30 years.

For patients with atypical symptoms, such as pain at rest or at night, or with a history of significant trauma, AP and lateral radiographs are indicated, as these may help to identify or rule out infection, bone tumor (visualize up to T10), fracture, or spondylolisthesis.

Differential diagnosis

Drug-seeking behavior

Extraspinal causes (ovarian cyst, nephrolithiasis, pancreatitis, ulcer disease)

Fracture of the vertebral body (major trauma or minimal trauma with osteoporosis)

Herniated nucleus pulposus or ruptured disk (unilateral symptoms in the leg that are equal to or greater than back)

Infection (fever, chills, sweats, elevated erythrocyte sedimentation rate or WBC)

Inflammatory conditions (family history, morning stiffness)

Myeloma (night sweats)

Adverse outcomes of the disease

Functional impairment is the primary disability. This may be of great significance for young adults whose normal activities are strenuous and whose general health is otherwise excellent. Many patients have difficulty accepting a condition that can impair function for several weeks. However, most symptoms are self-limiting.

Eighty-five percent of patients improve within 1 month. The 4% of patients whose symptoms persist longer than 6 months generate 85% to 90% of the costs to society for treating low back pain. This is the most frequent cause of time lost and disability in adults younger than age 45 years.

Treatment

Treatment focuses on relieving symptoms, with a short period (1 to 2 days) of bed rest, 10 days of NSAIDs or other non-narcotic pain medications, and reassurance. While muscle relaxants may be helpful in the first week or two, avoid prescribing narcotic analgesics and sedatives.

Physical therapy and spinal manipulation may be helpful in the first 3 to 4 weeks. Once the acute pain has diminished, emphasize exercise, aerobic conditioning, and strengthening regimens. Try to assist the patient in returning to normal activity patterns within 4 weeks.

Adverse outcomes of treatment

NSAIDs may cause gastric, renal, or hepatic complications. Although rare, manipulation may produce herniation of the intervertebral disk.

Referral decisions/Red flags

Neurologic abnormalities, unresponsive pain syndromes, or an unusual cause for the pain indicates the need for further evaluation.

Lumbar Degenerative Disk Disease

Definition

Synonyms

Chronic low back pain

Chronic low back pain is the condition of axial lumbar spine pain in which symptoms have lasted longer than 3 months. Symptoms may be recurrent and episodic, or in some patients, unremitting. Improvement in functional and occupational disability is the most important goal in treating this condition.

Degeneration of the intervertebral disk is a physiologic event of aging modified by such events as trauma, infection, heredity, and tobacco use. As the hydrophilic properties of the nucleus pulposus degrade, the disk loses height; formerly tight ligaments become loose, and newfound motions such as sliding and twisting create tears in the annulus fibrosus contributing to patterns of chronic, recurring low back pain.

Clinical symptoms

Lumbar pain that may radiate to one or both buttocks is the hallmark symptom. Often, the pain is "mechanical" in that it is aggravated by mechanical activities such as bending, lifting, stooping, or twisting. Patients may have a history of intermittent sciatica (ie, pain radiating down the back of the leg), but exacerbation of back pain is the predominant symptom. Thus, patients are usually always aware of some back discomfort or limited motion. Symptoms may persist for years, typically presenting between the third and sixth decades, and may interfere with the patient's vocation.

Some pain is relieved with a night's rest, but patients may be troubled with nighttime awakenings or sleep disturbances. Chronic pain often causes mood disturbances and difficulty concentrating on tasks.

Tests

Exam
Patients usually report stiffness when rising from a seated position, and there may be tenderness about the lumbar spine and sacroiliac joints (Figure 1). While not characteristic, nonorganic findings, such as widespread sensitivity to light touch, nonanatomic localization of symptoms, inappropriate grimacing, inconsistent actions, and exaggerated pain behaviors, are not infrequent. Usually these embellished symptoms are self-limited.

spine

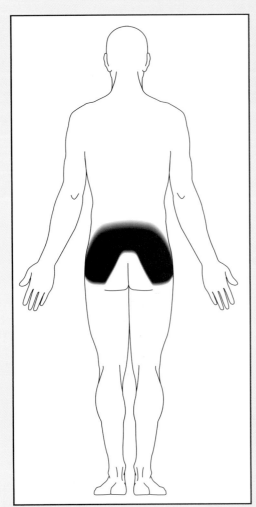

Figure 1
Typical pain diagram for a patient with lumbar degenerative disk disease

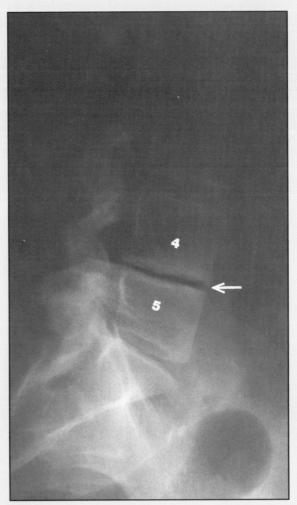

Figure 2
Lateral radiograph showing marked degenerative changes affecting the disk between L4 and L5

Diagnostic
AP and lateral radiographs may show age-appropriate changes, such as anterior osteophytes and reduced height of the intervertebral disks on the lateral view. Often there is a "vacuum sign" with apparent air (nitrogen) in the disk space (Figure 2). The Beck Depression Inventory is a simple self-administered test for depression. While depression is usually not the cause of chronic low back pain, it may complicate treatment.

Differential diagnosis

Depression

Drug-seeking behavior

Extraspinal causes (ovarian cyst, nephrolithiasis, pancreatitis, abdominal aortic aneurysm, ulcer disease)

Inflammatory arthritides (morning stiffness, abnormal laboratory studies)

Intervertebral disk infection or vertebral osteomyelitis (history of excruciating pain, fever, history of IV drug use)

Osteoporosis with compression fractures (females, previous fracture)

Metastatic tumors, myeloma, lymphoma (pathologic fractures, severe night pain)

Spinal tuberculosis (lower socioeconomic groups and patients with AIDS)

Workplace dissatisfaction

Adverse outcomes of the disease

The more severe symptoms may limit vocational and avocational activities, recreation, and sleep and may adversely affect mood, sexuality, and concentration. Deconditioning may come from reduced activities, making both symptoms and any occupational dysfunction worse.

Treatment

This is a chronic pain management problem. Advise the use of NSAIDs or other non-narcotic pain medications only, antidepressants if appropriate, and reassure the patient after ruling out the more serious causes of back pain. Physical activity, exercise, and weight reduction are important. These patients should not smoke.

Adverse outcomes of treatment

NSAIDs may cause gastric, renal, or hepatic complications. Do not label patients "disabled." Suggest a modified work schedule and more recreation. Narcotic abuse or dependency can be a problem for some that resort to narcotic analgesics.

Referral decisions/Red flags

Further evaluation is needed for patients who have fever, chills, unexplained weight loss, a history of cancer, significant nighttime pain, or a history of pain for more than 6 to 12 months. Other indications for additional evaluation include the presence of pathologic fractures (fractures occurring with minimal trauma, such as sneezing or bending), obvious deformity, saddle anesthesia, loss of major motor function, bowel or bladder dysfunction, abdominal pain, or visceral dysfunction.

spine

Lumbar Radiculopathy

Synonyms

Sciatica

Herniated nucleus pulposus

Lumbar root compression syndrome

Referred leg pain

Neurogenic leg pain

Definition

Lumbar radiculopathy (commonly called sciatica) is a condition with some degree of referred neurogenic dysfunction in the leg. Sciatica is most commonly an irritation of the fifth lumbar or first sacral nerve roots, usually the result of a herniated nucleus pulposus. In part, the root pain results from direct mechanical compression of the root, and in part from chemical irritation by substances in the nucleus pulposus. This chemical irritation may be ameliorated by NSAIDs and/or corticosteroids. Disk herniation will occur in about 2% of the population, and 10% to 25% of those patients will have symptoms that persist longer than 6 weeks. Only a small number of patients require surgery.

Clinical symptoms

The onset of symptoms is usually abrupt and often accompanied by low back pain. Some patients report that their preexisting back pain disappears when the leg pain begins.

Pain is often severe and is exaggerated by sitting, coughing, and sneezing. Patients have a difficult time finding a position of comfort. If they have to assume a knee-chest position, their disk fragment is so large they can only find comfort by increasing the interior diameter of the spinal canal with maximal spinal flexion.

Typically, the pain travels from the buttock down the posterior or posterolateral leg to the ankle or foot (Figure 1).

Midlumbar (L1 to L3) radiculopathies refer pain to the anterior aspect of the thigh and often do not radiate below the knee. Herniations at these levels constitute only 5% of all disk herniations.

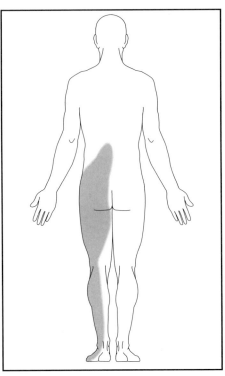

Figure 1
Pain diagram of patient with lumbar radiculopathy

Tests

Exam

With the patient standing, look for a list (trunk shifts to one side) (Figure 2). With the patient sitting, observe for pain and spinal extension (leaning back) when the leg is raised (flip sign), as it is highly reliable for a herniated disk, especially when coupled with back pain produced with supine straight leg raising at less than 45° of leg elevation. With the patient prone, observe for pain with hip extension (femoral stretch test), as it usually indicates involvement of the L3 root, but may indicate L2 or L4 involvement as well.

Classic findings include the following (Figure 3):

- L3-4 disk (L4 nerve root) may produce weakness in the ankle dorsiflexors, numbness in the shin, thigh pain, and an asymmetrical knee reflex. About 5% of disk ruptures occur at this level.

- L4-5 disk (L5 nerve root) may produce weakness in the great toe extensor, numbness on the top of the foot and first web space, and posterolateral thigh and calf pain.

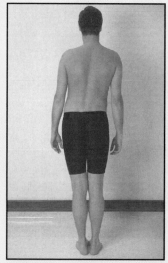

Figure 2
Trunk shift, or "list," to right

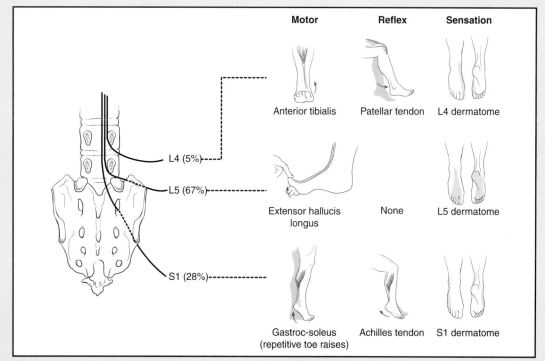

	Motor	Reflex	Sensation
L4 (5%)	Anterior tibialis	Patellar tendon	L4 dermatome
L5 (67%)	Extensor hallucis longus	None	L5 dermatome
S1 (28%)	Gastroc-soleus (repetitive toe raises)	Achilles tendon	S1 dermatome

Figure 3
Typical motor, sensory, and reflex findings with common lumbar radiculopathies

spine

- L5-S1 disk (S1 nerve root) may produce weakness in the gastrocnemius with inability to sustain tiptoe walking, numbness in the lateral foot, posterior calf pain and ache, and an asymmetric ankle reflex.

Diagnostic

Plain radiographs usually demonstrate age-appropriate changes. Electromyography (EMGs) should not be considered prior to 21 days, and these may not always detect clinically significant radiculopathy. An MRI scan may be useful if there is a diagnostic dilemma, or as a part of preoperative planning (Figure 4).

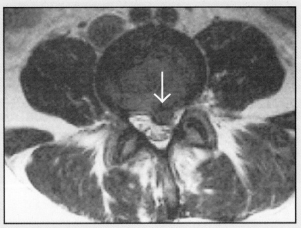

Figure 4
Axial MRI scan showing herniated lumbar disk

Differential diagnosis

Cauda equina syndrome (perianal numbness, urinary overflow incontinence or retention, reduced anal sphincter tone, bilateral involvement)

Demyelinating conditions

Extraspinal nerve entrapment (abdominal or pelvic mass)

Lateral femoral cutaneous nerve entrapment (sensory only, lateral thigh)

Spinal stenosis (older population)

Thoracic disk herniation (signs of myelopathy)

Trochanteric bursitis (no tension signs, pain down lateral thigh and leg, exquisite tenderness over trochanter)

Adverse outcomes of the disease

Cauda equina syndrome with permanent motor loss, urinary incontinence, and sensory numbness can develop. Other findings include motor or sensory loss in a specific root (drop foot in L5 root lesions).

Treatment

The use of NSAIDs or aspirin in the acute phase coupled with the judicious use of short-term oral steroids and/or epidural steroids may address the chemical neuritis. Patients should sit only for toilet needs and should eat standing or lying down. Bed rest for 1 or 2 days may help relieve the initial symptoms. Reassure the patient that most disk herniations resolve without residual problems. Oral pain medication and muscle relaxants are of value in the first few days, but should not be used for long-term pain control.

Many ruptures with a significant inflammatory component will improve within 2 weeks; patients who have persistent, disabling symptoms may have a large fragment that requires surgical excision.

spine

Adverse outcomes of treatment

NSAIDs may cause gastric, renal, or hepatic complications. Progression of neurologic deficit, or persistent numbness and weakness may occur despite treatment. It is rare to see significant complications from the use of a 6-day rapidly tapered low dose oral steroid or epidural steroids.

Referral decisions/Red flags

Patients with any of the following conditions need further evaluation: cauda equina symptoms; urinary retention; perianal numbness; motor loss; severe single nerve root paralysis; progressive neurologic deficit; radicular symptoms that last longer than 6 to 12 weeks; intractable leg pain; or recurrent episodes of moderately severe sciatica.

Acknowledgements

Figure 3 is reproduced with permission from Kasser JR (ed): *Orthopaedic Knowledge Update 5.* Rosemont, IL, American Academy of Orthopaedic Surgeons, 1996, pp 609–624.

Midback Sprain

Synonyms

Pulled upper back

Upper back strain

Thoracic sprain

Definition

A back sprain is an acute soft-tissue injury accompanied by pain in the thoracic and thoracolumbar areas. Symptoms may begin following an episode of mild trauma, especially twisting, and may eventually become chronic. An episode usually represents an incomplete tear in the annulus fibrosus of the intervertebral disk or an incomplete tear of supporting posterior ligamentous, muscular, or capsular structures.

Clinical symptoms

Patients report a history of constant midback pain for which they have tried simple remedies such as massage and NSAIDs without much relief (Figure 1). These patients rarely have neurologic symptoms and are frustrated with the chronic nature of the pain. They are often young adults having their first encounter with a health problem that does not resolve spontaneously.

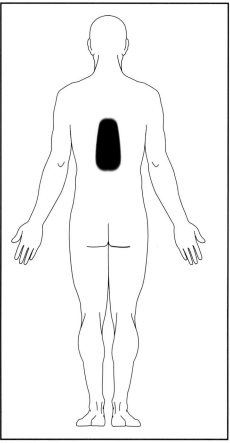

Figure 1
Typical pain diagram for a patient with midback sprain

Tests

Exam

Examination typically reveals a discrete tender spot, and patients may report that the pain is aggravated by motion or weightbearing. Reflexes are normal; sensation to light touch or pinprick may be diminished in a dermatomal distribution if there is a unilateral thoracic disk rupture.

Diagnostic

AP and lateral plain radiographs may show osteophytes at a specific intervertebral disk level. These radiographs are usually sufficient to rule out infection, bone tumor, or fracture.

Differential diagnosis

Fracture of the vertebral body (major trauma)

Herpetic neuralgia prodrome (await onset of vesicles)

Osteoporosis (fractures, percussion tenderness, in postmenopausal women)

Thoracic disk rupture (long-tract signs of clonus, spasticity, uncoordinated gait, or bilateral leg numbness)

Tumor (intense, constant pain and night pain sufficient to make the patient sit up)

Visceral pain (pancreatic, cardiac, or renal disease)

Adverse outcomes of the disease

Pain is the principal symptom and may be of great significance in young adults who normally participate in strenuous activities. Thoracic disk herniation may occur and produce unilateral sensory complaints, or even cord compression symptoms.

Treatment

Aerobic exercise (walking, bicycling) and thoracic extension exercises will often help if maintained for an extended period (12 months or longer). Spinal manipulation is of little help when the condition is chronic. When the pain is acute, a short period of rest and NSAIDs or other non-narcotic pain medication is appropriate. Bracing is rarely needed. Massage often provides temporary relief and can be done effectively by a family member.

Adverse outcomes of treatment

NSAIDs may cause gastric, renal, or hepatic complications.

Referral decisions/Red flags

Spasticity, clonus, or leg weakness suggests cord compression and indicates the need for further evaluation. Suspicion of associated serious diseases such as tumor and infection also requires further evaluation.

spine

Spinal Stenosis

Synonyms

Neurogenic claudication

Pseudoclaudication

Spinal claudication

Definition

Stenosis is a congenital or acquired narrowing of the spinal canal, usually worsening with age, that presents with insidious onset of neurogenic claudication (poor walking tolerance due to calf fatigue) and occasionally neurogenic (referred) leg pain. The age-related narrowing and inward buckling of the intervertebral disk, coupled with forward buckling of the interlaminar ligament and enlargement of arthritic facets is the common cause.

Clinical symptoms

Low back pain and leg symptoms are worsened by extension of the spine, as when walking, standing, or supine (Figure 1). Leg symptoms include weakness, pain, and sometimes inability to actually "find the legs." The pseudoclaudication differs from vascular claudication in that it usually recovers quickly with sitting or spinal flexion. Circulation may seem sufficient to the legs. With repeated exercise, the patient can usually walk the same distance each time after a short rest. The distance traveled before claudication occurs may vary daily.

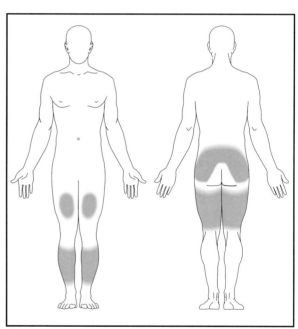

Figure 1
Typical pain diagrams for a patient with spinal stenosis

Patients usually experience relief with flexion, such as sitting or stooping, or with bending forward. Patients with spinal stenosis can typically walk relatively well with the lumbar spine flexed forward.

Tests

Exam

Examination reveals diminished knee and/or ankle reflexes. Have the patient stand with the spine extended (leaning backward); pain reproduction indicates spinal stenosis. Romberg's test is often positive. Strength testing after walking on a treadmill often reveals weakness in toe or heel walking, or in great toe dorsiflexion strength. Pinprick and temperature sensation are often abnormal while light touch sensitivity is frequently normal. Often, more than one nerve root is involved, presenting a confused or inconsistent picture.

Diagnostic

Radiographs may reveal degenerative spondylolisthesis, marked narrowing of the intervertebral disk at one or more levels, and degenerative scoliosis. Other patients simply have end plate changes about the disk space and marginal osteophyte formation about the facet joints; foraminal narrowing may also be seen. Electromyography may be positive. Special radiologic imaging is best reserved for preoperative planning or to resolve a diagnostic dilemma (Figure 2).

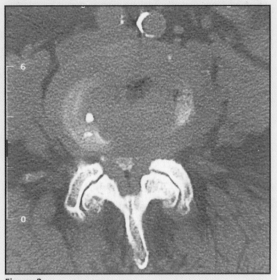

Figure 2
Axial CT scan with intrathecal contrast showing severe lumbar spinal stenosis in an older patient

Differential diagnosis

Diabetes mellitus with neuropathy

Herniated intervertebral disk (in younger patients, single root involvement)

Osteoarthritis of the hip or knee (painful range of motion in the affected joint)

Pathologic vertebral fracture with bony stenosis (osteoporosis)

Stroke (upper motor neuron findings)

Vascular claudication (check posterior tibial pulse, history)

Vitamin B_{12} or folic acid deficiency

Adverse outcomes of the disease

Patients may experience mild to marked functional impairment in ambulation, and/or variable degrees of back and leg pain. In rare instances, the bowel or bladder may be involved.

spine

Treatment

Explain the postural relationship between spinal flexion (leaning forward, sitting down) and relief of leg pain. NSAIDs or a low-dose, rapidly tapered oral steroid, or an epidural steroid injection should be used if postural treatment is not effective. If conservative treatment fails, surgery is often effective in relieving leg symptoms and improving walking distance.

Adverse outcomes of treatment

NSAIDs may cause gastric, renal, or hepatic complications. In addition, use of NSAIDs in elderly patients may lead to dyspeptic complications and adverse interactions with various medications.

Fluid retention, flushing of the skin, and shakiness can occur as a result of taking oral steroids.

Infection from injected steroid is also a risk, but can be largely avoided by use of careful sterile technique.

Referral decisions/Red flags

Further evaluation is needed if the patient has cauda equina symptoms (bowel or bladder incontinence, perianal numbness). Other indications for additional evaluation include failure of conservative treatment, diagnostic dilemma, and/or pseudoclaudication at less than two blocks of walking.

Thoracolumbar Fracture

Synonyms

None

Definition

Fracture of one or more components of the vertebrae in the thoracic or lumbar spine usually occurs due to high-energy trauma. Associated nerve root or spinal cord deficit is possible, and some fractures are highly unstable. Other associated injuries and fractures are also common.

Fractures from minimal trauma indicate underlying conditions such as osteoporosis or tumor.

Clinical symptoms

Pain and limited motion or inability to stand and walk are the most common symptoms. Injuries related to automobile accidents or falls and industrial injuries are acute in onset, whereas those related to minor trauma begin as a "pop," followed a few hours later by increasing local pain.

Tests

Exam

Swelling and hematoma formation, with or without bruising, are common over the back, especially with flexion injuries where the spinous processes or interspinous ligaments are pulled apart.

Nerve root findings, such as isolated reflex loss, dermatomal sensory pattern, or specific muscle loss indicate injury to a nerve root and may also indicate underlying instability in the fracture pattern.

Abnormal neurologic findings of spinal cord dysfunction, such as a positive Babinski's sign, motor paralysis, and spasticity all indicate a major injury, which is likely accompanied by an unstable fracture.

Diagnostic

AP and lateral radiographs are standard. A spot lateral view is sometimes necessary to visualize the L5-S1 junction well. Oblique or flexion-extension views may be needed to evaluate less obvious trauma or instability. Frequently, more than one level of the spine is injured.

With compression fractures, the actual fracture occurs anteriorly in the vertebral body with loss of height, but the height of the posterior portion appears normal in the lateral view, and the interpedicular distance is less than or equal to the vertebra below (Figure 1). Those fractures compressed by more than half of normal height may lead to chronic pain syndromes.

spine

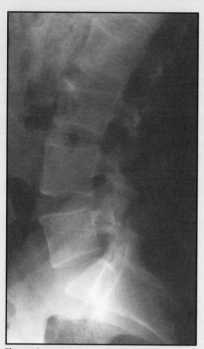

Figure 1
Lateral radiograph showing a minimal compression fracture of the lumbar spine

Axial burst fractures are often characterized by loss of height of both the anterior and posterior cortex of the vertebral body in the lateral view. Bony repulsion may be seen (Figure 2), and there is often widening of the interpedicular distance or an associated fracture of the lamina (Figure 3).

Chance fractures result from distraction, such as when the upper body is thrown forward as a result of a motor vehicle accident while the pelvis is stabilized by a seat belt. The vertebra is literally pulled in half with a superior and inferior part. The fracture line often runs posteriorly through the pedicle and into the lamina and spinous process.

Transverse process fractures result from rotation or extreme lateral bending and may be associated with significant retroperitoneal bleeding and muscle injury, even though the bony fragments are small.

Fracture-dislocations occur with varying degrees of displacement of one vertebra onto another, and these injuries are unstable.

Further imaging with CT or MRI scan helps classify the fracture pattern and demonstrates the degree of spinal canal compromise (Figures 4 and 5). These studies are usually obtained for preoperative evaluation.

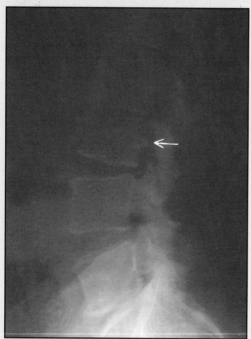

Figure 2
Lateral radiograph showing a burst fracture of the lumbar spine with bony retropulsion

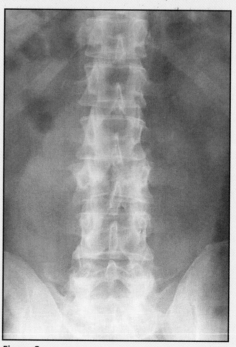

Figure 3
AP radiograph showing a burst fracture of the lumbar spine with widening of the interpedicular distance

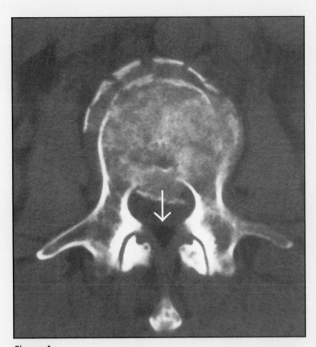

Figure 4
Axial CT scan of a lumbar spine burst fracture showing bony retropulsion into the spinal canal

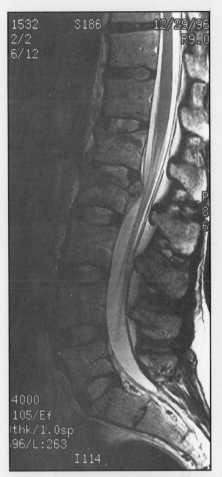

Figure 5
Sagittal MRI image of a lumbar spine burst fracture showing canal compromise

Differential diagnosis

Herniated thoracic or lumbar disk (nerve root findings; MRI or CT scan may show herniation)

Thoracic or lumbar sprain (no neurologic findings or fractures)

Visceral injuries (abnormal abdominal exam and ultrasound)

Adverse outcomes of the disease

Chronic pain syndromes with spinal instability and functional or occupational impairment is possible. Involvement of the nerve root or spinal cord can lead to paraplegia or significant paralysis. In some cases, sacral decubitus and loss of bowel and bladder function develop.

spine

Treatment

Immobilize the patient on a spine board with a sacral pad until the completion of the physical and neurologic examinations. If radiographs suggest the absence of significant injury, immobilization should be discontinued.

Compression fractures with anterior wedging of less than half the normal vertebral height, an intact posterior portion of the vertebral body, and normal interpedicular distance may be managed with pain medication and a Jewett-type orthosis or body cast for 6 to 12 weeks. Emphasize walking and exercise while the fracture is healing. Patients with workers' compensation claims may require job modifications or job changes, but they should return to work as soon as possible even if their duties are very limited.

Transverse process fractures should be evaluated for nerve root injury and may be treated in a Jewett-type spinal orthosis or a thoracolumbar corset along with an aerobic walking program for 6 weeks, followed by trunk flexion and extension exercises.

Adverse outcomes of treatment

Further collapse and deformity at the fracture, increasing neurologic deficit, and chronic pain syndrome with or without spinal instability and degenerative arthritis may all develop.

Referral decisions/Red flags

Patients who experience any neurologic deficit as a result of spinal fracture need further evaluation. Other conditions that require additional evaluation include burst fracture, fracture-dislocation, and/or any fracture involving more than just the anterior cortex of the vertebral body. Patients who require current spinal cord management protocols for medication and transportation should also be evaluated further.

Section Editor

Walter Greene, MD
Chairman, and J. Vernon Luck Sr, MD, Distinguished Professor
of Orthopaedic Surgery
University of Missouri Hospital
Columbia, Missouri

David D. Aronsson, MD
University of Vermont
College of Medicine
Department of Orthopaedics and Rehabilitation
McClure Musculoskeletal Research Center
Burlington, Vermont

Robert M. Campbell, Jr, MD
Associate Professor, Orthopaedics
University of Texas Medical School
San Antonio, Texas

Frederick R. Dietz, MD
Department of Orthopaedics
University of Iowa
Iowa City, Iowa

James C. Drennan, MD
Professor, Orthopaedics and Pediatrics
University of New Mexico School of Medicine
Medical Director/CEO
Carrie Tingley Hospital
University of New Mexico Health Sciences Center
Albuquerque, New Mexico

Brad Olney, MD
Associate Professor and Chairman
Department of Orthopaedic Surgery
University of Kansas Medical Center
Kansas City, Kansas

William L. Oppenheim, MD
Professor and Head, Pediatric Orthopaedics
UCLA Medical Center
Los Angeles, California

Peter D. Pizzutillo, MD
St. Christopher's Hospital for Children
Philadelphia, Pennsylvania

Thomas L. Schmidt, MD
Professor, Orthopaedic Surgery
University of Missouri
Kansas City, Missouri
Chief, Orthopaedic Surgery
The Children's Mercy Hospital
Kansas City, Missouri

George Sotiropoulos, MD
Clinical Assistant Professor
Department of Child Health
University of Missouri
Columbia, Missouri

Michael D. Sussman, MD
Chief of Staff
Shriners Hospital for Children
Portland, Oregon

George H. Thompson, MD
Rainbow Babies and Childrens Hospital
Case Western Reserve University
Cleveland, Ohio

Hugh Watts, MD
Clinical Professor of Orthopaedic Surgery
University of California at Los Angeles
Los Angeles, California

section nine

pediatric orthopaedics

pediatric orthopaedics— an overview

Musculoskeletal disorders in children are often unique and always have different considerations compared with similar problems observed in adults. For that reason, *Essentials of Musculoskeletal Care* includes this separate section on pediatric orthopaedics. In this overview, common presenting complaints and musculoskeletal concepts unique to children are discussed.

The common presenting complaints associated with pediatric disorders are pain, swelling, refusal to use a limb or limping, and deformity. In addition, age and gender must also be considered when evaluating children. For example, a boy between the ages of 4 and 8 years who has proximal thigh pain and a limp most likely has Legg-Calvé-Perthes disease. Similar symptoms in adolescent boys suggest a slipped capital femoral epiphysis. Female infants are more likely to have hip dysplasia, whereas clubfoot is more common in male infants. Adolescent girls are more likely to have scoliosis, but kyphosis of the spine is more common in adolescent boys.

Pain

Fractures and dislocations in children, as in adults, are characterized by an acute onset of severe pain. Severe pain that develops over a 1- to 4-day period is typical in osteomyelitis or septic arthritis in children. Other inflammatory arthritides such as juvenile rheumatoid arthritis have a more indolent onset, and the pain is typically worse on awakening. Night pain commonly accompanies tumors originating from bone (ie, osteosarcoma or osteoid osteoma) or bone marrow (ie, leukemia). Degenerative joint disease typically requires years to develop; therefore, deformities in children typically do not cause severe pain.

Swelling

The four common conditions that cause swelling or masses are tumors, tumorous conditions (eg, osteochondromas), infection (in either the bone or joint), and the various arthritides. The metaphyseal region of the bone has the greatest metabolic and vascular activity, and therefore is the osseous site of most diseases.

Refusal to use a limb or limping

Young children may not verbalize pain, but will express the problem by limping or refusing to use the extremity. The different diagnoses to be considered when evaluating this clinical problem are listed in the chapter called Evaluation of the Limping Child. Neuromuscular disorders causing muscle weakness may also cause a child to limp.

Deformity

Some deformities in children are "normal for the age." For example, genu varum or bowlegs is normal at birth, but by the time a child reaches age 3 years, normal alignment is a "knock-knee" or relatively large amount of genu valgum. Because pain is uncommon during childhood, understanding the natural history of these pediatric deformities is critical to knowing whether treatment is needed.

Special considerations

The oft-quoted statement *"Children are not just small adults,"* is particularly germane for musculoskeletal disorders. Growth, neurologic disorders, infections, injuries and fractures, and child abuse have unique impacts on musculoskeletal conditions in children.

Growth

Musculoskeletal growth may be a significant ally in treatment. For example, a newborn with a dislocated hip and a shallow acetabulum may develop a stable, normal joint after simple splinting of the hips for only a few months. Spontaneous correction of some deformities, such as metatarsus adductus, often occurs. Growth, however, may also exacerbate pediatric disorders. Fractures that damage the physis or growth plate cause either uneven growth (partial growth plate closure) with resulting angular deformity or cessation of growth (complete growth plate closure) with resulting shortening of the extremity. Excessive compression on one side of the physis secondary to obesity or weakening of the bone retards growth on that side of the growth plate. The result is progressive angulation. Examples include progressive bowleg deformity in infantile tibia vara (occurring in obese children who start walking at an early age) and rickets (weakening of the bony structure at the growth plate).

Neurologic disorders

Muscle imbalance, such as that observed in some neuromuscular disorders, including cerebral palsy, myelomeningocele, and muscular dystrophy, also causes unique problems in children. The underlying principle is that bones grow and muscles have to "catch up." With muscle imbalance plus growth, the result is contracture of the muscle and possible bony deformity. For example, in a child with cerebral palsy, the hip adductor and flexor muscles are often more spastic and therefore stronger than the opposing hip abductors and extensors. With growth, this muscle imbalance may cause progressive contracture of the adductors and flexors, abnormal torsion of the femur, deficient growth of the acetabulum, and subluxation or dislocation of the hip.

Variables in soft-tissue injuries

As children grow, their coordination and psyche are also developing. As a result, competitive sports require adaptation of the game to the age and size of the child.

Overuse injuries in the adult primarily manifest themselves as either microscopic tears or complete rupture of the musculotendinous junction or within the substance of the tendon. In children, the bone-tendon junction is the weak link. The result is different types of overuse syndromes. For example, Osgood-Schlatter disease results from microscopic avulsion fractures at the insertion of the patellar tendon during adolescence when the child is relatively big and active and when the relatively weak secondary ossification center of the proximal tibia is developing.

The Essentials of Musculoskeletal Care

Variables in infection and arthritis

The higher incidence of hematogeneous osteomyelitis in children is related to the unique anatomy of metaphyseal circulation in children. Direct trauma from frequent falls is a contributing factor to the increased incidence of septic arthritis in children. Chronic arthritides such as juvenile rheumatoid arthritis are also different in children. The etiology of this difference is less clear but may be related to a developing immune system causing a different response to triggering agents.

Pediatric fractures

Fractures are more common in children, related in part to their rambunctious play, and in part to the different characteristics of bone in children. Bone strength gradually increases as a child grows, but at most ages, the force required to fracture a bone in a child is less than that in an adult. Bone, therefore, is a weak link, and as a result, ligamentous injuries are uncommon until late adolescence. The bone of a child, however, is more plastic and less rigid than that of an adult. It may actually bend and deform, but not break. The resulting injury is a torus fracture, a compression or buckle fracture on one side of the bone. If the force continues, the bony cortex may break on the distraction or tension side, and remain bent on the compression side, producing a greenstick fracture.

Bone healing is more rapid in children. For example, a fracture of the femur in an adult requires 16 to 20 weeks of immobilization if treated by closed means. By comparison, the same fracture in a neonate heals in 2 weeks and during early childhood after 4 to 6 weeks of immobilization. Because of more rapid healing, nonunion is also rare in children.

Bone remodeling is greater in children, and fracture malalignment may spontaneously correct. Remodeling is greater in younger children, in fractures close to the physis, and in fractures angulated in the plane of motion. For example, a fracture of the distal radius with 35° of volar angulation in a 5-year-old child will completely remodel in 1 to 2 years, even if no reduction is performed. In this case, the principal reason for reduction is to relieve the pressure on adjacent soft tissues. Rotational deformities, however, do not correct.

Fractures involving the growth plate or physis are unique to children and account for 15% to 20% of pediatric fractures. These injuries are more common during adolescence, probably because the surrounding perichondral ring that supports the growth plate becomes thinner and weaker at this time. The most common pattern of physeal injury is often referred to as a Salter-Harris type II fracture. This injury includes a metaphyseal fragment on the compression side of the injury and accounts for approximately 75% of physeal fractures. Displaced type III and IV fractures require open reduction and are more likely to be complicated by premature closure (partial or complete) of the physis.

Closed reduction and casting is the most common treatment of pediatric fractures. The thick periosteum of the growing child aids in keeping the fracture reduced during healing. Prolonged bed rest is not associated with the increased risks seen in adults and is often the treatment of choice for complex fractures of the spine, pelvis, and extremities.

pediatric orthopaedics— physical exam

1

Figure 1
Hip internal rotation

With the patient prone, flex the knees to 90° and then move the feet apart. With restricted unilateral motion, one hip will stop moving and the other will continue to move. With internal femoral torsion (increased femoral anteversion), internal rotation will exceed external rotation by 30° or more.

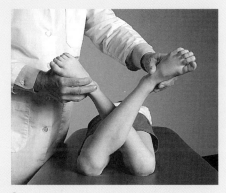

2

Figure 2
Hip external rotation

With the child prone and knees together, rotate the feet toward each other and cross the lower legs. It is easiest to view the patient from the bottom of the examining table. Compare the two sides for maximal rotation.

pediatric orthopaedics

3

Figure 3
Hip abduction

With the child supine and legs together, place dots on the anterior superior iliac spines as reference markers for the plane of the pelvis. Bring the feet apart and stop abduction on one side if the dot on that side begins to rise, indicating pelvic tilt rather than hip abduction. Compare the maximum abduction of the two sides.

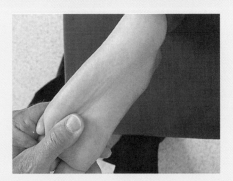

4

Figure 4
Tibial torsion

With the patient prone and the knee flexed, measure the angle formed by the axis of the thigh and the axis of the foot (thigh-foot angle).

5

Figure 5
Frontal view

With the patient standing, look at alignment of the leg for knock-knee (genu valgum), bowleg (genu varum), internal femoral torsion (patellae point toward each other), external femoral torsion (patellae face away from each other and not straight ahead), and leg length inequality (pelvic brims not level). Also, look for increased angulation at the elbow or extreme shoulder height asymmetry. Look for asymmetry in the angle formed by the humerus and forearm (elbow carrying angle). Unilateral deformity may indicate an acute injury or deformity from a congenital or traumatic growth disturbance.

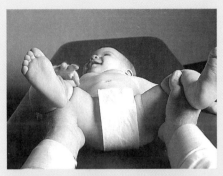

6

Figure 6
Abduction (neonate)

Place your thumbs on the patient's inner thighs, proximal to the knees, with fingers on the greater trochanters. Abduct the patient's legs and look for asymmetric motion. Hip abduction should exceed 45° bilaterally. Limited abduction suggests developmental dislocation of the hip.

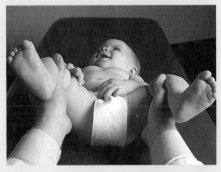

7

Figure 7
Barlow Sign

Place your thumbs on the patient's inner thighs, toward the hips, with fingers on the greater trochanters. Adduct the hips and lightly and rotate the femur to displace the proximal femur away from the pelvis. A sensation of slippage indicates subluxation. This test requires very little force.

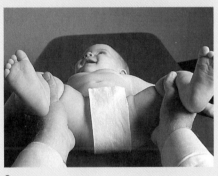

8

Figure 8
Ortolani's "reduction" sign

Place your thumbs on the patient's inner thighs, near the knees, with fingers on the greater trochanters. Abduct the legs, rotating the proximal femur toward the pelvis. Check for a sensation of reduction as the hips slip over the rim of the pelvis. This test requires very little force. A sense of reduction (a clunk) is perceived as the femoral head relocates into the actabulum.

pediatric orthopaedics

9

Figure 9
Scoliosis, bending

Ask the patient to bend forward with the knees straight and both arms hanging free, as you stand either in front of or behind the patient and sight along the upper spine. Rotation of the vertebrae will throw the ribs upward on the convex side and downward on the concave side. Forward bending brings this rotation into profile and makes the diagnosis easier. Measure the amount of rib rotation with a scoliometer.

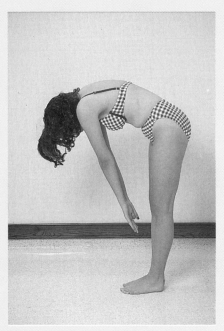

10

Figure 10
Kyphosis

Ask the patient to bend forward with arms hanging free. View the patient from the side. If the patient is not able to touch the distal shin or toes, the reason may be tight hamstrings secondary to a spondylolisthesis in the lower lumbar spine. Look for an excessive round back in the thoracic spine.

pediatric orthopaedics

Accessory Tarsal Navicular

Synonyms

None

Definition

Accessory tarsal navicular is a normal anatomic variant in which a secondary center of ossification forms in the medial portion of the tarsal navicular at the attachment of the posterior tibialis tendon. During adolescence, this becomes prominent and symptomatic by virtue of its size, or from repetitive sprains of the fibrous attachment of the ossicle to the navicular. The disorder is fairly common with one study observing a 14% incidence of symptoms during adolescence.

Clinical symptoms

Patients with an accessory tarsal navicular have pain and swelling on the medial side of the foot. The pain is exacerbated with activity or pressure from overlying shoes. Severe pain is uncommon.

Tests

Exam
Examination reveals tenderness on palpation over the navicular and commonly includes swelling (a bony mass) and redness at the insertion of the posterior tibialis tendon. Inversion of the foot against resistance may be painful. A flexible pes planus may be present.

Diagnostic
Radiographs are not necessary with a typical exam and mild symptoms. With persistent or severe symptoms, radiographs of the foot will document the disorder and exclude other possibilities (Figure 1). To demonstrate pes planus or other alignment problems, request AP and lateral radiographs as standing or weightbearing views. Oblique views often provide the best profile of the accessory ossicle. Some patients have a cornuated navicular (shaped like a horn of plenty), resulting from fusion of the accessory ossicle.

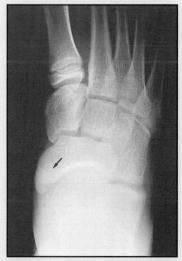

Figure 1
Oblique radiograph showing accessory navicular

pediatric orthopaedics

Differential diagnosis

Flexible pes planovalgus (developmental or acquired)

Posterior tibial tendinitis

Tarsal coalition

Adverse outcomes of the disease

Patients may experience pain or a limp.

Treatment

Most patients can be treated with short-term activity restrictions and shoe modifications (soft material medial to the bump and/or stretching of the shoe) to relieve pressure over the prominent navicular. In most instances, the bump is not large and symptoms resolve with cessation of growth. In patients who have an associated pes planus, use of a soft orthotic device may help medial arch pain.

Surgical excision of the prominent portion of the navicular is the preferred treatment for patients with persistent, disabling symptoms.

Adverse outcomes of treatment

Postoperative infection and a tender medial scar that is irritated by shoe wear are both possible.

Referral decisions/Red flags

Persistent pain signals the need for further evaluation.

Acknowledgements

Figure 1 is reproduced with permission from Davidson RS: Miscellaneous foot disorders, in Richards BS (ed): *Orthopaedic Knowledge Update: Pediatrics.* Rosemont, IL, American Academy of Orthopaedic Surgeons, 1996, pp 219–225.

Back Pain

Synonyms

Backache

Definition

Pain in the thoracic and lumbar spine or cervical spine during childhood, although unusual, is often due to organic causes (sometimes serious) when present for more than a few weeks.

Clinical symptoms

Pain is the most common symptom for a wide variety of pediatric spinal disorders. It is important to determine the nature of onset, as well as the location, character, and radiation of the pain.

Back pain accompanied by neurologic signs and symptoms (radicular pain, muscle weakness, gait abnormalities, sensory changes, bowel and bladder abnormalities), or systemic symptoms (fever, malaise, and weight loss) suggest serious problems. Night pain often signals a serious problem.

Tests

Exam
A differential diagnosis of back pain in children requires a complete musculoskeletal and neurologic evaluation. Examine the spine for deformities, loss of motion, muscle spasm, and areas of tenderness.

Neurologic evaluation should include assessment of gait, muscle strength, sensory testing, pain with straight leg raising, and deep tendon reflexes, including abdominal reflexes and pathologic reflexes (Babinski's sign). Abnormal neurologic signs may be a late manifestation of tumor. Therefore, a normal neurologic exam may not eliminate serious problems.

Diagnostic
The extent of the evaluation depends on the duration and degree of symptoms as well as the physical exam. No laboratory or radiographic studies are necessary if the exam is normal and symptoms are mild and of limited duration.

For symptoms in the thoracic and lumbar regions, obtain standing AP and lateral radiographs of the entire spine. Oblique views may be helpful for patients who have symptoms in the lumbar spine.

If patients have neck pain, obtain AP, odontoid, lateral, and oblique radiographs of the cervical spine. Flexion and extension views may also be necessary.

pediatric orthopaedics

Additional imaging studies (MRI scans, tomograms, bone scans, CT scans) will depend on the suspected etiology. Because these studies are expensive, the need for and sequence of these tests will depend on the differential diagnosis.

Differential diagnosis

Differential diagnosis of pediatric back pain are shown in Table 1. Scheuermann's disease is the most common cause of pain in the thoracic and thoracolumbar regions, while spondylolysis and spondylolisthesis are the most common causes in the lumbar and lumbosacral regions.

Table 1
Differential diagnosis of back pain in children

Congenital
- Diastematomyelia
- Congenital spine anomalies

Developmental
- Painful scoliosis
- Kyphosis (Scheuermann's disease)

Traumatic
- Fractures
- Muscle strain
- Spondylolysis and spondylolisthesis
- Herniated disk
- Left vertebral apophysis
- Upper cervical spine instability

Infections
- Diskitis
- Vertebral osteomyelitis
- Tuberculosis

Systemic diseases
- Chronic infection
- Storage diseases
- Juvenile osteoporosis

Juvenile arthritis
- Rheumatoid arthritis
- Ankylosing spondylitis

Neoplastic (benign and malignant)
- Benign
  - Osteoid osteoma
  - Osteoblastoma
  - Aneurysmal bone cyst
  - Eosinophilic granuloma
- Malignant
  - Osteogenic sarcoma
  - Spinal cord tumor
  - Metastatic
- Psychogenic

Adverse outcomes of the disease

When back pain is due to an organic cause, failure to diagnose may result in progression of the condition. In certain conditions, such as instability of the upper cervical spine, tumors, or infections, this may ultimately result in spinal cord or peripheral nerve injury.

Treatment

Treatment of back pain in children is diagnosis specific. Because of the extensive differential diagnoses, it is not possible to discuss all aspects of treatment here.

If there are no neurologic abnormalities, and the range of motion and exam of the spine are normal, then conservative treatment with activity modification and a mild analgesic is appropriate initially. More extensive studies will be necessary if the pain does not improve within 1 to 2 weeks, if the pain progresses, or new symptoms and findings appear.

Adverse outcomes of treatment

Adverse outcomes of treatment are also diagnosis specific. The major adverse outcome occurs when conservative treatment continues and there is no clinical improvement or the symptoms become worse. This indicates failure to recognize an underlying organic cause for the pain.

Referral decisions/Red flags

The following factors suggest a serious underlying etiology as the cause of back pain: 1) persistent or increasing pain; 2) pain accompanied by systemic symptoms such as fever, malaise, or weight loss; 3) neurologic symptoms or findings; 4) bowel or bladder dysfunction; 5) onset of symptoms at a young age, especially 4 years of age or younger (tumors should be suspected); and 6) a painful left thoracic scoliosis. If these findings occur, further evaluation is necessary.

pediatric orthopaedics

Calcaneal Apophysitis

Definition

Calcaneal apophysitis is characterized by pain in the posterior aspect of the heel after play and sports activities in prepubertal children. Sports that require cleated, hard-soled shoes are especially likely to cause trouble. This condition is due to overuse and repetitive microtrauma from pull on the calcaneal apophysis by the Achilles tendon. This weak link is obliterated when the apophysis fuses to the main body of the calcaneus, a process that occurs by age 9 years in girls and age 11 years in boys.

Clinical symptoms

Patients have posterior heel pain and a limp.

Tests

Exam
Examination reveals tenderness, especially with compression, at the junction of the Achilles tendon and the calcaneus.

Diagnostic
With bilateral involvement and a typical history, radiographs are probably not necessary. With unilateral involvement, obtain a lateral radiograph of the heel to rule out unicameral bone cyst or tumors. Radiographs are not diagnostic. Sclerosis at the secondary ossification center is normal (Figure 1).

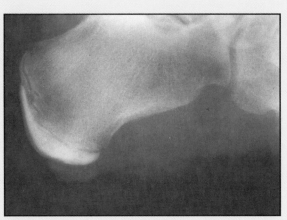

Figure 1
Lateral radiograph of the heel in a child (note sclerosis in the secondary ossification center of the calcaneus is normal)

Differential diagnosis

Achilles tendinitis (can be associated with Reiter's syndrome or other seronegative spondyloarthropathies)

Infection (unilateral, elevated sedimentation rate)

Tumor (unilateral)

Adverse outcomes of the disease

Pain, a limp, and activity modifications are possible, but there are no long-term sequelae.

Treatment

Treatment includes short-term modification or restriction of the precipitating activity. Shoe modifications using a 1/4″ heel lift or heel cushion and Achilles tendon stretching may be helpful. Casting is rarely needed, but may be used for 2 to 4 weeks if the pain and limp are severe. Neither surgery nor cortisone injection is indicated.

Adverse outcomes of treatment

There are no long-term sequelae.

Referral decisions/Red flags

Suspicion of tumor indicates the need for further evaluation.

Acknowledgements

Figure 1 is reproduced with permission from Wilkins KE: The painful foot in the child, in Bassett FH (ed): *Instructional Course Lectures 38.* Park Ridge, IL, American Academy of Orthopaedic Surgeons, 1988, pp 77–85.

pediatric orthopaedics

Cavus Foot Deformity

Synonyms

High-arched foot

Pes cavus

Definition

Cavus foot deformity is characterized by a high-arched foot that takes one of two forms: cavovarus (inverted calcaneus with tightness of the heel cord) or calcaneo-cavus (high arch with normal heel alignment, usually from weakness of the calf muscles resulting in increased ankle dorsiflexion and increased plantar flexion of the forefoot) (Figure 1). There is often a family history, and this condition may be associated with progressive neurologic diseases.

Clinical symptoms

Idiopathic cavus foot typically follows a static course. These patients often prefer western boots or shoes with elevated heels, as these relieve stress on the tight heel cord and plantar fascia.

Cavus deformities secondary to neurologic disorders are usually progressive and commonly lead to a stumbling or falling pattern of gait. Patients experience frequent ankle sprains and marked restriction in activity.

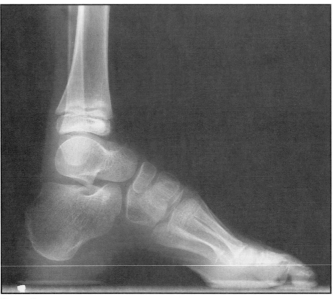

Figure 1
Standing lateral radiograph showing the exaggerated arch in a cavus foot

Tests

Exam
The deformity is present both with and without weightbearing. With the patient standing, look for asymmetrical muscle bulk in the calf, and with walking, look for foot drop. Test the heel cord for tightness (inability to dorsiflex beyond neutral). Assess muscle strength and sensation of the foot, and examine the lower spine as well.

Diagnostic
Obtain weightbearing AP and lateral radiographs of the foot and an AP radiograph of the ankle. AP and lateral radiographs of the spine should also be obtained to rule out spinal dysraphism. An MRI scan of the spine may be necessary.

Electromyography helps differentiate myopathy and neuropathy and confirms neurologic involvement. Nerve conduction studies are important to rule out Charcot-Marie-Tooth disease.

Muscle biopsies are sometimes required.

Differential diagnosis

Charcot-Marie-Tooth, hypertrophic form (in teenage years)

Charcot-Marie-Tooth, neuronal form (in late teens or 20s; "stork leg")

Club foot residual

Déjerine-Sottas disease (in infancy, childhood)

Diastematomyelia (bony bar through spinal cord; symptoms may progress with growth)

Friedreich's ataxia

Poliomyelitis

Spinal cord tumor

Spinal dysraphism (myelomeningocele, lipomyelomeningocele)

Traumatic tendon laceration

Adverse outcomes of the disease

Adverse outcomes of the disease include the following: progressive neurologic disease with increasing weakness and difficulty in walking; plantar callosities; claw toes; and abnormal shoe wear. Plantar ulcers associated with an insensate foot are also possible.

Treatment

Use drop-foot bracing as needed. Most patients need surgery to improve alignment, which ultimately improves function and prevents plantar ulceration. Some patients require stabilization of the hindfoot or ankle, and others require tendon transfers to augment weakened muscle groups.

Adverse outcomes of treatment

Postoperative infection and failure of fusion are both possible.

Referral decisions/Red flags

Because patients with cavus foot deformity often have insensate feet, further evaluation upon recognition is appropriate, especially if there are plantar callosities.

Acknowledgements

Figure 1 is reproduced with permission from Davidson RS: Miscellaneous foot disorders, in Richards BS (ed): *Orthopaedic Knowledge Update: Pediatrics.* Rosemont, IL, American Academy of Orthopaedic Surgeons, 1996, pp 219–225.

Child Abuse

Definition

Approximately 2.4 million reports of child abuse are filed annually in the United States, and 4,000 of these children die every year as a result of this abuse or neglect. Firstborn children, premature infants, stepchildren, and handicapped children are at greater risk for child abuse than other children, and most cases of abuse involve children younger than 3 years of age. Failure to recognize injuries due to child abuse results in a child being returned to an abusive environment where there is a 25% risk of serious reinjury and a 5% risk of death.

Fractures are seen in almost 50% of child abuse cases, and soft-tissue injuries such as bruising, burns, or scars are seen in most patients. Toddlers typically have "normal" bruises over the chin, the brow, elbows, knees, and shins, but bruises on the back of the head, the neck, buttocks, abdomen, legs, arms, cheeks, or genitalia are suspicious for abuse. The critical issue in situations of suspected child abuse is whether the history given by the family adequately explains the child's injuries.

The investigative interview

The most important guiding principle in conducting and documenting an interview in a situation of suspected abuse is to remain objective, as there is a strong likelihood that the medical record will be used later as evidence in court. Other guidelines include the following:

- interview individual family members in private;
- carefully document the given history of injury *verbatim,* as well as its source;
- establish a scenario for the injury from each witness, noting carefully any inconsistencies;
- note any delay in seeking medical attention for injuries; and
- ensure that you are attentive, nonjudgmental, and that you avoid leading questions during the history.

If family members later alter the story about how the injury occurred, carefully record this change as an addendum in the record, noting the date in which the revision occurred.

Obtain a social history as well to identify unusual stresses on the family, such as a recent loss of a job, separation or divorce, death in the family, housing problems, or inadequate funds for food. Alcohol abuse in the home is a risk factor for child abuse, and maternal cocaine use increases the risk of abuse fivefold.

Tests

Exam

The first step in the examination is to check for any acute extremity fractures, then carefully conduct a head-to-toe examination to evaluate any suspicious soft-tissue injuries. The head-to-toe examination is important because in half of the confirmed abuse cases, there is evidence of prior abuse. Obtain a CT scan of the abdomen if the head-to-toe examination shows abdominal tenderness or if results of liver function tests are elevated.

If sexual abuse is suspected, look for physical signs of sexual assault such as bruising and chafing of the genitalia, as the physical examination of the genitalia is often normal. If the child's mental status is abnormal, consult neurology and ophthalmology to evaluate for subdural hematoma and retinal hemorrhage secondary to violent shaking. Check bleeding studies when there is bruising, and order a toxicology screening if there is a history of substance abuse in the family.

Diagnostic

While there is no predominant fracture pattern seen in child abuse, fractures of some bones seem to be more suspicious for child abuse than others. Fractures considered highly specific for child abuse include posterior rib fractures, scapula fractures, posterior process fractures of the spine, and fractures of the sternum. One type of fracture unique to child abuse is the "corner" or "chip" fracture of the metaphysis (Figure 1). With this type of fracture, the edge of the metaphysis is avulsed from the epiphysis of the long bone due to downward traction or pull on the extremity. Transverse or oblique fractures due to child abuse are usually caused by direct trauma or violent pulling on the extremities, while spiral fractures are caused by rotational injury. Rib fractures are also common and when healed may appear only as fusiform thickening of the ribs. A bone scan may be helpful in detecting rib fractures, but it may not show skull fractures or long bone fractures near the epiphyseal growth plates.

Fractures considered to be of moderate specificity for abuse include multiple fractures, especially bilateral ones, fractures of different ages, epiphyseal separations, vertebral body fractures, fractures of the fingers, and complex skull fractures. Multiple fractures at various stages of healing without explanation strongly suggest a history of child abuse (Figure 2). The age of fractures can be estimated by their appearance. Ten to 14 days after the injury, new periosteal bone can be seen; by 14 to 21 days after the initial injury, there is loss of definition of the

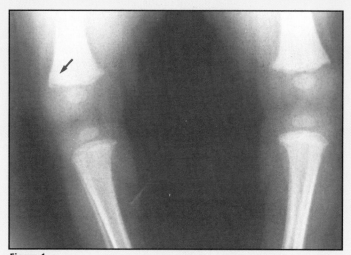

Figure 1
AP radiograph of the distal femur with a "corner" or "chip" fracture

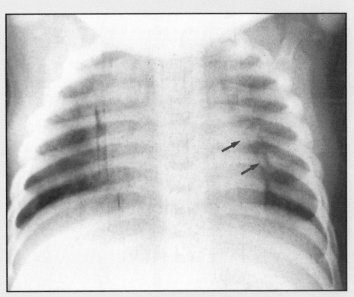

Figure 2
Rib fractures at various stages of healing

fracture line. Periosteal bone scar, called soft callus, also appears 14 to 21 days after fracture. More dense callus is then seen 21 to 42 days after injury. Fractures older than 6 weeks tend to reshape to a more normal configuration through a process called remodeling. These older fractures are distinguished by subtle fusiform sclerotic thickening, which is best seen if compared to a normal contralateral bone.

Fractures of low specificity for abuse include clavicle fractures and linear skull fractures. Note, though, that a vague history of injury or other risk factors for child abuse outweigh fracture specificity in determining child abuse.

Therefore, it is important to order a skeletal survey to check for additional fractures when child abuse is suspected, including both AP and lateral views of all long bones, the hands, feet, spine, and the chest, as well as a skull series. Do not order a "babygram," as the detail in these radiographs is poor and may miss subtle fractures.

Differential diagnosis

It is important to avoid overdiagnosing child abuse. Spontaneous fractures can occur in diseases such as osteogenesis imperfecta. The presence of osteopenia, a family history of osteogenesis imperfecta, as well as the presence of blue sclera would suggest this disease. Spontaneous pathologic fractures are also seen in osteomyelitis, tumor, rickets, neuromuscular disease, and other metabolic diseases.

The final diagnosis of child abuse

Once pathologic fractures are excluded, determine whether the child's fracture is due to either accidental trauma or to inflicted trauma. A diagnosis of accidental trauma can only be made when an acute injury is brought promptly to medical attention, has a plausible mechanism of injury, and lacks other risk factors for child abuse. Suspicious injuries must be reported as child abuse.

Reporting child abuse

In the United States, any physician who reports suspected child abuse in good faith is protected from both civil and criminal liability, but failure to report suspected abuse exposes the physician to liability. Sexual abuse is a criminal offense and must always be reported. In addition to making notes in the medical record, the physician may be asked by child protective services to complete a notarized affidavit summarizing findings in the abuse case and stating that the patient may be at risk for injury or loss of life if returned to the home environment. The child may then be placed in a foster home until an investigation is completed. The physician must be prepared to defend his or her findings in custodial hearings if the family challenges the actions of child protective services.

Acknowledgements

Figures 1 and 2 are reproduced with permission from Thompson JD: Child abuse, in Richards BS (ed): *Orthopaedic Knowledge Update: Pediatrics.* Rosemont, IL, American Academy of Orthopaedic Surgeons, 1996, pp 285–287.

Clubfoot

Synonyms

Congenital clubfoot

Talipes equinovarus

Talipes equinovarus congenita

Definition

Clubfoot is a congenital foot deformity characterized by four distinct components: plantar flexion (equinus) of the ankle, adduction (varus) of the heel (hindfoot), high arch (cavus) at the midfoot, and adduction of the forefoot (Figure 1). While there are many theories about the causes of clubfoot, none have been proven. Most cases are idiopathic and occur in an otherwise normal infant. Neuromuscular clubfoot is also a congenital deformity associated with disorders such as myelomeningocele, arthrogryposis, and congenital constriction band syndrome. Only idiopathic clubfoot is discussed here.

The incidence of clubfoot in males is twice that of females. If a family has one child with a clubfoot, the risk in subsequent siblings is 3% to 4%. If one parent and one child in a family have clubfoot, subsequent children have a 25% chance of having clubfoot.

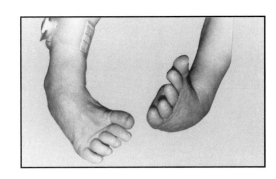

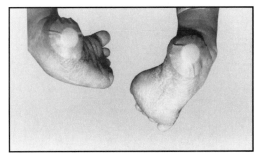

Figure 1
Newborn with bilateral congenital clubfoot

Clinical symptoms

Typically, infants with clubfoot appear as if they could walk on the top or dorsolateral aspect of the foot. In most instances, all four components of a clubfoot are present to some degree, but different aspects of the deformity are more striking in some patients. Plantar flexion (equinus) is usually most severe and is characterized by a drawn up position of the heel and the inability to pull the calcaneus down on attempts at dorsiflexion of the foot. The high arch (cavus) may be difficult to see, but in patients with a severe clubfoot, it is indicated by a transverse crease across the sole of the foot. Occasionally, forefoot adduction may be relatively severe while the other components appear less severe.

Tests

Exam

Examination should rule out neuromuscular disorders and determine whether muscle function and sensation are intact. In rare instances, congenital absence of the anterior compartment muscles will occur in association with rigid clubfoot. Absence of muscle function and altered sensation may indicate a spinal cord disorder such as lipomyelomeningocele.

A true idiopathic clubfoot is not fully correctable by passive manipulation. A foot that can be placed in a normal position by manipulation is not considered a clubfoot. Rather, this condition is talipes equinovarus due to intrauterine molding and typically resolves without therapy.

Diagnostic

Radiographs are not usually necessary to confirm the diagnosis or to begin treatment. However, they should be obtained if the diagnosis is unclear, or if conservative treatment is failing. Standardized stress views provide the most information about the rigidity of deformity and differential diagnosis.

Differential diagnosis

Arthrogryposis

Congenital constriction band syndrome

Myelomeningocele

Adverse outcomes of the disease

Untreated, children with clubfoot will have a severe disability. Not only are walking and shoe wear severely impaired, but there is the psychological trauma of living with what many perceive as a grotesque deformity.

Even with successful treatment, the affected foot will be smaller and less mobile than a normal foot. The difference is not as obvious with bilateral involvement, but is readily apparent with unilateral involvement. The shoe size of the affected foot is typically 1 to $1^1/_2$ sizes smaller than the normal foot. In addition, the calf muscles are smaller in the affected leg and present a cosmetic problem for which there is no good solution.

Treatment

Manipulation and casting should be started immediately upon diagnosis; with passage of time, the untreated deformity only becomes more resistant to conservative treatment. Two to 4 months of manipulation and casting are required to correct a clubfoot. The success rate of casting is dependent on many variables, including the rigidity of the deformity and the experience and skill of the physician. Correction by manipulation usually requires prolonged splinting to minimize recurrence of deformity. Recurrence after casting is most common within the first 2 to 3 years of life, but may happen up to age 5 to 7 years.

Surgery is required when conservative treatment fails, usually after 3 to 4 months of treatment. Initial surgery can range from heel cord lengthening to a complete posterior, medial, and lateral release of the foot. The purpose of surgery is to lengthen or release contracted tendons and ligaments so that the bones can be positioned in normal alignment. Casting is done after surgery to allow healing of the tendon and remodeling of the tarsal bones. Although less common, recurrence may also occur after surgical treatment.

Treatment is successful if the foot is in a good weightbearing position and the child can run and play without pain.

Adverse outcomes of treatment

Recurrent deformity is more likely to occur if the treatment program did not completely correct the deformity.

Overcorrection may occur after surgical release and results in a severe flatfoot with lateral translation of the heel. Flatfoot may also result from casting if the hindfoot varus is not fully corrected before the foot is manipulated into dorsiflexion.

Referral decisions/Red flags

Immediate diagnosis and referral in the first week of life provides the best chance for successful correction by casting and manipulation. Neurologic abnormalities or stiff joints suggest an underlying disorder associated with the clubfoot deformity.

pediatric orthopaedics

Congenital Absence of the Fibula

Synonyms

Fibula hemimelia

Definition

Fibula hemimelia is characterized by congenital partial or complete deficiency of the fibula (Figure 1). The fibula is the most commonly absent long bone. Usually, there is no family history. Leg length discrepancy, ranging from mild to severe, is universal. Frequently, there are associated defects, including absence of lateral ray(s) of the foot, anomalous bony fusions of the tarsal bones, hypoplasia of the lateral femoral condyle, and mild shortening of the femur.

Clinical symptoms

The deformity is readily apparent.

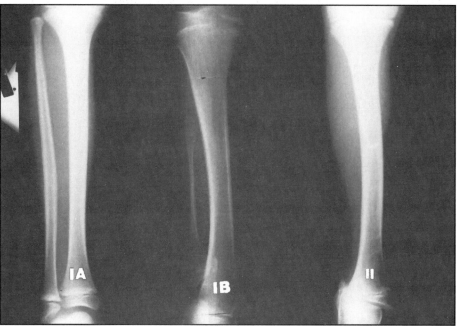

Figure 1
Classification of fibular hypoplasia from minimal shortening of the fibula with a ball and socket ankle (IA) to complete absence of the fibula (II)

pediatric orthopaedics

Tests

Exam
The hindfoot is frequently in valgus, and inversion and eversion movement of the hindfoot is often restricted. An equinus contracture of the ankle may be present. Assess for leg length discrepancy and lateral ray deficiencies.

Diagnostic
Obtain full-length radiographs of the lower extremities.

Differential diagnosis

Congenital vertical talus

Flexible calcaneovalgus

Adverse outcomes of the disease

Leg length discrepancy, ankle instability, severe gait disturbances, and possible need for amputation are all possible.

Treatment

Surgical treatment should be kept at a minimum. The goal of treatment is function, not a cosmetic radiograph. The leg length discrepancy will increase as the child grows and frequently will be in the range of 8 to 16 cm when the child reaches maturity. This degree of anticipated discrepancy, particularly when associated with abnormal hindfoot function, is best managed by ankle disarticulation and prosthetic fitting when the patient is between ages 1 and 2 years. This amputation level is quite functional and allows the child to run and play in a relatively normal manner. Milder deformities can be managed by other modalities.

Adverse outcomes of treatment

Failure of reconstructive surgery to stabilize the ankle, along with infection, painful scars, or neuromas, is possible.

pediatric orthopaedics

Referral decisions/Red flags

Although amputation is usually not done until a patient is ready to walk, it is helpful for a surgeon to evaluate the patient at an early age. This gives the treating surgeon an opportunity to gain the parent's trust, as the concept of amputation is not easy to accept when the amount of discrepancy may not seem severe in a small infant. Support and counseling of the parents by the primary care physician may be critical.

Acknowledgements

Figure 1 is reproduced with permission from Kasser JR (ed): *Orthopaedic Knowledge Update 5.* Rosemont, IL, American Academy of Orthopaedic Surgeons, 1996, pp 437–451.

Congenital Dislocation of the Knee

Synonyms

None

Definition

Congenital hyperextension of the knee ranges from a mild positional deformity that readily responds to short-term splinting of the knee in flexion to frank dislocation of the tibia on the femur, which is complicated and difficult to treat. Fortunately, the latter condition is rare. Approximately 80% of patients who have true subluxation or dislocation will have other orthopaedic abnormalities, such as a dislocated hip or clubfoot (Figure 1). This condition often occurs in association with breech deliveries. In patients with dislocations of the knee, the pathology includes fibrotic quadriceps, flattened, rather than rounded joint surfaces, absent or attenuated anterior cruciate ligaments, and displaced hamstring tendons.

Clinical symptoms

Patients will have obvious deformity and loss of motion.

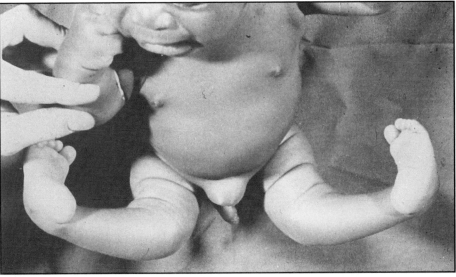

Figure 1
Newborn with a transverse anterior knee crease, posterior prominence of the femoral condyles, and associated talipes equinovarus

Tests

Exam
Test the compliance of the fibrotic quadriceps tendon by flexing the knee gently to its maximum extent. Some knees will reduce and are stable in flexion.

Diagnostic
Obtain AP and lateral radiographs of the knee in flexion and extension.

Differential diagnosis

None

Adverse outcomes of the disease

Limited knee motion, gait disturbance, and premature arthritis are all possible.

Treatment

Knees that easily flex should be splinted in flexion for 2 to 3 weeks. An aluminum finger splint is adequate for these patients. More rigid deformities require serial casting in the first few months of life. Open reduction is indicated if initial treatment fails or if the diagnosis has been delayed.

Adverse outcomes of treatment

Patients may experience redislocation and stiffness, or have a weak quadriceps muscle following treatment. Failure to recognize associated hip dislocation is also a possibility.

Referral decisions/Red flags

These patients should be sent for further evaluation immediately upon diagnosis.

Acknowledgements

Figure 1 is reproduced with permission from Drennan JC: Congenital dislocation of the knee and patella, in Heckman JD (ed): *Instructional Course Lectures 42*. Rosemont, IL, American Academy of Orthopaedic Surgeons, 1993, pp 517–524.

Congenital Limb Deletions of the Upper Extremity

Synonyms

Thalidomide infant

Congenital constriction band amputation (Streeter dysplasia)

Congenital amputations

Aplasia and hypoplasia

Hemimelia

Phocomelia

Definition

Children may be born missing all or part of an upper extremity. In most instances, the cause is unknown. With a terminal defect, the limb is absent distal to a particular level. The most common terminal transverse lesion is a congenital below-elbow amputation. The most common terminal partial deletion is radial hemimelia with absence of the thumb. With an intercalary lesion, the distal part is intact and an intermediate segment is missing. Phocomelia is a severe intercalary defect where the hand is attached to the shoulder.

Clinical symptoms

There are no symptoms other than the obvious deformity.

Tests

Exam
The deformity is obvious.

Diagnostic
No radiographs are needed except when used with prosthetic design or fitting.

Differential diagnosis

Note that specific anomalities affecting other organ systems are often associated with radial hemimelia or congenital deficiency of the thumb. Some anomalies, such as VATER syndrome, and chromosomal and craniofacial abnormalities, are obvious. Others, such as Holt-Oram syndrome (atrial septa defect and radial hemimelia), become apparent on complete examination; some, such as Diamond-Blackfan anemia, Fanconi anemia, and thrombocytopenia, may not become apparent for several months. The possibility of these associated hematopoietic disorders should be evaluated before any surgery is considered.

Adverse outcomes of the disease

Impaired function is the principal finding.

Treatment

When any congenital anomaly is discovered, the presence of other congenital anomalies should be investigated. To understand what has happened, families do better with genetic counseling. They also need to understand that function can often be improved by prosthetic fitting or reconstructive surgery. During adolescence, the patient also may need counseling. An understanding and supportive family and physician will minimize psychological difficulties.

Adverse outcomes of treatment

These vary due to the multiple types of congenital limb deletions.

Referral decisions/Red flags

Children with congenital deletions of the upper extremity may need to be evaluated at specialized facilities. Fitting of upper extremity prosthetics is commonly started between ages 4 and 6 months.

Congenital Muscular Torticollis

Synonyms

None

Definition

Congenital muscular torticollis is a unilateral contracture of the sternocleidomastoid muscle. Scar tissue in the muscle impedes growth of the sternocleidomastoid muscle. As a result, the involved muscle develops a contracture that causes the head to tilt toward the affected side and rotate toward the unaffected side (Figure 1). Facial asymmetry also develops.

Clinical symptoms

Parental concern about the head posture is the usual reason for the visit. At age 4 to 6 weeks, the parents may also note a lump or swelling in the muscle. If present, this mass disappears in a few weeks.

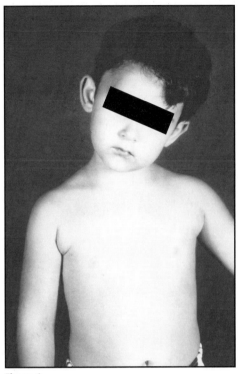

Figure 1
Torticollis secondary to sternocleidomastoid contracture. Note lateral tilt to the left and rotation to the right

Tests

Exam
The head is tilted in a "cock robin" position. Contracture of the right sternocleidomastoid muscle causes limitation of lateral tilt to the left and rotation to the right. Flattening of the face is noted on the involved side.

Exam should exclude ocular or neurogenic disorders. Check the hips to rule out an associated developmental hip dysplasia.

Diagnostic
Before surgery, radiographs of the cervical spine should be obtained to rule out underlying congenital bony anomalies.

pediatric orthopaedics

Differential diagnosis

Cervical cord neoplasm

Congenital anomalies of the base of the skull

Klippel-Feil syndrome (congenital cervical spine anomalies)

Ocular disorders

Posterior fossa tumor

Rotary subluxation of C1 and C2 vertebrae

Sprengel's deformity (congenital elevation of the scapula)

Adverse outcomes of the disease

Facial asymmetry and persistent limitation of motion are possible.

Treatment

Conservative treatment consists of frequent stretching exercises that tilt and rotate the head. Supervision by a therapist is helpful. If the problem persists after 12 to 18 months, surgical treatment should be considered.

Adverse outcomes of treatment

Failure of stretching programs and persistence of deformity are possible. The contracture may recur after surgery.

Referral decisions/Red flags

Failure to improve following 2 to 3 months of conservative treatment, the presence of anomalies of the skull or cervical spine, or abnormal results of neurologic or eye examination all indicate the need for further evaluation.

Congenital Radial-Ulnar Synostosis

Synonyms

"One bone" forearm

Definition

Congenital radial-ulnar synostosis is characterized by a failure of development. The proximal ends of the radius and ulna fail to separate, resulting in an inability to pronate and supinate the forearm.

Clinical symptoms

If the condition is bilateral, the child may have to substitute shoulder motion to place the hand in a pronated or supinated position. When the condition is unilateral, there are few symptoms, since the child can compensate well with the normal opposite extremity.

Tests

Exam
The limitation of pronation and supination of the forearm can be readily detected, even during examination of the newborn.

Diagnostic
After ossification occurs, radiographs demonstrate bony union of the proximal radius and ulna (Figure 1).

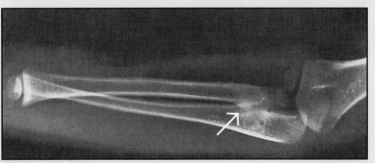

Figure 1
Radiograph showing cross synostosis of the proximal radius and ulna

Differential diagnosis

Congenital dislocation of the radial head

Adverse outcomes of the disease

Loss of pronation and supination of the elbow are characteristic.

Treatment

No treatment is needed if the condition is unilateral, and the forearm is in a satisfactory position. Surgery to divide the synostosis and restore motion has not been successful. Surgical treatment is reserved for patients with disability secondary to a forearm positioned in either extreme pronation or supination. Aligning the forearm in a neutral position improves function. This procedure is usually not done until the child is at least age 4 years and may be delayed to the adolescent or adult years if disability is questionable.

Adverse outcomes of treatment

Compartment syndrome and infection may develop with surgical treatment.

Referral decisions/Red flags

Extreme supination or pronation, or bilateral synostoses with malposition indicate the need for further evaluation.

Acknowledgements

Figure 1 is reproduced with permission from Bayne LG, Costas BL: Malformations of the upper limb, in Morrissy RT (ed): *Lovell and Winter's Pediatric Orthopaedics,* ed 3. Philadelphia, PA, JB Lippincott, 1990, vol 2, pp 574–575.

Developmental Dysplasia of the Hip

Definition

Synonyms

Congenital dysplasia
of the hip

Congenital dislocation
of the hip

Congenital subluxation
of the hip

In a neurologically normal child, developmental dysplasia of the hip (DDH), or insta-bility of the hip, is usually present at birth, but it may also occur later. DDH is associ-ated with ligamentous laxity, almost always by age 18 to 24 months. It most commonly affects the left hip (3:1 ratio), and most often affects girls, infants born in breech presentation, Caucasians of northern European ancestry, Native Americans, or families with a history of DDH.

Clinical symptoms

At birth, children are asymptomatic. Hip instability prior to walking is usually not noticed by the parents; however, as the child gets older, apparent shortening of the leg or altered walking may be the first sign.

Tests

Exam

At birth, the hip can usually be reduced by flexion and abduction. Therefore, provocative reduction diagnostic maneuvers are critical in the newborn exam. In older children, subsequent contracture of the muscles make these tests impossible. While the exact age at which the provocative reduction diagnostic maneuvers can no longer be performed depends on various factors, such as the degree of ligamentous laxity, positive responses to provacative diagnostic maneuvers included in the new-born exam are generally impossible to elicit by age 1 to 3 months.

From birth to age 3 months, two provocative diagnostic maneuvers are helpful in diagnosing hip instability: the Barlow test and the Ortolani maneuver. The Barlow test detects hips that are dislo-catable, but in a resting position, are reduced. The Barlow (dislocation) test should be performed first. For this test, it is important that the child is fully undressed, warm, and comfortable. With the long finger over the greater trochanter and the thumb in the region of the lesser trochanter, flex the hips to 90°. With the hip in flexion and abduction (Figure 1), proceed to bring the hip to the mid-line with gentle posterior pressure. If the hip is dislocatable, the femoral head will fall out of the acetabulum or socket.

The Ortolani maneuver is useful for newborns with hips that are dislocated in a resting position. With this maneuver, you reduce the hip by abducting the hip and pushing the femoral head anteri-orly, with your fingers over the greater trochanter (Figure 2). For the Ortolani (relocation) test, flex the hips to 90°. Beginning with the hips in neutral, apply pressure with the long finger against the

pediatric orthopaedics

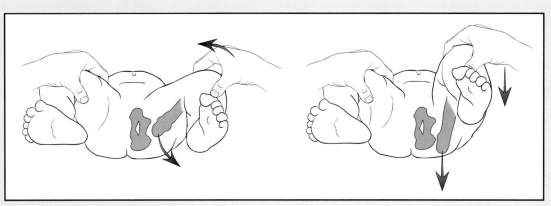

Figure 1
Barlow test

greater trochanter as you move the affected hip into abduction. With this maneuver, the femoral head will relocate with a "clunk" into the acetabulum.

In an older child with unilateral DDH, asymmetrical hip abduction and apparent shortening of the thigh (both with hips in flexion) are the positive signs. Because hip instability is not always present or apparent at birth, it is appropriate to perform these diagnostic maneuvers on "well baby" exams up to age 18 months.

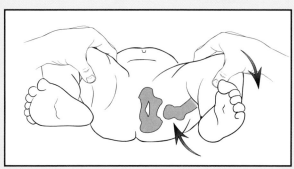

Figure 2
Ortolani maneuver

In an older child with bilateral DDH, the diagnosis is more difficult. The thigh lengths are equal, and hip abduction is symmetrical. Because of the ligamentous laxity, hip abduction is not as restricted as one might anticipate. In one study of children with bilateral DDH diagnosed after birth, hip abduction averaged 45°. Most children younger than age 2 years have 70° of hip abduction with the legs held in a flexed position.

Diagnostic
Radiographs and ultrasound are not needed when the clinical exam is positive, but these tests may be helpful if the exam is equivocal. Ultrasound is a good adjunct to the clinical exam in an infant whose femoral head has not started to ossify.

Differential diagnosis

Fracture of the upper femur

Proximal femoral focal deficiency

pediatric orthopaedics

Adverse outcomes of the disease

Prolonged dislocation of the hip makes closed reduction less likely. Persistent dislocation results in secondary osteoarthritis.

Treatment

The goal of treatment is to contain the femoral head within the confines of the acetabulum. This stimulates growth and deepening of the acetabulum and development of joint stability. An abduction brace such as a Pavlik harness is usually sufficient for newborns and small infants. The use of a brace should be continued until joint stability and acetabular development have occurred.

Adverse outcomes of treatment

Osteonecrosis of the femoral head and persistent instability are possible.

Referral decisions/Red flags

Demonstrated instability or dislocation indicate the need for further evaluation.

Acknowledgements

Figures 1 and 2 are reproduced with permission from Wenger DR: Developmental dysplasia of the hip, in Wenger DR, Rang M (eds): *The Art and Practice of Children's Orthopaedics.* New York, NY, Raven Press, 1993, pp 256–296.

pediatric orthopaedics

Diskitis

pediatric orthopaedics

Synonyms

None

Definition

Diskitis is an infection occurring in or around the intervertebral disk space. MRI studies have documented that diskitis is associated with inflammation of one or both adjacent vertebrae (osteomyelitis) (Figure 1). The cause is thought to be a bacterial infection of hematogenous origin, but aspiration of the disk retrieves an infecting organism less than 50% of the time. Diskitis can occur anywhere in the spine, but most commonly occurs in the low thoracic and lumbar regions and tends to affect children from toddler age to young adolescence.

Clinical symptoms

Children who are able to communicate may be able to localize the pain to the back, but they may also perceive their pain as abdominal discomfort. Therefore, pyelonephritis, mesenteric adenitis, appendicitis, and inflammatory bowel disorders are often suspected. Toddlers are frequently first seen when they refuse to walk.

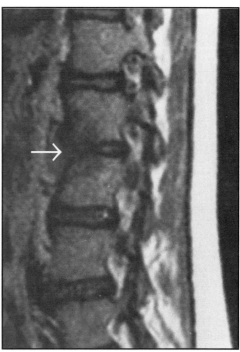

Figure 1
T1 MRI showing the characteristic signal changes of diskitis at L2–3

Tests

Exam
Children rarely appear to be systemically ill, and fever, if present, may only be low grade. Toddlers may refuse to walk and/or even sit unsupported. Those who walk often lean forward and place their hands on their thighs for support (the psoas sign). Percussion of the spinous processes with the child prone may help localize the pain. The straight leg raising test may be positive.

Diagnostic
WBC is often within normal limits, but the erythrocyte sedimentation rate and C-reactive protein are usually elevated. Blood cultures are positive in about 50% of patients with proven cases of infection.

Irregularity of the vertebral end plate and narrowing of the disk may not be apparent on plain radiographs until 2 or 3 weeks after onset of symptoms, but AP and lateral radiographs of the spine should be obtained as part of the initial assessment. The lateral view is most helpful.

A bone scan is usually, but not always, positive early in the disease process. For atypical cases, an MRI scan will confirm the diagnosis.

Due to the low yield and morbidity of the procedure, aspiration of the disk is usually not done. However, studies have demonstrated that *Staphylococcus aureus* is the most common organism when cultures are positive.

A PPD should be performed, unless done recently.

Differential diagnosis

Epidural abscess

Herniated disk (in teenagers)

Pyelonephritis

Retrocecal appendicitis

Scheuermann's disease (erythrocyte sedimentation rate normal, no fever)

Septic arthritis of the hip

Spinal tuberculosis (spares the disk, involves bone, usually thoracolumbar junction area)

Spine tumor

Spondylolisthesis (in teenagers)

Adverse outcomes of the disease

There are minimal residual problems unless chronic osteomyelitis occurs. There may be a spontaneous fusion of the vertebrae adjacent to the infected disk space. In young children, this could potentially produce a kyphosis if the posterior elements continue to grow while the anterior elements are fused.

Treatment

With better understanding of the pathophysiology of diskitis, antibiotics (usually a penicillinase-resistant antibiotic or a cephalosporin) are used more routinely. A common method is to administer the antibiotic for 2 to 4 days IV and then continue oral antibiotics for 4 to 6 weeks.

Immobilization in a body cast is used less frequently, but may be quite successful in reducing severe pain. Surgical debridement is almost never necessary.

Adverse outcomes of treatment

Antibiotic allergies or secondary gastrointestinal problems can develop.

Referral decisions/Red flags

Persistent fever, vertebral osteomyelitis and/or collapse, diagnostic dilemma, presence of a psoas abscess, and any neurologic deficit or suspicion of epidural abscess indicate the need for further evaluation.

Acknowledgements

Figure 1 reproduced with permission from Kasser JR (ed): *Orthopaedic Knowledge Update 5.* Rosemont, IL, American Academy of Orthopaedic Surgeons, 1996, pp 643–655.

pediatric orthopaedics

Evaluation of the Limping Child

Definition

Many conditions cause a child to limp, and making the correct diagnosis can be a challenge. The differential diagnosis is extensive and includes various disease categories and anatomic sites, ranging from the spine to the foot (Table 1). In addition, the history is often vague, the complaints nonfocal, and the examination underwhelming despite diagnostic possibilities that carry significant implications.

History—Key questions

1. *Is the problem acute or chronic?*

Recent onset makes infectious or traumatic conditions more likely. A traumatic etiology, however, is not as common as you might expect. Young children often fall and parents often attribute one of these episodes as the cause of the limp. Therefore, it is important to ask how long the parents have noticed the limp rather than what caused the child to limp. Acute problems, unless an infection or other serious process is suspected, may be observed; chronic problems need further evaluation with laboratory and radiographic studies if the diagnosis is not clear on physical examination.

2. *What time of the day is the limp worst?*

Most musculoskeletal conditions are exacerbated by activity and relieved by rest. Therefore, parents often report that the pain or limp is worst in the afternoon. However, in some conditions, such as transient synovitis of the hip or juvenile rheumatoid arthritis, the limp is more pronounced in the morning. Pain at night suggests leukemia or other neoplasms.

3. *Are there systemic symptoms?*

Questions about endurance, malaise, swelling, and fever are important to ask, even with an acute onset. With chronic symptoms, also ask about the child's appetite, and bowel and bladder function.

pediatric orthopaedics

Table 1

The principal causes of limping in childhood

	With trauma	With fever or systemic illness	Without trauma, fever, or systemic illness

Bones	Fracture
	Toddler's stress
	Child abuse
	Periostitis
		Osteomyelitis
		Hemoglobinopathy
		   Hand-foot syndrome
		   Sickle cell crises
		   Gaucher crises
		Neuroblastoma
		Leukemia
		Ewing's sarcoma
			Osteonecrosis
			Leg length discrepancy
			Aneurysmal bone cyst
			Chondroblastoma
			Chondromyxoid fibroma
			Fibrous dysplasia
			Langerhans cell histiocytosis
			Unicameral bone cyst
			Osteogenic sarcoma

Joints	Hemarthrosis
	Sprain
	Dislocation
	Seronegative spondylo-
	   arthropathies
	   Ankylosing spondylitis
	   Reiter's disease
	   Psoriatic arthritis
	   Chronic inflammatory
	      bowel disease
		Septic arthritis
		Juvenile rheumatoid
		   arthritis (systemic)
		Systemic lupus
		   erythematosus or
		   other collagen vascular
		   disease
		Hemoglobinopathy
		Neuroblastoma
		Leukemia
			Acute rheumatic fever
			Postinfectious reactive arthritis
			Pigmented villonodular synovitis
			Sarcoidosis
			Reactive arthritis of sickle cell
			   disease

Soft tissue	Tendinitis Bursitis Contusion Laceration Muscle strain Muscle hematoma Nerve injury	Abscess Cellulitis Viral myositis Trichinosis Dermatomyositis/ polymyositis Insect/spider bite	Muscular dystrophy Postinfectious myositis Charcot-Marie-Tooth disease Spinal cord tumor Diastematomyelia Rhabdomyosarcoma Overuse syndrome
Spine and pelvis	Avulsion fracture vertebral body Disk herniation	Ankylosing spondylitis Diskitis Sacroiliac septic arthritis	Spondylolysis/spondylolisthesis Disk herniation
Hip	Slipped capital femoral epiphysis	Septic arthritis	Developmental dysplasia of the hip Transient synovitis Legg-Calvé-Perthes disease Slipped capital femoral epiphysis
Knee	Referred hip pain	Referred hip pain Septic arthritis	Referred hip pain Popliteal cyst Discoid meniscus Torn meniscus Sinding-Larsen-Johansson disease Osgood-Schlatter disease Osteochondritis dissecans Subluxating patella
Foot and ankle		Foreign body	Köhler disease Sever disease Tarsal coalition Accessory tarsal navicular Freiberg's infraction

pediatric orthopaedics

Examination—Pertinent screening tips

Record the patient's temperature. A fever of > 37.5°C (99.5°F) increases the possibility of an inflammatory or neoplastic process. Therefore, order a CBC, erythrocyte sedimentation rate, and appropriate radiographs.

Ask older children to "place one finger on the one spot that hurts the most." This maneuver localizes the anatomic site and obviously greatly narrows the differential diagnosis.

Compare findings on the affected side with those on the contralateral side.

Young children are unable to cooperate for measurement of muscle strength. Muscle atrophy, however, is a sentinel sign of a significant process. Muscle weakness or atrophy in a young child is best determined by measuring and comparing limb girths at symmetric locations. For example, asymmetric thigh girths ≥ 1 cm measured 6 to 8 cm above the patella indicate a significant knee problem that deserves further laboratory and radiographic examinations.

Radiographs and special imaging studies

Radiographs are indicated when a fracture or a chronic process is suspected. AP and lateral views of the suspected area are routine. Different diagnostic possibilities dictate special views. For example, in an older child who presents with a several week history of limping and knee pain after sports activity, a tunnel view, in addition to routine AP and lateral views of the knee, is indicated to rule out osteochondritis dissecans.

A bone scan often localizes the site of disease causing an occult limp that cannot be diagnosed by routine studies. However, this test is not always positive even with a chronic process. For example, leukemia in children may cause increased, decreased, or even normal uptake on bone scans. A CT scan is generally best for benign bony lesions. An MRI scan provides better information on soft-tissue lesions and sarcomas; however, before proceeding to these tests, consult with an appropriate specialist to prevent ordering studies that provide limited information.

Flatfoot

Synonyms

Flexible flatfoot

Valgus foot

Pronated foot

Peroneal spastic flatfoot

Pes planus

Definition

A flatfoot is defined as an abnormally low or absent longitudinal arch and is characterized as either flexible or rigid. Flexible flatfoot is very common and often a variant of normal. Rigid flatfoot is discussed under tarsal coalition.

Flexible flatfoot is the norm in infants, common in children, and within normal limits in adults. Most children will continue to develop an arch until at least age 5 to 6 years; approximately 10% to 20% of adults have flexible flatfeet.

Clinical symptoms

A flexible flatfoot is usually asymptomatic. If pain is present, it usually does not develop until the patient is an adolescent or an adult. Aching occurs along the plantar aspect of the midfoot.

Tests

Exam
Examination reveals loss of the longitudinal arch on weightbearing, often with heel valgus. When the child is not weightbearing or is standing on tiptoes, the arch is recreated. Range of motion is usually normal, but a secondary contracture of the Achilles tendon may develop in older children or adults.

Diagnostic
Radiographs are not necessary in asymptomatic children. Standing lateral radiographs of the foot show a loss of the longitudinal arch, with disruption of the line that normally connects the midaxis of the talus and first metatarsal (Figure 1). On the standing AP view, there may be lateral shift (subluxation) of the navicular on the head of the talus.

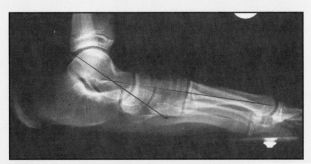

Figure 1
Standing lateral radiograph of flexible flatfoot

Differential diagnosis

None

Adverse outcomes of the disease

Excessive shoe wear (roll over or "breaking down" of the medial counter) and mild pain may be reported.

Treatment

Asymptomatic flexible flatfoot does not require any treatment. Shoe modifications and orthotic devices have not been shown to alter the shape of the foot. For symptomatic patients with a flexible flatfoot, soft longitudinal arch supports and heel cord stretching exercises often relieve discomfort.

Operative realignment is only occasionally needed for adolescents with persistent pain.

Adverse outcomes of treatment

Postoperative infection, failure of fusion, and a medial scar are all possible.

Referral decisions/Red flags

While rare, persistent pain is possible and warrants further evaluation.

Acknowledgements

Figure 1 is reproduced with permission from Kasser JR (ed): *Orthopaedic Knowledge Update 5.* Rosemont, IL, American Academy of Orthopaedic Surgeons, 1996, pp 503–514.

Fracture of the Condyle of the Distal Humerus

Synonyms

Elbow fracture

Definition

Fractures of the condyle are serious because the fracture typically passes through the condyle and physis of the distal humerus (Figure 1). Therefore, the growth plate and articular surface may be mal-aligned. Fracture of the lateral condyle is more common.

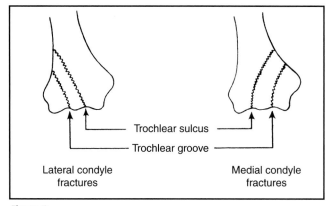

Figure 1
Typical fracture patterns of lateral and medial condyles of distal humerus in children

Clinical symptoms

Patients have obvious pain and do not use the arm. Parents often report a history of a fall.

Tests

Exam
Examination reveals tenderness over the injured condyle of the distal humerus, along with swelling and/or deformity of the elbow.

Diagnostic
This fracture may be missed, particularly in very young children who have limited ossification of the distal humerus. In these children, only a thin wafer of metaphyseal fragment will be apparent with the rest of the fracture crossing the cartilaginous physis and trochlea. In addition to AP and lateral radiographs of the elbow, an oblique view may be helpful. A fat pad sign is usually present.

Differential diagnosis

Capitellum fracture

Monteggia fracture (fracture of the proximal ulna with dislocation of the radial head)

Supracondylar fracture of the distal humerus

Adverse outcomes of the disease

Delayed union, nonunion, malunion (rotational or angular), growth arrest of the distal humerus, and compartment syndrome are all possible.

Treatment

Undisplaced fractures can be treated with cast immobilization. Initial immobilization in a well-made posterior splint, followed by radiographs in 3 to 5 days, is appropriate. If the radiographs show that the alignment has been maintained, immobilization in a long arm cast for another 3 to 5 weeks can follow. Delayed union usually responds to continued immobilization. Active range of motion after casting is indicated, but physical therapy is not necessary.

Fractures displaced 2 mm or more need either closed reduction and pin fixation or open reduction and pin fixation.

Adverse outcomes of treatment

Displacement of the fragment during treatment, malunion, nonunion, and osteonecrosis of the condylar fragment are all possible.

Referral decisions/Red flags

Any displacement or angulation, inability to extend all fingers of the affected hand, failure of over-the-counter medication to provide pain relief, or absent or diminished radial pulse are all indications for further evaluation.

Acknowledgements

Figure 1 is reproduced with permission from Milch H: Fractures and fracture-dislocations of the humeral condyles. *J Trauma* 1963;3:592–607.

pediatric orthopaedics

Fracture of the Distal Radius

Synonyms

Torus fracture

Both-bone fracture of the forearm

Greenstick fracture

Galeazzi's fracture

Impaction fracture

Definition

The distal third of the forearm is the most common location for fractures in children and in one series accounted for 23% of all pediatric fractures. The pattern of injury is age related. Most of these fractures heal without incident and can be treated by closed means. Residual angulation, particularly in the volar or dorsal plane, will typically remodel completely if 1 to 2 years of growth remain.

Clinical symptoms

Patients have acute pain, tenderness, swelling, and deformity, and most often a history of falling on an outstretched arm.

Tests

Exam
Evaluate the status of the median, ulnar, and radial nerves distal to the fracture. Also assess circulation at the fingertips.

Diagnostic
Obtain AP and lateral radiographs of the forearm to include the wrist and elbow joints; comparison views are rarely needed with these injuries.

Differential diagnosis

Osteogenesis imperfecta

Pathologic fracture through a bone cyst

Adverse outcomes of the disease

Malunion, loss of pronation or supination, compartment syndrome (uncommon), and osteomyelitis in open fractures are all possible.

Treatment

In treatment of any open fractures, irrigation and debridement is critical.

Fractures of the distal radius alone, greenstick fractures of the distal radius (with or without involvement of the ulna), or complete fractures of both bones usually affect children between the ages of 6 and 12 years. Closed reduction is indicated for fractures with angulation of greater than 15°, followed by immobilization for 6 to 10 weeks. Even with this treatment, reangulation can occur in up to 35% of patients. Loss of rotational alignment is the only absolute indication for remanipulation.

Physeal fractures of the distal radius usually occur in adolescents and are typically Salter type II fractures (Figure 1). More than 15° of angulation warrants closed reduction and immobilization in a sugar tong splint or long arm cast with the forearm in neutral position.

A torus fracture is buckling of the cortex on one side of the bone, usually the dorsal cortex of the distal radius. Short-term immobilization for 3 to 4 weeks in a volar splint or short arm cast is adequate.

Galeazzi's fracture is a displaced fracture of the distal radius with an injury of the distal radioulnar joint. This requires closed reduction and immobilization for 6 weeks to permit healing of the disrupted joint capsule and fracture.

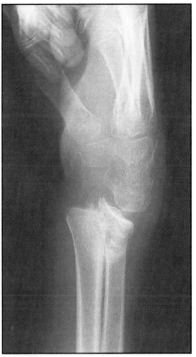

Figure 1
Radiograph of an 11-year-old boy who presented late with a significantly displaced Salter type II distal radial physeal fracture

Adverse outcomes of treatment

Recurrence of deformity, malunion, loss of pronation or supination, and compartment syndrome are all possible. Physeal fractures in children ages 5 to 10 years may be complicated by an arrest of physeal growth and resulting deformity at the wrist.

Referral decisions/Red flags

Inability to extend all fingers, pain not relieved by over-the-counter pain medications, dislocation of the distal radioulnar joint, and/or angulation of greater than 15° indicate the need for further evaluation.

Acknowledgements

Figure 1 is reproduced with permission from Waters PM: Forearm and foot fractures, in Richards BS (ed): *Orthopaedic Knowledge Update: Pediatrics.* Rosemont, IL, American Academy of Orthopaedic Surgeons, 1996, pp 251–257.

pediatric orhtopaedics

Fracture of the Proximal Forearm

Synonyms

Monteggia fracture-dislocation

Both-bone fracture of the forearm

Definition

Fractures of the proximal or midportion of the forearm typically involves disruption of both the radius and the ulna (Figure 1). A Monteggia fracture-dislocation is a dislocation of the radial head associated with a fracture of the ulna. In children, if these injuries are recognized early, then closed reduction is usually successful.

Clinical symptoms

Patients have acute pain, tenderness, and swelling, usually in association with a fall on the outstretched arm.

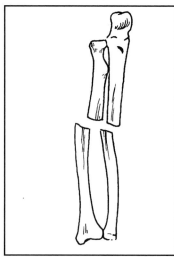

Figure 1
Fracture of the middle third of the forearm

Tests

Exam
Assess median, ulnar, and radial nerve function distal to the fracture site. Inordinate pain on passive extension of the fingers should raise concern for possible compartment syndrome.

Diagnostic
Obtain full-length AP and lateral radiographs of the forearm that include the wrist and elbow. The radial head should line up with the capitellum on both views. In a Monteggia injury, dislocation of the radial head is usually anterior, but may be posterior or lateral.

Differential diagnosis

Osteogenesis imperfecta

Pathologic fracture through a bone cyst

Adverse outcomes of the disease

If recognized early, a Monteggia fracture-dislocation in children can be treated by closed reduction. Failure to diagnose the injury in a timely fashion will necessitate either open reduction, reconstructive surgery, or acceptance of the deformity. Proximal forearm fractures are more likely to cause complications such as malunion, compartment syndrome, or loss of forearm rotation.

Treatment

Irrigation and debridement of open fractures are indicated.

Both-bone forearm fractures angulated more than 15° and all Monteggia fracture-dislocations require closed reduction and long arm casting for 6 to 10 weeks. In adolescents, internal fixation may be required.

Adverse outcomes of treatment

Adverse outcomes include recurrence of deformity, malunion, loss of forearm rotation, and compartment syndrome. Both-bone fractures may develop a synostosis or bony bar connecting the radius and ulna.

Referral decisions/Red flags

Inability to extend all fingers, pain not relieved by over-the-counter pain medications, and angulation greater than 15° all indicate the need for further evaluation.

Acknowledgements

Figure 1 is adapted with permission from Müller ME, Nazarian S, Koch P, et al (eds): *The Comprehensive Classification of Fractures of Long Bones.* Berlin, Germany, Springer-Verlag, 1990, pp 96–105.

Fracture of the Radial Neck

Synonyms

None

Definition

A fracture of the radial head or neck occurs through the physeal plate or neck of the proximal radius. The incidence of this fracture peaks in children between ages 9 and 10 years.

Clinical symptoms

Patients have pain and restricted motion, particularly forearm rotation.

Tests

Exam
Examination reveals tenderness over the radial head that is exacerbated by attempts at pronation and supination. In addition, patients will not be able to extend the elbow and will have an elbow effusion.

Diagnostic
AP and lateral radiographs of the elbow reveal the fracture and allow measurement of angulatory displacement of the radial head (Figure 1).

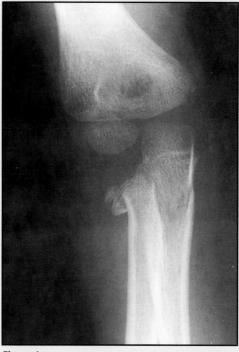

Figure 1
Radial neck fracture (the coexisting fracture of the olecranon is a clue to the valgus mechanism)

Differential diagnosis

Lateral condyle fracture of the distal humerus

Radial head subluxation (pulled elbow)

Adverse outcomes of the disease

Limited pronation or supination, or minor loss of elbow extension are possible.

Osteonecrosis may occur, but because the radial head is small, it usually revascularizes.

Treatment

For fractures with less than 30° of angulation, place the extremity in a long arm cast for 3 weeks with the elbow flexed 90°. Closed reduction is indicated for patients with 30° to 60° of angulation, followed by immobilization for 3 to 4 weeks. Closed reduction is performed with the elbow extended and the forearm supinated. While an assistant applies a varus (adduction) stress, push medially on the proximal lateral portion of the radial head and restore to less than 30° angulation on the AP radiograph. Either closed or open reduction in the operating room is indicated for patients with more than 60° of angulation.

Adverse outcomes of treatment

Failure of reduction and postoperative infection are possible.

Referral decisions/Red flags

Failure of reduction, loss of position, or angulation of greater than 60° indicate the need for further evaluation. Patients who have less than 50° to 60° of pronation or supination following reduction also require additional evaluation.

Acknowledgements

Figure 1 is reproduced with permission from Sponseller PD: Injuries of the humerus and elbow, in Richards BS (ed): *Orthopaedic Knowledge Update: Pediatrics.* Rosemont, IL, American Academy of Orthopaedic Surgeons, 1996, pp 239–250.

pediatric orhtopaedics

Freiberg's Infraction

Synonyms

Osteonecrosis of the metatarsal head

Definition

Freiberg's infraction is osteonecrosis of the head of the metatarsal bone, and typically occurs in adolescence (mostly females) (Figure 1). The etiology is likely related to trauma. The second (longest) metatarsal is most commonly involved, but occasionally, the head of the third metatarsal may be affected.

Clinical symptoms

Patients have pain in the forefoot that is exacerbated by activity.

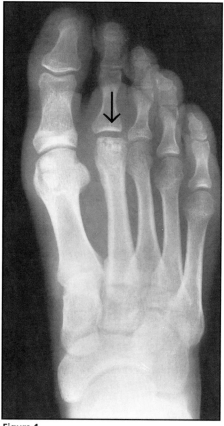

Figure 1
Radiograph showing Freiberg's infraction. Note flattening and fragmentation at the head of the second metatarsal

Tests

Exam
Examination reveals tenderness under the second metatarsal head, sometimes swelling on the dorsal aspect of the metatarsal head, and pain at the extremes of dorsiflexion and plantar flexion.

Diagnostic
Radiographs usually demonstrate the condition within 2 to 3 weeks of onset of symptoms.

pediatric orthopaedics

Differential diagnosis

Infection (elevated erythrocyte sedimentation rate, redness)

Juvenile rheumatoid arthritis

Metatarsal stress fracture (tenderness over the shaft of the metatarsal)

Morton's neuroma

Synovial conditions (of the joint and tendon sheath)

Adverse outcomes of the disease

Deformity of the metatarsal head and adult arthritis are possible. Occasionally, part of the necrotic metatarsal head will separate, creating a loose body that is painful and clicks.

Treatment

Restricted activities and use of a metatarsal pad or bar is indicated, but this treatment may take several months. With persistent symptoms, use a short leg cast for 4 to 6 weeks. Occasionally, surgery is required to remove loose bodies or to realign the metatarsal head.

Adverse outcomes of treatment

Failure of pain relief and transfer lesion with pain under the adjacent third metatarsal head following surgery are possible.

Referral decisions/Red flags

Inability to relieve pain or development of osteoarthritis indicates the need for further evaluation.

Genu Valgum

Synonyms

Knock-knees

Definition

Genu valgum is characterized by alignment of the knee with the tibia abducted (valgus) in relation to the femur. At birth, the child is bowlegged, with a genu varum of 10° to 15°. The bowing gradually straightens to 0° by 12 to 18 months of age. Continued growth results in maximum genu valgum that averages 10° to 15° at age 3 to 4 years. With subsequent growth, the genu valgum decreases, and normal "adult" alignment of 5° to 10° genu valgum occurs by early adolescence. Young children have a fairly wide range of normal knee alignment (two standard deviations is typically ± 10°).

Clinical symptoms

Parental concern is the usual reason for the visit.

Tests

Exam

Measure the child's height and then plot it on a height nomogram. Measure the tibiofemoral angle using a goniometer or by measuring the intermalleolar (IM) distance. The IM distance is the distance between the medial malleoli with the medial femoral condyles touching; this measurement quantifies the genu valgum. The IM distance has the disadvantage of being a relative rather than an absolute measurement that is dependent on leg length.

Diagnostic

Radiographs are not usually necessary, but should be considered when the valgus is more than 15° to 20°, or if the child is short statured.

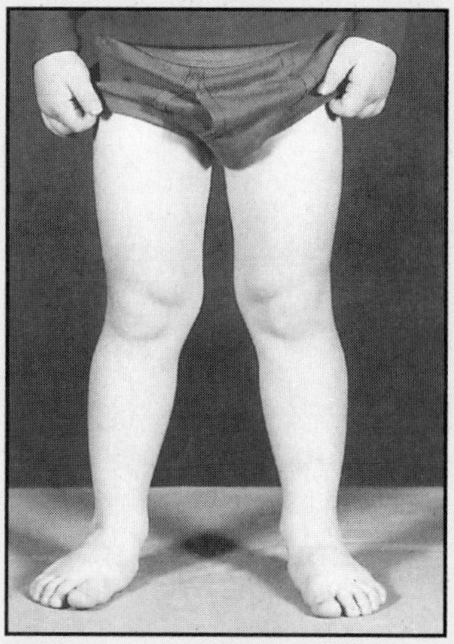

Figure 1
Boy referred for evaluation of "knock-knees"

A standing AP radiograph on a 36- × 43-cm cassette with the film centered at the knees and the feet pointing straight ahead provides screening for a possible skeletal dysplasia, as well as accurate measurement of the tibiofemoral angle (genu valgum). Nonweightbearing lateral radiographs should also be obtained.

The child shown in Figure 1 is a 3-year, 6-month-old boy referred for evaluation of "knock-knees." His height was in the 70th percentile, and development was normal. Genu valgum measured 18° by clinical examination. The patient was considered to be within normal limits and an explanation was provided to the mother. No further follow-up was required.

Differential diagnosis

Hypophosphatemic rickets (short stature, wide growth plate, low serum phosphorus)

Multiple epiphyseal dysplasia (short stature, multiple joint involvement)

Pseudoachondroplasia (short stature, early arthritic change)

Adverse outcomes of the disease

None

Treatment

Observation is the treatment of choice for the otherwise normal 3- to 4-year-old child with marked genu valgum. Shoe modifications are ineffective, and long leg braces are not indicated because this condition spontaneously corrects more than 99% of the time.

Adverse outcomes of treatment

None

Referral decisions/Red flags

Patients of short stature and those with asymmetric or excessive genu valgum need further evaluation (greater than two standard deviations from normal for age).

Acknowledgements

Figure 1 is reproduced with permission from Greene WB: Genu varum and genu valgum in children, in Schafer M (ed): *Instructional Course Lectures 43*. Rosemont, IL, American Academy of Orthopaedic Surgeons, 1994, pp 151–159.

Genu Varum

Synonyms

Bow legs

Definition

Genu varum is an angular deformity at the knee with the tibia adducted (varus) in relation to the femur. At birth, genu varum of 10° to 15° is normal. The bowing gradually straightens to 0° by age 12 to 18 months. Continued growth results in maximum genu valgum of 10° to 15° at age 3 to 4 years. Young children have a fairly wide range of normal knee alignment (two standard deviations is ± 10°).

Clinical symptoms

Parental concern is the usual reason for the visit.

Tests

Exam

Measure the child's height and then plot it on a height nomogram. Measure the tibiofemoral angle using a goniometer or quantify the genu varum by measuring the intercondylar (IC) distance, which is the distance between the medial femoral condyles when the lower extremities are positioned with the medial malleoli touching. This measurement has the disadvantage of being relative rather than absolute since it is affected by the length of the legs. Also assess tibial torsion.

Diagnostic

Radiographs are appropriate if the child is under the 25th percentile for height, the varus is relatively severe for the child's age, there is excessive internal tibial torsion, the varus is increasing after age 16 months, or there is significant asymmetry of the two sides. Obtain standing AP radiographs of the lower extremities on a 36- × 43-cm cassette with the film centered at the knees

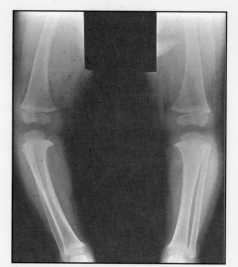

Figure 1
1-year, 6-month-old girl with genu varum

and the feet pointing straight ahead (Figure 1). Assess radiographs for the tibiofemoral angle, the slope of the proximal tibia (medial aspect depressed in infantile tibia vara), and the width of the growth plate (widened in rickets and certain skeletal dysplasia).

The child shown in Figure 1 has a metaphyseal-diaphyseal angle 20° on the right, and the medial aspect of the right proximal tibial physis has an inferior slope. The radiographic picture and abnormal metaphyseal-diaphyseal angle indicate that this patient has a high probability of having infantile tibia vara on the right side. A knee-ankle-foot orthosis was prescribed.

Infantile tibia vara and normal development may be difficult to differentiate in children under age 3 years.

Differential diagnosis

Hypophosphatemic rickets (vitamin D-resistant rickets)

Infantile tibia vara

Metaphyseal chondrodysplasia (short stature)

Physiologic bow legs (spontaneous correction by age 3 years)

Adverse outcomes of the disease

The principal adverse outcome is persistent varus deformity.

Treatment

For young children with infantile tibia vara, bracing may be successful. After age 30 to 36 months, tibial osteotomy may be required for these children.

Adverse outcomes of treatment

The deformity can recur.

Referral decisions/Red flags

The presence of disorders other than physiologic genu varum indicates the need for further evaluation.

Acknowledgements

Figure 1 is reproduced with permission from Greene WB: Infantile tibia vara, in Heckman JD (ed): *Instructional Course Lectures 42.* Rosemont, IL, American Academy of Orthopaedic Surgeons, 1993, pp 525–538.

Intoeing

Synonyms

Pigeon-toed

Femoral torsion

Femoral anteversion

Internal tibial torsion

Definition

Intoeing means that the foot turns in more than expected during walking and running activities. By age 2 years, children typically walk with the foot turned out relative to the line of progression. Intoeing may be secondary to deformities in the foot or may be due to inward rotation of the femur or tibia, or a combination of the two.

Clinical symptoms

In children, intoeing usually does not cause pain or interfere with development or stability in gait. With severe intoeing, children may stumble as they catch their toes on the back of the trailing leg.

Tests

Exam

Observe the child walking and quantify the degree of intoeing by estimating the angle of the foot relative to the line of progression. Normally, the foot is turned out 10° to 20°. Note whether the deviation appears to be in the femur (with the patella pointing inward) or through the tibia or foot (with the patella pointing straight ahead). Also, examine the foot for metatarsus adductus.

Although not an absolute measurement, the easiest way to assess femoral anteversion is to measure hip rotation, and the easiest way to assess tibial torsion is to measure the thigh-foot angle. Both measurements are done with the child in a prone position, the hips fully extended, and the knees flexed 90°. To measure hip rotation, use the lower leg as a pointer and rotate the legs through the axis of the hip joint (Figure 1). The average range for a child older than age 2 years is approximately 50° of internal rotation and 40° of external rotation. An excessive amount of internal rotation (greater than 65°) coupled with a limited degree of external rotation indicates increased femoral anteversion.

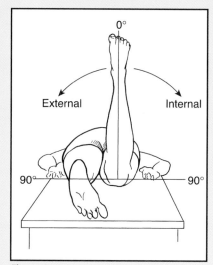

Figure 1
Rotation: Zero starting position

Tibial torsion is most easily assessed by measuring the thigh-foot angle (the axis of the foot relative to the axis of the thigh with the knee flexed to 90° and the hip and foot in a neutral or zero starting point alignment) (Figure 2). A neutral or internal thigh-foot angle indicates internal tibial torsion.

It is important to realize that clinical measurement of hip rotation does not profile the degree of bony femoral torsion until the child is age 1 to 2 years. Due to intrauterine position, a child is born with tightness of the hip joint capsule that creates an external contracture of the hip even though femoral torsion is maximum at birth. The effect of this on measurement of hip rotation does not resolve until the child is at least age 1 year and sometimes even up to age 2 years.

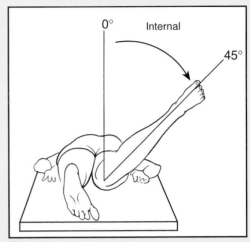

Figure 2
Internal rotation

Differential diagnosis

Mild cerebral palsy

Other neuromuscular disorders may cause intoeing because of dynamic muscle imbalance and associated bony abnormalities.

Adverse outcomes of the disease

If intoeing due to internal tibial torsion or increased femoral anteversion persists, the child may have a cosmetically unpleasant gait pattern and may trip more frequently, but this condition has not been linked to degenerative arthritis in adulthood.

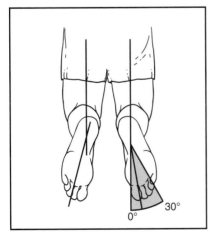

Figure 3
Thigh foot angle: Normal range

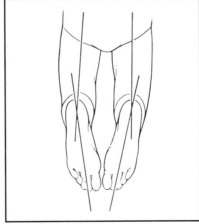

Figure 4
Bilateral internal tibial torsion

The Essentials of Musculoskeletal Care   611

pediatric orthopaedics

Treatment

Tibial torsion—In almost all children with normal neuromuscular function, internal tibial torsion resolves spontaneously by the time the child is 4 years old. Although a variety of splints, special shoes, and exercise programs have been recommended, these are unnecessary because they do not change the natural history of the deformity. Surgical correction is rarely necessary for a child with normal neuromuscular function. Rather, surgery should be reserved for children older than 8 to 10 years who have persistent internal tibial torsion that causes significant gait dysfunction.

Femoral torsion—This is the most common cause of intoeing in children. Slow correction of anteversion to an acceptable degree is typical. An associated increase in external tibial torsion will diminish the degree of intoeing. Shoe modifications, braces, and exercises do not alter developmental changes in femoral or tibial torsion.

In very severe cases in which children have persistent tripping or an unsightly gait, surgical correction by rotational osteotomy of the femur may be indicated. Note that surgery is generally done after age 6 years to allow for spontaneous resolution to occur. If surgery is delayed until adolescence, the patient is likely to have external tibial torsion that must also be corrected.

In children with cerebral palsy, both internal tibial torsion and increased femoral anteversion persist in a much higher percentage of patients and are more likely to require surgical correction due to their impact on gait.

Adverse outcomes of treatment

Risks of casting for metatarsus adductus are minimal. Risks of surgery include failure to heal, infection, over- or undercorrection of the deformity, or postoperative angular deformity.

Referral decisions/Red flags

The primary care practitioner who is familiar with examination for this problem can adequately reassure parents regarding the resolution of these deformities. When metatarsus adductus persists in a child older than age 6 months, or internal tibial torsion or abnormal femoral anteversion persists past age 4 years and the deformity is significant, further evaluation may be warranted if the family is concerned.

Acknowledgements

Figures 1 and 2 are reproduced with permission from Greene WB, Heckman JD (eds): *The Clinical Measurement of Joint Motion.* Rosemont, IL, American Academy of Orthopaedic Surgeons, 1994, pp 106, 107.
Figures 3 and 4 are adapted with permission from Alexander IJ: *The Foot: Examination and Diagnosis,* ed 1. New York, NY, Churchill Livingstone, 1990, p 115.

pediatric orthopaedics

Juvenile Rheumatoid Arthritis

Synonyms

JRA

Still's disease

Definition

The American Rheumatic Association lists four criteria for diagnosing juvenile rheumatoid arthritis (JRA): 1) chronic synovial inflammation of unknown cause; 2) onset in children younger than age 16 years; 3) objective evidence of arthritis in one or more joints for 6 consecutive weeks; and 4) exclusion of other diseases. Pauciarticular JRA involves four or fewer joints after 6 months of symptoms, whereas polyarticular JRA involves five or more joints. Systemic JRA is the subgroup characterized by an illness beginning with high spiking fevers—temperatures above 102.7°F (39.3°C). The number of joints involved is not included in the definition of systemic onset JRA but most patients in this group have several affected joints.

Clinical symptoms

Pain, swelling, and stiffness are less severe in juvenile rheumatoid arthritis than in the adult form.

Pauciarticular JRA is the most common pattern. Age at onset is typically younger than 4 years, and girls are affected four times more often than boys. The parents will comment that the child is irritable, lethargic, or shows a reluctance to play. The disease commonly begins in a single joint, most frequently the knee, ankle, fingers, or wrist.

The polyarticular form is characterized by symmetrical involvement of the knees, wrists, fingers, and ankles and is more common in girls. The onset of the seronegative form typically occurs in children from ages 1 to 3 years, and the seropositive form usually begins in adolescence and is virtually indistinguishable from adult rheumatoid arthritis.

Systemic onset commonly occurs in children between the ages of 4 and 9 years and is associated with polyarthralgias, myalgias, an evanescent maculopapular rash with central clearing, and a high erythrocyte sedimentation rate.

pediatric orthopaedics

Tests

Exam
Evaluate the joints for jointline tenderness, effusion, range of motion, warmth, and adjacent muscle atrophy.

Diagnostic
The erythrocyte sedimentation rate is often normal. A positive ANA test indicates a tendency for uveitis.

Differential diagnosis

Leukemia (night pain)

Osteomyelitis (abnormal radiographs)

Septic arthritis (fever, single joint involvement unless overwhelming sepsis)

Adverse outcomes of the disease

Spontaneous remission is common. Blindness may develop with untreated uveitis.

Treatment

NSAIDs are the mainstay of treatment. Other medications, such as methotrexate, are used if synovitis is persistent. Splinting and orthotics help maintain functional joint alignment. Physical therapy is helpful in maintaining joint motion and strength when the pain has diminished. Regular ophthalmologic slit lamp examinations for uveitis are necessary in the pauciarticular form.

Adverse outcomes of treatment

NSAIDs may cause gastric, renal, or hepatic complications.

Referral decisions/Red flags

Failure of NSAIDs, splinting, and therapy to relieve symptoms indicates the need for further evaluation. Persistent active synovitis is also an indication that the patient needs additional evaluation.

Klippel-Feil Syndrome

Synonyms

None

Definition

Patients with Klippel-Feil syndrome have congenital cervical abnormalities that result from failure of segmentation. The spectrum of involvement ranges from fusion of two vertebrae (block vertebrae) to massive fusion of the entire spine.

Clinical symptoms

Children with this condition are asymptomatic. Many bony anomalies are not discovered until the patient is involved in an automobile accident and radiographs show anomalies. With severe involvement, shortening and webbing of the neck are observed.

Tests

Exam

Most patients with congenital anomalies of the cervical spine appear normal. Patients with multiple bony anomalies have a short neck, low hairline, and limited range of neck motion, especially in sidebending.

Patients may also exhibit scoliosis, congenital deformity of the thumb, heart murmur, mirror motion of the fingers and thumb, diminished hearing acuity, and renal anomalies.

Diagnostic

AP and lateral radiographs of the cervical spine will confirm the presence or absence of congenital anomalies of the cervical spine, and standing PA radiographs of the thoracic and lumbar spine identify an associated scoliosis.

For patients with Klippel-Feil syndrome, rule out renal anomalies with an ultrasound examination. Because these patients often have hearing problems, obtain an audiogram.

Differential diagnosis

Acquired fusion of the cervical spine (due to juvenile rheumatoid arthritis)

Associated visceral anomalies

Neoplasm of the cervical spine

Rotary subluxation of the atlas and axis

pediatric orthopaedics

The Essentials of Musculoskeletal Care   615

Adverse outcomes of the disease

Secondary instability of the cervical spine with spinal cord or nerve root compression may develop, as may disk degeneration and arthritis of the cervical spine. Patients may also complain of chronic pain. Facial asymmetry may be apparent with congenital torticollis.

Treatment

Most patients with congenital anomalies of the cervical spine do not have clinical problems (Figure 1). If disk degeneration and arthritis of the cervical spine develop, most patients will respond to symptomatic treatment. When cervical instability develops, surgical fusion of the unstable segments may be indicated to protect neural elements. Surgical treatment is usually not required. Treatment may be indicated for associated clinical problems such as renal anomalies, cardiac anomalies, hearing deficits, or scoliosis.

Adverse outcomes of treatment

Symptomatic treatment may be unsuccessful in relieving pain, and if the patient undergoes surgery, there may be perioperative complications.

Referral decisions/Red flags

The presence of associated conditions, such as hearing deficits, cardiac anomalies, renal anomalies, and scoliosis indicate the need for further evaluation. Patients with an abnormal neurologic examination or instability of the cervical spine on radiographic examination also need additional evaluation.

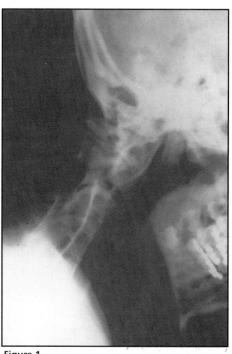

Figure 1
Lateral radiograph of a patient with Klippel-Feil syndrome

Acknowledgements

Figure 1 is reproduced with permission from Pizzutillo PD: Spinal considerations in the young athlete, in Heckman JD (ed): *Instructional Course Lectures 42*. Rosemont, IL, American Academy of Orthopaedic Surgeons, 1993, pp 463–472.

Legg-Calvé-Perthes Disease

Synonyms

Idiopathic osteonecrosis of the femoral head

Avascular necrosis of the femoral head

Aseptic necrosis of the femoral head

Definition

Legg-Calvé-Perthes disease (LCPD) is idiopathic osteonecrosis of the femoral head in children. It typically affects children between the ages of 4 and 8 years (the range is 2 to 12 years). It is four times more common in boys than in girls, is uncommon in African-Americans, and is unilateral in 90% of patients. After the bone dies and loses structural integrity, the articular surface of the femoral head may collapse, leading to deformity and arthritis. The greater the percentage of involvement of the femoral head and the older the child at onset, the worse the prognosis.

Clinical symptoms

Typically, the child has been limping for 2 to 3 weeks, sometimes longer, at the initial visit. Activity worsens the limp, making symptoms more noticeable at the end of the day. If the child complains of pain, it is typically an aching in the groin or proximal thigh.

Tests

Exam

Examination reveals mild to moderate restriction of hip motion. Abduction, in particular, is limited and typically measures 20° to 30° at presentation. The easiest way to recognize mild restriction of abduction is to spread both extremities at the same time. This maneuver prevents the pelvis from tilting. Another technique is to place one hand on the opposite pelvis and use the other hand to abduct the extremity. The degree of abduction is recorded when the pelvis starts to move or tilt.

pediatric orthopaedics

Diagnostic

Obtain AP and frog lateral radiographs of the pelvis. Increased density of the femoral head is an early sign of LCPD. The crescent sign indicates that a shear fracture has occurred in the subchondral bone (Figure 1). If plain radiographs are normal, a technetium bone scan or MRI scan may demonstrate LCPD.

In a child with bilateral involvement, obtain a screening AP radiograph of the hand and knee to rule out epiphyseal dysplasia or thyroid disease.

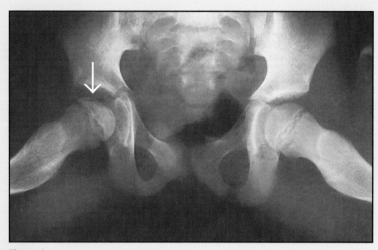

Figure 1

A 7-year-old boy with a 2-month history of right hip pain. The subchondral fracture is through 75% of the femoral head

Differential diagnosis

Atypical septic arthritis (increasing pain and constitutional symptoms)

Gaucher's disease

Hypothyroidism

Multiple epiphyseal dysplasia (bilateral, mild short stature, autosomal dominant)

Sickle cell anemia

Spondyloepiphyseal dysplasia, Stickler syndrome variant (bilateral, short stature)

Transient synovitis (pain more noticeable in the morning)

Adverse outcomes of the disease

Osteoarthritis of the hip can result in severe symptoms at an early age (adolescence to 40s). Other problems include pain, a limp, and the need to modify activities beginning at an early age.

Treatment

Healing of LCPD involves revascularization of the femoral head, removal of necrotic bone, and replacement with live bone that is initially relatively weak (woven bone as contrasted to strong lamellar bone). While this process occurs more rapidly and with great consistency in children, it is still a biologic phenomenon that requires many months. Unfortunately, no current interventions accelerate this process.

Observation may be acceptable for children who are unlikely to develop a femoral head deformity that will progress to early arthritis. Therefore, observation is commonly indicated for children younger than age 6 years who do not exhibit significant subluxation and who maintain at least 40° to 45° of abduction. Older children who have no involvement of the lateral portion of the femoral head may also be observed.

The principal reason for brace treatment is to "contain" the femoral head within the acetabulum, thus maintaining the most spherical (and least flattened) head possible. As a general rule, osteotomy is reserved for older children. Note that none of these treatment modalities completely contains the femoral head during all phases of gait.

Before any treatment is initiated, motion must be regained, and bed rest and traction are commonly required. The abduction brace is generally worn all day and is discontinued when the lateral portion of the femoral head has regenerated, a process that generally takes 12 to 18 months.

Adverse outcomes of treatment

None

Referral decisions/Red flags

None

Acknowledgements

Figure 1 is reproduced with permission from Beaty JH: Legg-Calvé-Perthes disease: Diagnostic and prognostic techniques, in Barr JS (ed): *Instructional Course Lectures 38.* Park Ridge, IL, American Academy of Orthopaedic Surgeons, 1989, pp 291–296.

pediatric orthopaedics

Metatarsus Adductus

Synonyms

Metatarsus varus

Definition

Metatarsus adductus is a common congenital deformity characterized by medial deviation (adduction or varus deformity) of the forefoot (Figure 1). Objective data concerning the incidence of metatarsus adductus are sparse, but in one prospective study, the deformity was observed in 25% of preterm infants (especially twins) and in 13% of term infants. The forefoot deformity often resolves spontaneously, but it may persist.

Clinical symptoms

Symptoms are absent during infancy. Parental concern typically initiates evaluation.

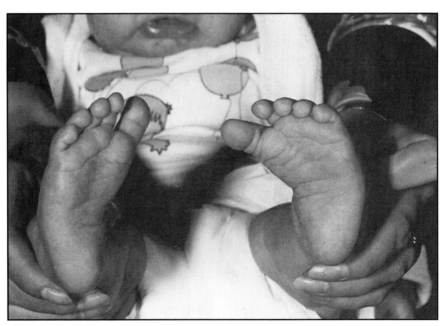

Figure 1
Three-month-old infant with obvious metatarsus adductus. The foot appears to be supinated; however, when the foot was placed in a weightbearing position, the hindfoot alignment was normal

pediatric orthopaedics

Tests

Exam

The most striking feature is the convex lateral border of the foot with a palpable prominence at the base of the fifth metatarsal. The hindfoot is in neutral or increased valgus and never demonstrates a varus posture. Normal ankle dorsiflexion is also present. These latter two findings are important, because some children with severe metatarsus adductus may, at first inspection, appear to have a clubfoot deformity.

Diagnostic

Making serial photocopies of the foot is a low-cost, no-risk method of charting the progression of metatarsus adductus (Figure 2). Although somewhat subjective, it is useful to rate the severity of metatarsus adductus by clinical criteria.

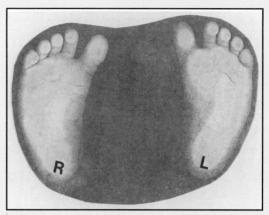

Figure 2
Photocopy of feet in a 9-month-old infant with bilateral metatarsus adductus

A commonly used system is based on the heel bisector line (Figure 3). Normally, a line bisecting the heel crosses the forefoot between the second and third toes. Metatarsus adductus is considered mild when the heel bisector crosses the third toe, moderate when the heel bisector goes between the third and fourth toes, and severe when the heel bisector crosses between the fourth and fifth toes. Flexibility of the forefoot should also be assessed, and perhaps the simplest criteria is to define a flexible foot as one in which the second toe can easily be brought in line with or past the heel bisector.

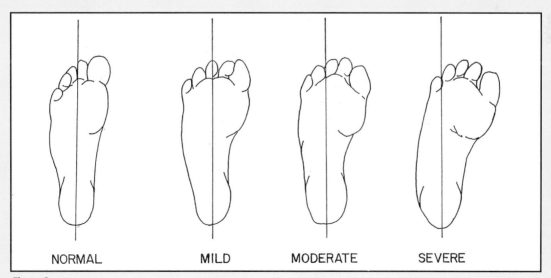

NORMAL   MILD   MODERATE   SEVERE

Figure 3
Classification of metatarsus adductus as described by Bleck

pediatric orthopaedics

Differential diagnosis

Cavus foot (high arch, heel in varus)

Clubfoot (heel in varus, ankle in equinus)

Hyperactive abductor hallucis ("monkey toe"; lateral border of foot straight)

Internal tibial torsion (foot rotated in; lateral border of foot straight)

Skeletal dysplasia (most common in diastrophic dwarfism)

Skewfoot (Z or serpentine foot; valgus hindfoot and midfoot, severe forefoot adduction)

Adverse outcomes of the disease

Abnormal shoe wear and subsequent discomfort with prolonged standing are possible. Some children are emotionally stressed when teased about their foot alignment.

Treatment

Most newborns are not treated, although parents are advised to avoid putting the infant in the knee-chest position, which accentuates metatarsus adductus. A supine sleeping position is better and is also important for preventing sudden infant death syndrome (SIDS). Because metatarsus adductus corrects spontaneously in most children, the decision to begin treatment may be delayed until the child is 6 months old; however, it is reasonable to initiate treatment earlier if the deformity is severe and inflexible.

There is no evidence that stretching exercises done by the parents, the use of nighttime splints, or outflare shoes is effective. The limited studies of these modalities have included younger patients whose metatarsus adductus may have resolved spontaneously.

Serial casting is the gold standard of conservative treatment, and if started at age 6 to 9 months, has a high degree of success. Casts are applied for 2-week periods, and usually three or four casts will suffice. Casting may even be successful in children ages 1 to 3 years, but at this age, casting is certainly more difficult and is associated with a higher rate of recurrent deformity. The cast should be applied in a manner that molds the forefoot into abduction without accentuating heel valgus. A long leg cast is probably more effective, but this concept has not been proven.

Surgical treatment is limited.

Adverse outcomes of treatment

Recurrent deformity after serial casting may suggest a skewfoot deformity. Exacerbating heel valgus while casting may result in development of a severe flatfoot or skewfoot deformity. Degenerative changes at the tarsometatarsal joints have been noted after complete surgical tarsometatarsal capsulotomy. Damage to the physis of the first metatarsal and subsequent shortening of that bone is the unique complication of correcting metatarsus adductus by metatarsal osteotomies.

pediatric orthopaedics

Referral decisions/Red flags

Rigid metatarsus adductus that is not responsive to treatment, especially in children older than age 3 months, signals the need for further evaluation.

Acknowledgements

Figures 1 and 2 are reproduced with permission from Greene WB: Metatarsus adductus and skewfoot, in Schafer M (ed): *Instructional Course Lectures 43*. Rosemont, IL, American Academy of Orthopaedic Surgeons, 1994, pp 161–177.

Figure 3 is reproduced with permission from Bleck EE: Metatarsus adductus: Classification and relationship to all kinds of treatment. *J Pediatr Orthop* 1983;3:2–9.

pediatric orthopaedics

Osgood-Schlatter Disease
Sinding-Larsen-Johansson Disease

Synonyms

Osteochondritis of the tibial tuberosity

Osteochondritis of the inferior patella

Tibial tubercle traction apophysitis

Definition

Osgood-Schlatter disease or condition results from repetitive injury and small avulsion injuries at the bone-tendon junction where the patellar tendon inserts into the secondary ossification center of the tibial tuberosity. The onset of the disease during early adolescence coincides with development of this secondary ossification center, which is a weak link to repetitive quadriceps contraction. A traumatic or overuse etiology also explains the fivefold greater incidence in adolescents who are active in sports and the two to three times greater incidence in males. Sinding-Larsen-Johansson disease is a similar disorder that occurs at the junction of the patellar tendon and the distal pole of the patella.

Clinical symptoms

Patients report pain that is exacerbated by running, jumping, and kneeling activities. Pain may also occur after prolonged sitting with the knees flexed. Most patients continue all activities.

Tests

Exam

Examination reveals tenderness and swelling at the insertion of the patellar tendon into the tibial tubercle in Osgood-Schlatter disease and at the inferior pole of the patella in Sinding-Larsen-Johansson disease. These conditions are often bilateral, but one side may be more symptomatic than the other. Although knee motion is usually not restricted, kneeling is painful during the acute phase. The patellofemoral joint is stable.

Diagnostic

When a patient has bilateral symptoms, radiographs are rarely needed. Radiographs typically show soft-tissue swelling. In Osgood-Schlatter disease, small spicules of heterotopic ossification may be seen anterior to the tibial tuberosity (Figure 1). In Sinding-Larsen-Johansson disease, there may be elongation of the inferior pole of the patella with "fragmentation" of the bone in that area.

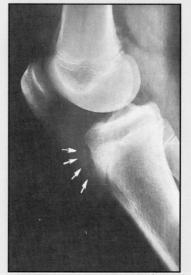

Figure 1
Symptomatic tibial tubercle ossicle in a 13-year-old boy

Differential diagnosis

Infection (rare, elevated erythrocyte sedimentation rate)

Neoplasm (rare, unilateral)

Adverse outcomes of the disease

The long-term prognosis for patients who undergo conservative treatment is good, with minimal adult disability. Some residual prominence of the tibial tubercle is common, particularly in patients who have fragmentation of the epiphysis and heterotopic ossification during the active phase of the disease. In one long-term study, 76% of patients reported no limitation of activity, 18% had tenderness that was similar to but not as painful as the original condition, and the remaining 6% had anterior knee pain but no symptoms referable to the tibial tuberosity. On directed questioning, 60% of patients reported discomfort when kneeling, but this was not typically perceived as a disability.

Treatment

Symptoms are often controlled adequately with intermittent use of ice after sports, coupled with occasional use of aspirin, and use of a protective knee pad. Decreased activity to permit healing of the microscopic avulsion fractures is the key to treating patients with severe symptoms.

Occasionally, immobilization is needed for severe or recalcitrant symptoms. This usually can be accomplished with a prefabricated knee immobilizer that is removed once a day for bathing and range of motion exercises.

Another important aspect of the management process is helping the parents and the patient understand the anticipated duration of restricted activity. This is especially true if the child is a competitive athlete, in which case patients with Osgood-Schlatter disease may need to stop training for an average of 2 to 3 months. The disease may interfere with fully effective training for an average of 6 to 7 months.

Surgical treatment is not commonly needed, and with rare exceptions should be restricted to patients who have persistent symptoms after growth is completed. At that time, simple excision of heterotopic ossification in the patellar tendon will usually relieve discomfort and aching after running activities.

Adverse outcomes of treatment

These are the same as the adverse outcomes of the disease.

Referral decisions/Red flags

Unilateral pain at rest or pain not directly over the tibial tubercle should raise concerns of neoplastic or another process.

Acknowledgements

Figure 1 is reproduced with permission from Stanitski CL: Knee overuse disorders in the pediatric and adolescent athlete, in Heckman JD (ed): *Instructional Course Lectures 42*. Rosemont, IL, American Academy of Orthopaedic Surgeons, 1993, pp 483–495.

pediatric orthopaedics

Osteomyelitis

Synonyms

None

Definition

Osteomyelitis is an infection in bone, usually bacterial in origin, and of acute, sub-acute, or chronic duration. In children, the bacterial inoculation is usually by hematogenous spread, but may be secondary to direct contamination (open fracture, nail puncture wound). Acute osteomyelitis presents within 2 weeks of disease onset. Subacute osteomyelitis is a more balanced response between host and organism that results in a quasi-contained lesion in the bone. Children with subacute osteomyelitis typically present after 1 to several months. Chronic osteomyelitis is also typically of 1 to several months duration and is characterized by sequestrum of necrotic bone harboring bacteria.

The metaphysis of long bones is the most common location of osteomyelitis in children. This is due to the sluggish circulation at the metaphyseal-physeal barrier where the vessels are required to make a "U" turn.

Clinical symptoms

Pain, swelling, tenderness, erythema, increased localized warmth, and generalized malaise can all be associated with osteomyelitis. When the lower extremity, pelvis, or spine is involved, refusal to walk or a limp is an early sign, particularly in a young child. Pseudoparalysis (failure to use a limb despite normal neuromuscular structures) is observed in the upper extremity, particularly in small children.

Tests

Exam

Fever with a temperature above 100.4°F (38°C) is typical, but not always present. Tenderness is noted in the involved region. Movement of the adjacent joint is limited, but not to the degree observed in septic arthritis. Limping or refusal to walk is seen with osteomyelitis in the lower extremity, pelvis, or spine. Also, look for other foci or infection.

Diagnostic

Obtain AP and lateral radiographs of the suspected area. Early in the disease process, the radiographs are normal or will only show soft-tissue swelling. Osseous changes appear 7 to 10 days after onset of symptoms and include periosteal elevation and/or destruction of bone with radiolucency, fading cortical margins, and no surrounding reactive bone.

pediatric orthopaedics

Although not often needed, consider an MRI or bone scan in unclear situations. The MRI scan is more definitive, but more expensive. Ultrasound may show an area of periosteal elevation and provide direction for aspiration that will confirm infection and provide a specimen for Gram stain and culture.

Blood studies should include a CBC with differential, erythrocyte sedimentation rate, and C-reactive protein, as these confirm an inflammatory process and are parameters in monitoring response. Neonates and children who are seen early in the disease process may not have an abnormal WBC, elevated erythrocyte sedimentation rate, or significant fever. A blood culture should also be obtained as it will often (in 40% to 50% of patients) identify the infecting organism.

Subacute osteomyelitis has an indolent onset of several weeks duration. Diagnosis is often delayed because symptoms are often vague, and other findings such as fever and abnormal laboratory studies are not remarkable. A bone scan in this instance may help pinpoint the site of infection. Two categories of subacute osteomyelitis are observed. Cavitary osteomyelitis occurs in the epiphysis or metaphysis and is characterized by a small, localized lucency surrounded by reactive, dense-appearing bone (Brodie abscess). In cavitary osteomyelitis, the diagnosis is usually clear; the cavity is usually filled with granulation tissue, and cultures are often negative. Cavitary subacute osteomyelitis may be treated by a trial of oral antibiotics.

The second category of subacute osteomyelitis simulates a neoplastic process and, therefore, mandates a biopsy with or without debridement. These lesions are typically in the diaphysis but may also be metaphyseal. Periosteal elevation and cortical thickening are the early radiographic signs.

Differential diagnosis

Acute leukemia

Acute rheumatic fever

Acute rheumatoid arthritis

Cellulitis

Malignant bone tumors, in particular, Ewing's sarcoma and osteosarcoma

Septic arthritis

Adverse outcomes of the disease

Adverse outcomes include growth disturbance and leg length discrepancy, destruction of adjacent joints, pathologic fracture, and massive bone defects leading to limb dysfunction. Chronic draining of the sinuses that persists into adulthood may result in squamous cell carcinoma.

Treatment

Immobilization of the affected part, hydration, fever control, and IV antibiotics guided by the culture and sensitivity, when available, are indicated. Change to oral antibiotics after improvement of clinical and appropriate laboratory parameters.

Surgical drainage is needed when the child is seen late in the process (approximately 4 to 7 days after onset), or if symptoms and fever persist despite antibiotic therapy. Surgery for acute osteomyelitis consists primarily of drainage of the subperiosteal abscess, but in chronic osteomyelitis also includes resection of necrotic and avascular tissue.

Staphylococcus aureus is the most common pathogen for patients of all ages, but group B streptococcus and enteric rod organisms should also be covered in neonates. In children ages 6 months to 4 years, *Haemophilus influenzae* is also covered, although the vaccination will eventually make that recommendation obsolete.

The typical course of antibiotics is 6 weeks, but it may need to be longer, depending on clinical response and laboratory values.

Immobilization should continue for 3 weeks or more to assist the healing process and protect against pathologic fracture if osseous destruction has weakened the bone.

Osteomyelitis secondary to nail puncture wounds is usually caused by *Pseudomonas aeruginosa.* This infection requires surgical debridement and antibiotic therapy for 7 to 10 days. Failure to debride this infection adequately can lead to a chronic infection that is difficult to eradicate.

Adverse outcomes of treatment

Recurrence of infection with progression to chronic osteomyelitis and/or inadequate drainage may occur.

Referral decisions/Red flags

Even when surgical drainage is not necessary, consultation is indicated for assistance with splinting or casting to prevent pathologic fracture. Early consultation aids in subsequent decisions concerning the failure to respond to antibiotic therapy.

Osteonecrosis of the Tarsal Navicular

Synonyms

Köhler disease

Osteonecrosis of the tarsal navicular

Definition

Idiopathic osteonecrosis of the tarsal navicular affects children ages 4 to 8 years. It is more common in boys.

Clinical symptoms

Patients have pain in the medial arch with weightbearing and a limp with an antalgic gait (short stance time on the painful leg).

Tests

Exam

Examination reveals tenderness over the medial arch at the tarsal navicular.

Diagnostic

Radiographs are diagnostic and show a dense, fragmented, thin navicular (Figure 1). Radiographs may be normal initially, but should be repeated after 6 weeks if patients remain symptomatic. The navicular bone appears dense, fragmented, and thin.

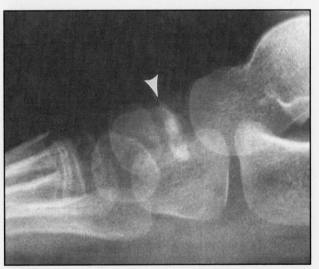

Figure 1
Lateral radiograph of the foot showing a shattered, fragmented navicular

Differential diagnosis

Rigid flatfoot

Stress fracture of the navicular

Tumor

Adverse outcomes of the disease

None

Treatment

Activity modifications or casting is indicated. A short leg walking cast for 4 to 8 weeks relieves pain, provides better walking (less pain), and may speed resolution of the osteonecrosis. However, the eventual outcome is good whether casting or activity modification is chosen.

Adverse outcomes of treatment

Cast sores may develop.

Referral decisions/Red flags

Persistent pain or limp that is not responsive to activity modifications indicates the need for further evaluation.

Acknowledgements

Figure 1 is reproduced with permission from Kasser JR (ed): *Orthopaedic Knowledge Update 5*. Rosemont, IL, American Academy of Orthopaedic Surgeons, 1996, pp 503–514.

Outtoeing

Synonyms

External rotation contracture of infancy

Femoral retroversion

External tibial torsion

External femoral torsion

Definition

Toeing out is normal in adults and older children. This is obvious when one observes footprints in the snow or sand and sees the foot turned out at an angle of 10° to 20° relative to the line of progression. Excessive outtoeing, although not as frequent a concern as intoeing, represents a variation from normal development. The excessive rotation may occur in the femur, the tibia, or both. Understanding normal development of femoral and tibial torsion, as well as the changes that occur in hip rotation, is critical.

Clinical symptoms

Parents or grandparents often notice excessive inward or outward toeing in infants, and they must be reassured and advised about the natural history of rotational variations in the femur and tibia. Inward femoral rotation (anteversion) is greatest at birth (approximately 40°) and gradually declines to adult values of 10° to 15° by age 8 years. After the age of 18 to 24 months, measurement of hip rotation is a good approximation of the degree of femoral rotation, averaging 40° to 45° in each direction. Children with outtoeing due to femoral retroversion have external rotation in the range of 60° to 80°, while internal rotation is limited to 10° to 30°.

The tibia normally twists outward between the knee and ankle. Tibial torsion is normally 0° at birth, gradually increasing to 20° or 30° of external rotation by age 3 to 4 years, with the left side slower to rotate than the right.

Tests

Exam

The physical examination should assess five factors, as follows: 1) the angle that the foot makes relative to the line of progression when walking (foot progression angle); 2) the degree of femoral rotation; 3) the degree of tibial torsion (thigh-foot angle); 4) the posture of the foot; and 5) the possibility of neuromuscular disorders. Neuromuscular disorders may present as problems of outtoeing and intoeing. Therefore, the physical examination should include evaluation for spasticity, muscle contractures, clonus, and a stiffness of gait. Appropriate motor development milestones are listed in Table 1.

Assessment of the foot progression angle does not identify the site of the problem; however, it does indicate the severity of the problem.

Table 1
Gross motor developmental screens during early childhood

Age	Activity
2 years	Stairs, one step at a time
2.5 years	Jumps
3 years	Stairs, alternating feet
4 years	Hops on one foot
5 years	Skips

Femoral rotation (hip rotation) should be measured with the hip joint in extension, which is the position of walking. Placing the child in a prone position is the easiest way to measure hip rotation, and it also facilitates measurement of the thigh-foot angle (Figures 1 and 2). Excessive external rotation of the hip (femoral retroversion) is accompanied by reduced internal rotation.

The degree of tibial torsion is estimated by the thigh-foot angle. This is most easily measured with the child prone, the knee flexed to 90°, and the hip and foot in a neutral or zero starting point alignment. The measured angle is the axis of the foot relative to the axis of the thigh. A thigh-foot angle of 40° or more indicates excessive external tibial torsion (Figures 3 and 4).

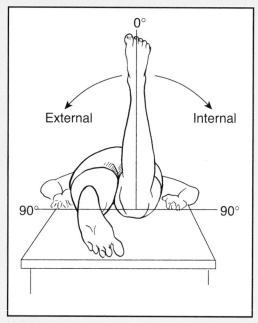

Figure 1
Starting point for measuring hip rotation with the hip extended (child prone)

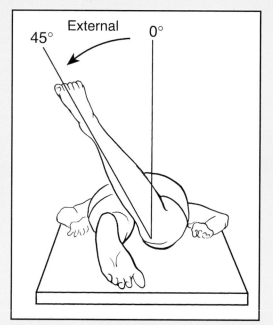

Figure 2
Measurement of hip external rotation

Foot posture is assessed for flatfoot, which increases outward posturing of the limb, or metatarsus adductus, which turns the foot inward.

Diagnostic
Radiographs are rarely needed unless the child has short stature.

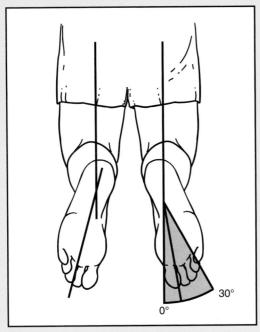

Figure 3
Thigh-foot angle: normal range

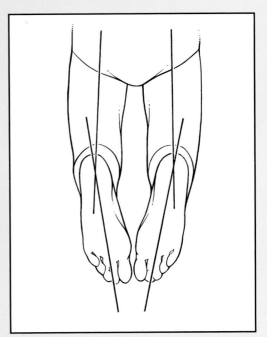

Figure 4
Bilateral internal tibial torsion

Differential diagnosis

Cerebral palsy (usually an internally rotated gait)

Meningomyelocele (externally rotated limb)

Adverse outcomes of the disease

Outtoeing is a benign disorder, often improving with age and rarely requiring treatment.

Treatment

Outtoeing or external rotation before the child walks is a common posture, especially in males and African-Americans, and especially in the right leg. In the absence of neuromuscular disease, observation and reevaluation when the child walks are the appropriate measures.

As the child walks, the parents may need reassurance that the condition is changing; charting the external rotation of the femur and the thigh-foot angle will help as the years pass. Nonsurgical treatment is not helpful, and only rarely does outtoeing persist to a degree that rotational osteotomy is required.

Excessive external tibial torsion may develop as a compensatory mechanism for persistent inward rotation of the femur (femoral anteversion). In the rare situation in which an adolescent has patellofemoral problems from the "malignant malalignment" syndrome (severe femoral anteversion and severe external tibial torsion), a rotational osteotomy of both the femur and tibia is required to place the patella straight ahead during walking.

Adverse outcomes of treatment

None

Referral decisions/Red flags

Further evaluation is not really needed, unless parental reassurance becomes an issue.

Acknowledgements

Figures 1 and 2 are reproduced with permission from Greene WB, Heckman JD (eds): *The Clinical Measurement of Joint Motion.* Rosemont, IL, American Academy of Orthopaedic Surgeons, 1994, pp 106, 108.

Figures 3 and 4 are adapted with permission from Alexander IJ: *The Foot: Examination and Diagnosis,* ed 1. New York, NY, Churchill Livingstone, 1990, p 115.

Table 1 is adapted from Paine RS, Oppe TE: Neurological examination of children, in *Clinics in Developmental Medicine,* Nos. 20/21. London, England, William Heinemann Medical Books.

Preparticipation Athletic Evaluation

Definition

Although it is not intended to substitute for a routine annual visit, the preparticipation evaluation is often the only contact a teenager or older child has with a physician. Therefore, this encounter should be as effective as possible.

Evaluation goals

Goals of this evaluation include the following:

- identify conditions that would predispose participants to serious injury or death, or worsening of current medical or psychological conditions;
- diagnose previously undetected conditions;
- assess general health and risk-taking behaviors;
- satisfy school, state, and insurance requirements; and
- assess fitness level and performance parameters (optional).

Optimal timing of this examination is 6 weeks before the beginning of the season. This schedule allows adequate time for further consultation, diagnostic testing, or rehabilitation of identified deficiencies. To supplement the information provided at the evaluation, the parents and the child (when appropriate) should provide as much medical history as possible.

Types of evaluation

Three types of exams are typically used: examination by a personal physician, the locker room examination, or the station method. Evaluation conducted by the child's personal physician is obviously preferred, but too often is not feasible. With the "locker room" examination, a large number of children parade past a single examiner, resulting in an impersonal and inadequate evaluation. With the "station" method, children move through several stations for different portions of the examination. One advantage of this approach is that physicians often volunteer their time, allowing for a large number of children to be examined in a short time at relatively little cost. Specialists such as cardiologists and orthopaedic surgeons are often available to expedite consultations, and coaches and athletic trainers are usually involved with this type of evaluation, which helps maintain order and facilitates communication. Disadvantages include the lack of a physician-patient relationship and difficulty in communicating with the parents.

The preparticipation physical evaluation form can be used with any of the methods of evaluation (Figure 1).

Preparticipation Physical Evaluation

HISTORY

DATE OF EXAM _____

Name_____ Sex_____ Age_____ Date of birth_____

Grade____ School_____ Sport(s)_____

Address_____ Phone_____

Personal physician_____

In case of emergency, contact

Name _____ Relationship _____ Phone (H) _____ (W) _____

Explain "Yes" answers below.
Circle questions you don't know the answers to.

Yes No

1. Have you had a medical illness or injury since your last check up or sports physical? ☐ ☐
   Do you have an ongoing or chronic illness? ☐ ☐
2. Have you ever been hospitalized overnight? ☐ ☐
   Have you ever had surgery? ☐ ☐
3. Are you currently taking any prescription or nonprescription (over-the-counter) medications or pills or using an inhaler? ☐ ☐
   Have you ever taken any supplements or vitamins to help you gain or lose weight or improve your performance? ☐ ☐
4. Do you have any allergies (for example, to pollen, medicine, food, or stinging insects)? ☐ ☐
   Have you ever had a rash or hives develop during or after exercise? ☐ ☐
5. Have you ever passed out during or after exercise? ☐ ☐
   Have you ever been dizzy during or after exercise? ☐ ☐
   Have you ever had chest pain during or after exercise? ☐ ☐
   Do you get tired more quickly than your friends do during exercise? ☐ ☐
   Have you ever had racing of your heart or skipped heartbeats? ☐ ☐
   Have you had high blood pressure or high cholesterol? ☐ ☐
   Have you ever been told you have a heart murmur? ☐ ☐
   Has any family member or relative died of heart problems or of sudden death before age 50? ☐ ☐
   Have you had a severe viral infection (for example, myocarditis or mononucleosis) within the last month? ☐ ☐
   Has a physician ever denied or restricted your participation in sports for any heart problems? ☐ ☐
6. Do you have any current skin problems (for example, itching, rashes, acne, warts, fungus, or blisters)? ☐ ☐
7. Have you ever had a head injury or concussion? ☐ ☐
   Have you ever been knocked out, become unconscious, or lost your memory? ☐ ☐
   Have you ever had a seizure? ☐ ☐
   Do you have frequent or severe headaches? ☐ ☐
   Have you ever had numbness or tingling in your arms, hands, legs, or feet? ☐ ☐
   Have you ever had a stinger, burner, or pinched nerve? ☐ ☐
8. Have you ever become ill from exercising in the heat? ☐ ☐
9. Do you cough, wheeze, or have trouble breathing during or after activity? ☐ ☐
   Do you have asthma? ☐ ☐
   Do you have seasonal allergies that require medical treatment? ☐ ☐

Yes No

10. Do you use any special protective or corrective equipment or devices that aren't usually used for your sport or position (for example, knee brace, special neck roll, foot orthotics, retainer on your teeth, hearing aid)? ☐ ☐
11. Have you had any problems with your eyes or vision? ☐ ☐
    Do you wear glasses, contacts, or protective eyewear? ☐ ☐
12. Have you ever had a sprain, strain, or swelling after injury? ☐ ☐
    Have you broken or fractured any bones or dislocated any joints? ☐ ☐
    Have you had any other problems with pain or swelling in muscles, tendons, bones, or joints? ☐ ☐
    If yes, check appropriate box and explain below.
    ☐ Head   ☐ Elbow   ☐ Hip
    ☐ Neck   ☐ Forearm   ☐ Thigh
    ☐ Back   ☐ Wrist   ☐ Knee
    ☐ Chest   ☐ Hand   ☐ Shin/calf
    ☐ Shoulder ☐ Finger   ☐ Ankle
    ☐ Upper arm   ☐ Foot
13. Do you want to weigh more or less than you do now? ☐ ☐
    Do you lose weight regularly to meet weight requirements for your sport? ☐ ☐
14. Do you feel stressed out? ☐ ☐
15. Record the dates of your most recent immunizations (shots) for:
    Tetanus _____   Measles _____
    Hepatitis B _____   Chickenpox _____

FEMALES ONLY
16. When was your first menstrual period? _____
    When was your most recent menstrual period? _____
    How much time do you usually have from the start of one period to the start of another? _____
    How many periods have you had in the last year? _____
    What was the longest time between periods in the last year? _____

Explain "Yes" answers here: _____
_____
_____
_____
_____
_____
_____
_____
_____

I hereby state that, to the best of my knowledge, my answers to the above questions are complete and correct.

Signature of athlete _____ Signature of parent/guardian _____ Date _____

Figure 1A

Preparticipation physical evaluation form: history questionnaire

Preparticipation Physical Evaluation

PHYSICAL EXAMINATION

Name _____ Date of birth _____

Height _____ Weight _____ % Body fat (optional) _____ Pulse _____ BP ___/___ (___/___ , ___/___)

Vision R 20/ _____ L 20/ _____ Corrected: Y N Pupils: Equal _____ Unequal _____

	NORMAL	ABNORMAL FINDINGS	INITIALS*
MEDICAL			
Appearance			
Eyes/Ears/Nose/Throat			
Lymph Nodes			
Heart			
Pulses			
Lungs			
Abdomen			
Genitalia (males only)			
Skin			
MUSCULOSKELETAL			
Neck			
Back			
Shoulder/arm			
Elbow/forearm			
Wrist/hand			
Hip/thigh			
Knee			
Leg/ankle			
Foot			

* Station-based examination only

CLEARANCE

❑ Cleared

❑ Cleared after completing evaluation/rehabilitation for: _____

_____

_____

_____

❑ Not cleared for: _____ Reason: _____

Recommendations: _____

_____

_____

_____

Name of physician (print/type) _____ Date _____

Address _____ Phone_____

Signature of physician _____, MD or DO

© 1997 *American Academy of Family Physicians, American Academy of Pediatrics, American Medical Society for Sports Medicine, American Orthopaedic Society for Sports Medicine, and American Osteopathic Academy of Sports Medicine.*

Figure 1B

Preparticipation physical evaluation form: physical examination record

pediatric orthopaedics

Referral decisions/Red flags

Cardiac events are the most common cause of sudden death during exercise and in young people; these events are most often secondary to hypertrophic obstructive cardiomyopathy, complications of Marfan syndrome, or other congenital abnormalities. Further evaluation is needed if questioning reveals a history of exercise intolerance, syncope, near-syncope, chest pain with exercise, a family cardiac history or unexplained death before age 50, or hyperlipidemia.

Note: For more information about the preparticipation physical evaluation, see the *Preparticipation Physical Evaluation,* written jointly by the American Academy of Family Physicians, the American Academy of Pediatrics, the American Medical Society for Sports Medicine, the American Orthopaedic Society for Sports Medicine, and the American Osteopathic Academy of Sports Medicine. This monograph was published by The Physician and Sportsmedicine, 4530 W 77th Street, Minneapolis, MN 55435. To obtain a copy of this monograph, contact AAP Publications, 1-800-433-9016.

Proximal Femoral Focal Deficiency

Synonyms

None

Definition

Proximal femoral focal deficiency (PFFD) is a rare, nonhereditary condition characterized by dysgenesis of the proximal femur associated with coxa vara or pseudarthrosis of the hip. The amount of leg length discrepancy and bony deficiency ranges from mild to severe.

Several different classifications exist, but the most widely used is that of Aitken (Figure 1). In type A, there is initially a gap between the proximal and distal femur, but it is eventually ossified (by maturity). The femur is short with varus of the neck and subtrochanteric region. In type B, the gap fails to unite, and the acetabulum and femoral head are hypoplastic. In type C, the femoral head is absent or hypoplastic, and the acetabulum is markedly dysplastic; the distal femur migrates progressively proximally. In type D, both the acetabulum and femoral head are absent, and the distal femur consists of little more than the condyles.

Clinical symptoms

Deformity and shortening of the limb are obvious at birth.

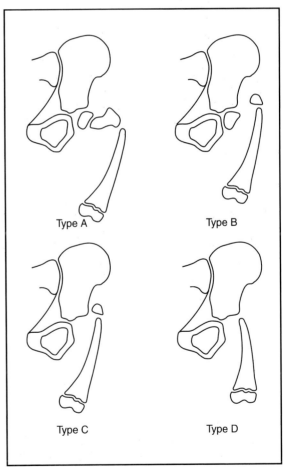

Type A

Type B

Type C

Type D

Figure 1

Aitken classification of PFFD

pediatric orthopaedics

Tests

Exam

With marked shortening, the knee may lie only inches from the pelvis. Adduction and internal rotation of the hip are limited. Aside from the severe bony involvement, there is a lack of musculature and associated joint contracture.

Diagnostic

Obtain an AP radiograph from the hip to the toes. The degree of shortening advances with the degree of dysplasia; however, as patients get older, an apparent pseudarthrosis of the femoral neck may resolve with advancing ossification.

Differential diagnosis

Coxa vara

Developmental dislocation of the hip

Adverse outcomes of the disease

A severe limp and severe leg length discrepancy are possible.

Treatment

Use of an appropriate prosthesis, along with surgical procedures tailored to the type and severity of the PFFD are indicated. Because the surgical procedures for this condition are complex, proper selection of the procedure is critical.

Adverse outcomes of treatment

Infection, failure of surgical fixation or of fusion, and/or painful scars or neuromas can develop.

Referral decisions/Red flags

Patients need further evaluation as soon as the initial diagnosis is made.

Acknowledgements

Figure 1 is reproduced with permission from Kasser JR (ed): *Orthopaedic Knowledge Update 5*. Rosemont, IL, American Academy of Orthopaedic Surgeons, 1996, pp 351–364.

Scheuermann's Disease

Synonyms

Juvenile kyphosis

Postural round back

Posterior convex curvature of the spine

Definition

Scheuermann's disease is an exaggeration of the normal posterior convex curvature of the thoracic spine associated with wedging of the vertebrae. This condition occurs primarily in boys in early to middle adolescence. Although the etiology is unknown, current theories include occult trauma to the developing spine, or hormonal or nutritional deficiencies.

Clinical symptoms

The child's posture typically initiates concern. Older adolescents with severe kyphosis may report pain. Patients may have tight hamstrings, but neurologic abnormalities are not common. Postural kyphosis is a flexible curve that has lesser degrees of deformity. Congenital kyphosis typically occurs earlier and may be associated with neurologic deficit or rapid progression.

Tests

Exam
Examine the patient from the side, first with the patient standing upright, then bending forward (flexion), and then either supine or bending back (hyperextension). The normal thoracic spine has 20° to 45° of kyphosis. Flexible kyphosis will correct in the supine or extended positions.

Diagnostic
PA and lateral shielded standing radiographs of the entire spine on a long (14″ × 36″) cassette is the standard for evaluating spinal deformity (Figure 1). The classic triad of findings includes the following: 1) three adjacent vertebrae with 5° or more of anterior wedging and a total thoracic kyphosis of greater than 45° measured from T5 to T12; 2) Schmorl's nodes; and 3) irregularity of the cartilaginous end plates.

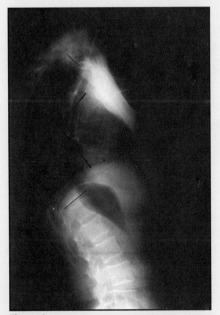

Figure 1
Lateral radiograph of the spine demonstrating kyphosis and the method of estimating its magnitude

Differential diagnosis

Congenital kyphosis (earlier presentation, rapid progression, congenital bony anomalies)

Inflammatory/infectious/related to tumor

Myelodysplasia (severe bony deficits in the spine)

Neurofibromatosis (vertebral notching, café-au-lait spots)

Postirradiation kyphosis (Wilm's tumor)

Postlaminectomy (obvious laminectomy)

Posttraumatic and postparalytic kyphosis (spinal fracture)

Postural round back (flexible curves, no radiographic changes)

Related to metabolic bone disease

Adverse outcomes of the disease

Progression of the deformity, especially in midadolescence is possible, as is persistent back pain, but this is difficult to predict. Other problems include the inability to perform heavy manual labor and possible psychological problems related to cosmesis.

Treatment

Hyperextension exercises and instruction on proper posture are appropriate for postural kyphosis, but these modalities do not affect the natural history of Scheuermann's disease. Alternatives are similar to treatment options for idiopathic scoliosis: observation, bracing, or surgery. Bracing is indicated when adequate spinal growth remains and the curve is less than 75°. A successful bracing program either stops progression (60%) or improves the curvature (40%). Surgical treatment is infrequent, usually requires anterior and posterior fusion, and is reserved for patients with severe kyphosis.

Adverse outcomes of treatment

Despite a bracing program, the curve may progress. Pseudarthrosis, paralysis, and other postsurgical complications are possible.

Referral decisions/Red flags

Patients with congenital kyphosis, rigid curves, painful kyphosis, and curves greater than 45° should be evaluated further.

Scoliosis

Definition

Scoliosis is a lateral curvature of the spine of greater than 10°, usually thoracic or lumbar, associated with rotation of the vertebrae and sometimes excessive kyphosis or lordosis. The most common etiology is idiopathic, but it may be secondary to other etiologies (Table 1).

Table 1
Etiologic classification of structural scoliosis

Idiopathic

Congenital

    Failure of formation—hemivertebra

    Failure of segmentation—bar joining one side of adjacent vertebrae

Neuromuscular scoliosis

    Cerebral palsy

    Muscular dystrophy

    Myelomeningocele

    Spinal muscular atrophy

    Friedreich's ataxia (spinocerebellar degeneration)

Scoliosis associated with vertebral disease

    Tumor

    Infection

    Metabolic bone disease

Scoliosis associated with spinal cord disease

    Tumor

    Syringomyelia

Neurofibromatosis

Marfan's syndrome

Clinical symptoms

Idiopathic scoliosis usually appears in early adolescence, with girls seven times more likely than boys to have a significant curvature that requires treatment. Progression typically occurs in girls between the ages of 10 and 16 years. Parents often notice that the child's clothes do not "hang" correctly, or the disorder can be discovered in school screening programs. Typically, there is no pain during the adolescent years. Significant pain suggests a possible bone tumor, tethered spinal cord, or other abnormalities.

Tests

Exam

Mild degrees of scoliosis may not be apparent when the patient is standing. In addition, unequal shoulder and pelvic height are unreliable in detecting mild degrees of scoliosis; besides, these parameters may be observed in normal subjects.

The forward bending test is the most sensitive clinical test for scoliosis, as it brings into profile the rotation of the vertebrae and ribs (Figure 1). To perform this test, stand behind the patient, ask him or her to bend forward with arms hanging free, feet together, and knees straight. Look for elevation of the rib cage on one side and depression of the ribs on the other. In the lumbar spine, look for elevation of the paravertebral muscle mass on one side. The amount of rib rotation can be quantified with an inclinometer. Inclinations of 5° to 7° suggest the need for a radiograph to quantify the curve.

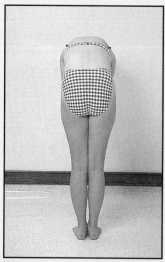

Figure 1

Diagnostic

Obtain PA and lateral radiographs to measure the curve. These should be done with the X-ray source 6′ away from the patient, preferably with a long (14″ × 36″) film and a grid. Film technique should shield X-ray exposure of the gonads and breasts.

The Cobb angle is a measure of curve magnitude and is the angle made by a line drawn along the superior end plate of the upper most tilted vertebra in the curve and a line drawn along the inferior end plate of the lowest vertebra in the curve (Figure 2).

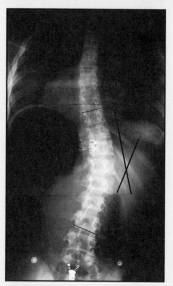

Figure 2

PA radiograph of the entire spine on a long cassette demonstrating a scoliotic deformity and the Cobb method of measuring its magnitude

The Essentials of Musculoskeletal Care

Differential diagnosis

See Table 1.

Adverse outcomes of the disease

Progression of the deformity is possible. Restricted respiratory function can develop, although cor pulmonale is very unlikely. Earlier data, which suggested a high rate of pulmonary problems with adolescent idiopathic scoliosis, included many patients with congenital, infantile, and neuromuscular disorders. Back pain is also possible.

Treatment

Many idiopathic curves do not progress to a degree that requires treatment. Observation for curve progression is appropriate while the child is growing. The frequency of evaluation depends on the degree of curvature and the amount of growth remaining.

Brace treatment is reserved for patients with progressive curves in the 20° to 40° range; however, bracing is not always effective, even with a well-made brace and good compliance. Surgery to fuse the involved vertebrae is appropriate for progressive curves of more than 50°. Scoliosis associated with neuromuscular disorders is more likely to progress and require surgical stabilization.

Adverse outcomes of treatment

The spinal curve may continue to progress despite brace treatment. Pseudarthrosis of the fusion and/or paralysis can occur following surgery.

Referral decisions/Red flags

Patients with an obvious curvature or a mild curvature with more than 5° to 7° inclination should be further evaluated for quantification of the scoliosis, an estimate of the likelihood of progression, and selection of a treatment regimen. Patients with left thoracic curves, those with significant pain, and those with abnormal neurologic exams are more likely to have a neuromuscular disorder or bony tumor and should be evaluated further.

Septic Arthritis

Synonyms

Pyarthrosis

Infected joint

Definition

Joint infections (septic arthritis) most commonly affect children and the hip, knee, and ankle are the most common sites involved. The route of infection may be hematogenous (secondary to upper respiratory infection, impetigo, etc), contiguous spread from an adjacent osteomyelitis (where the metaphysis of the involved bone is intra-articular, such as the proximal femur, proximal humerus, radial head, and lateral aspect of the distal tibia), and direct inoculation (via a puncture wound or needle passing through a cellulitis and into the joint).

Clinical symptoms

Pain, malaise, loss of appetite, and failure to use the affected joint or limb are common symptoms. A toddler may refuse to walk if the lower extremity is involved. Temperature is commonly above 102°F (38.8°C), but neonates may not have a fever, and their blood studies may be normal.

The hip is commonly affected in children younger than age 2 years and tends to sustain the most severe residual disability. The source is usually hematogenous but may be secondary to inoculation while attempting a femoral venipuncture in a septic neonate.

Tests

Exam
Swelling, tenderness, and increased localized warmth of the involved joint are characteristic findings. More than any other condition, the child resists any attempts to move the joint. The limb is held in a position allowing the greatest distention or comfort of the joint. Therefore, the hip is positioned in flexion, abduction, and external rotation, and the knee and elbow are in flexion. Pseudoparalysis (flaccid limb) is characteristic of infants with septic arthritis of the shoulder. All children, but particularly infants, should be examined for other sites of infection.

Diagnostic
Obtain baseline AP and lateral radiographs. Even if concomitant osteomyelitis is present, osseous changes will not be visible unless symptoms have been present for longer than 8 to 14 days. Ultrasound is not usually needed, but will show a joint effusion.

Laboratory studies include CBC with differential, erythrocyte sedimentation rate, C-reactive protein, and blood cultures.

Aspirate the joint to confirm infection and obtain a specimen for Gram stain and culture to direct antibiotic therapy. Fluoroscopic control may be necessary. Analysis of synovial fluid would be expected to show a WBC greater than 50,000 mm^3, polymorphonuclear cells of 90%, decreased sugar, and increased protein. Synovial fluid WBC is lower with *Neisseria gonorrhoeae* and typically ranges from 20,000 to 70,000 mm^3.

Differential diagnosis

Acute leukemia

Acute rheumatic fever

Cellulitis

Hemarthrosis from injury

Henoch-Schönlein purpura

Juvenile rheumatoid arthritis

Lyme disease

Osteomyelitis

Transient synovitis (temperature < 99.5°F {37.5°C})

Adverse outcomes of the disease

Septic arthritis can lead to destruction of the joint surface, secondary arthritis, scarring of the capsule with severe joint contractures, recurrent infections, and severe limb dysfunction (Figure 1).

Septic arthritis of the hip may interrupt the blood supply to the femoral head and cause osteonecrosis. For that reason, prompt surgical drainage is indicated for septic arthritis of the hip joint. Prognosis is good if surgical drainage and appropriate antibiotic therapy is initiated within 4 days of the onset of symptoms.

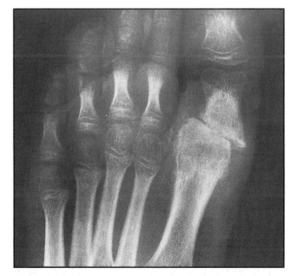

Figure 1
Destruction of the great toe metatarsophalangeal joint secondary to septic arthritis

pediatric orthopaedics

Treatment

Initiate IV antibiotics after obtaining blood and joint cultures. Drain the joint as appropriate (see below), and then splint the joint as needed to reduce pain and prevent dislocation. The parameters allowing the change to oral antibiotics are presently controversial. Certainly, a good clinical response is required, but whether studies measuring antibiotic titers are needed is controversial.

Staphylococcus aureus is the most common pathogen; group B streptococcus is common in infants younger than 1 year. *Haemophilus influenzae* is common in children from ages 6 months to 4 years, but the prevalence of this type may be waning with vaccination. *Neisseria gonorrhoeae* occurs in children between ages 12 and 18 years.

Continue antibiotic therapy for 2 to 3 weeks, or until there is complete resolution of the clinical infection. With concomitant osteomyelitis, the duration of antibiotic therapy is longer.

Joint drainage by repeated needle aspiration, arthroscopy, or arthrotomy is not always necessary, but commonly improves the clinical response by removing inflammatory products that destroy articular cartilage. Drainage also improves antibiotic effectiveness by reducing the size of the inoculum.

Adverse outcomes of treatment

Joint contractures, degenerative arthritis, and osteonecrosis of the femoral head are all possible.

Referral decisions/Red flags

Septic arthritis of the hip is a surgical emergency. Early consultation facilitates management of patients with involvement of other joints.

Acknowledgements

Figure 1 is reproduced with permission from Wilkins KE: The painful foot in the child, in Bassett FH (ed): *Instructional Course Lectures 38.* Park Ridge, IL, American Academy of Orthopaedic Surgeons, 1988, pp 77–85.

pediatric orthopaedics

Seronegative Spondyloarthropathies

Synonyms

Reiter's syndrome

Ankylosing spondylitis

Psoriatic arthritis

Definition

The seronegative spondyloarthropathies include ankylosing spondylitis, Reiter's syndrome, psoriatic arthritis, and arthritis of inflammatory bowel disease. They have the following characteristics in common: 1) inflammation of tendon, fascia, or joint capsule insertions (enthesitis); 2) pauciarticular arthritis usually involving the lower extremity; 3) extra-articular inflammation involving the eye, skin, mucous membranes, heart, and bowel; and 4) association with the presence of the HLA-B27 antigen.

Clinical symptoms

Unlike that in adults, the onset of ankylosing spondylitis in children is more likely to affect the joints of the lower extremities. Asymmetric pauciarticular arthritis involving the lower extremity in children age 9 years or older, particularly boys, should suggest the possibility of ankylosing spondylitis. The family history is often positive.

Reiter's syndrome, with its triad of conjunctivitis, enthesitis, and urethritis, may be triggered in young children by infectious diarrhea caused by *Yersinia, Campylobacter, Salmonella,* or *Shigella.* In adolescents, nongonococcal urethritis secondary to Chlamydia or trachoma may cause Reiter's syndrome. All three components of the disorder are not necessarily present in every patient, nor are they always present at the same time. The Achilles tendinitis or plantar fasciitis associated with Reiter's syndrome may be extremely painful.

Psoriatic arthritis is considered uncommon in children, but approximately one third have the onset of this disorder before age 15 years, especially girls. Arthritis frequently antedates skin problems when this disorder occurs in childhood. A family history of psoriasis is a helpful clue when joint symptoms occur first.

Arthritis of inflammatory bowel disease, either ulcerative colitis or Crohn's disease, typically causes symptoms before age 21 years, but only 15% of patients are diagnosed before age 15 years. Arthralgia without joint effusion is twice as common as arthritis with joint effusion.

Tests

Exam

A purplish discoloration may occur around the joint and is one of the distinguishing features of a juvenile spondyloarthropathy. Likewise, a child with ankylosing spondylitis may initially have inflammation of a tendon or ligament insertion site, such as patellar tendinitis, Achilles tendinitis, or plantar fasciitis. Although back pain may be absent, limited mobility of the spine may be present.

Mild conjunctivitis or an acute anterior uveitis causing painful red eyes and photophobia are also associated with Reiter's syndrome.

In psoriatic arthritis, monoarticular involvement of the knee is the most common presenting picture. Progression to other joints proceeds in an asymmetric fashion. Compared with other spondyloarthropathies, upper extremity involvement is more common in psoriatic arthritis. Tenosynovitis involving the digits and nail pits is also characteristic.

Pauciarticular arthritis of the lower extremity in inflammatory bowel disease is typically of short duration and either resolves spontaneously or with treatment of the bowel lesion. However, progressive ankylosing spondylitis may develop in some patients.

Diagnostic

The presence of the HLA-B27 antigen and a positive family history for spondyloarthropathy support the diagnosis of ankylosing spondylitis. Sterile pyuria supports the diagnosis of Reiter's syndrome.

Differential diagnosis

Juvenile rheumatoid arthritis

Various overuse syndromes

Adverse outcomes of the disease

Many lower extremity problems associated with the childhood spondyloarthropathies resolve spontaneously. Persistent erosive arthritis may occur. Ultimately, changes in the sacroiliac joint develop in children with ankylosing spondylitis.

Treatment

NSAIDs, muscle strengthening, orthotics for the painful joint, and counseling about activity modification are all indicated.

Adverse outcomes of treatment

NSAIDs can cause gastric, renal, or hepatic complications.

Referral decisions/Red flags

Loss of function or inability to gain pain control indicate the need for further evaluation.

Slipped Capital Femoral Epiphysis

Synonyms

Slipped epiphysis

Definition

With this condition, the proximal femoral epiphysis (head of the femur) literally falls off the neck of the femur. The slippage occurs through the physis, which is weaker as it begins to close. The mean age at which this condition occurs is 11 years in girls and 13 years in boys. Boys, African-Americans, athletically inclined children, and obese children are most commonly affected. Bilateral involvement eventually occurs in 30% to 40% of affected children. Endocrine disorders that weaken the physis are associated with slipped epiphysis and are particularly prevalent in preadolescent children.

Clinical symptoms

Pain and a limp, often of indolent onset or related to an injury, are the most common presenting symptoms. Patients typically have pain in the groin or anterior thigh area, but the pain may be referred to the distal thigh (knee).

Tests

Exam
Examination reveals restricted hip motion, particularly loss of internal rotation. The loss of internal rotation is most obvious when the hip is moved in flexion. Patients usually walk with an antalgic limp and the limb externally rotated (foot out). The affected limb may be 1 to 3 cm shorter than the normal limb.

Diagnostic

AP and frog lateral radiographs will confirm the diagnosis. The lateral view is particularly important. Look for posterior displacement in the lateral view and medial displacement of the femoral head on the AP view (Figures 1 and 2). In some patients, displacement is not obvious, but the physeal plate is widened. Obtain appropriate tests for endocrine disorders in preadolescent children and when clinically indicated.

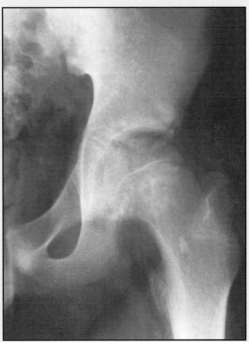

Figure 1
Chronic SCFE AP view

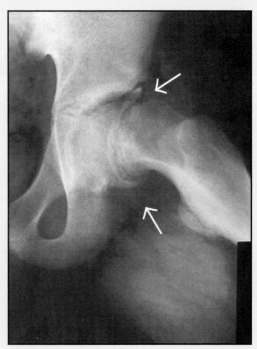

Figure 2
Chronic SCFE lateral view

Differential diagnosis

Femoral cutaneous nerve entrapment (more common in muscular girls)

Growth hormone deficiency

Hyperthyroidism

Hypothyroidism

Multiple endocrine neoplasia

Panhypopituitarism

Legg-Calvé-Perthes disease (in younger age range)

Adverse outcomes of the disease

Untreated, a slipped epiphysis is likely to progress, causing greater deformity of the proximal femur and a greater likelihood of symptomatic arthritis at an early age. Other adverse outcomes include osteonecrosis of the femoral head, secondary osteoarthritis (chondrolysis), and subsequent slip of the contralateral side (about 40%).

Treatment

Immediate cessation of weightbearing and surgical stabilization are indicated.

Adverse outcomes of treatment

Osteonecrosis of the femoral head, chondrolysis, or infection may develop.

Referral decisions/Red flags

Further evaluation for surgical treatment should occur immediately upon diagnosis.

Acknowledgements

Figures 1 and 2 are reproduced with permission from Weiner DS: Bone graft epiphysiodesis in the treatment of slipped capital femoral epiphysis, in Barr JS (ed): *Instructional Course Lectures 38.* Park Ridge, IL, American Academy of Orthopaedic Surgeons, 1989, pp 263–272.

pediatric orthopaedics

Sports Injuries of the Elbow in Children

Synonyms

Little League elbow

Pitcher's elbow

Panner disease

Definition

Excessive throwing and the subsequent abduction (valgus) stress causes most injuries of the elbow (Figure 1). The injury may affect either the medial (tension) or lateral (compression) side of the humerus. Medial injuries are either acute (avulsion fracture of the medial epicondyle) or gradual in onset (traction apophysitis of the medial epicondyle, better known as Little League elbow). Lateral involvement is secondary to osteonecrosis of the capitellum. When it affects children younger than age 10 years, it is typically called Panner disease and has a good prognosis. Osteonecrosis of the capitellum in adolescents, called osteochondritis dissecans, has a more guarded prognosis.

Figure 1
Throwing athletes impose valgus stress on the elbow

Clinical symptoms

The common history is aching pain over the involved area that is activity related. A sudden, forceful pitch may cause avulsion of the medial epicondyle, with acute swelling and pain, especially in children between the ages of 9 to 12 years. Osteonecrosis of the capitellum may result in an osteochondral loose body and cause a "locking" or "catching" sensation.

Tests

Exam
Watch for tenderness over the involved humeral condyle. Mild swelling and limitation of elbow motion may be present.

Diagnostic
AP and lateral radiographs of the elbow may be normal or may show the avulsion fracture or fragmentation and heterotopic ossification of the medial epicondyle (Little League elbow), or fragmentation and irregularity of the capitellum.

Differential diagnosis

Atypical tumor

Subluxating ulnar nerve

Ulnar neuritis

Adverse outcomes of the disease

These injuries are generally self-limited with the exception of osteonecrosis of the capitellum in adolescents, which may lead to deformity of the joint and later radiohumeral arthritis.

Treatment

Resting the arm, with no throwing for 3 to 6 weeks is indicated, followed by rehabilitation to restore elbow motion and upper extremity strength.

Adverse outcomes of treatment

Postoperative infection is possible.

Referral decisions/Red flags

Intra-articular loose body with locking indicates the need for further evaluation. Osteonecrosis of the capitellum in adolescents is more likely to cause residual symptoms and consideration of surgical intervention.

Acknowledgements

Figure 1 is adapted with permission from Hunter-Griffin LY (ed): *Athletic Training and Sports Medicine,* ed 2. Park Ridge, IL, American Academy of Orthopaedic Surgeons, 1991, p 943.

pediatric orthopaedics

The Essentials of Musculoskeletal Care   655

Subluxation of the Radial Head

Synonyms

Jerked elbow

Pulled elbow

Nursemaid's elbow

Temper tantrum elbow

Slipped elbow

Supermarket elbow

Definition

Subluxation of the radial head is the most common elbow injury in young children. Young children normally have greater joint laxity, and children who experience this condition have more joint laxity than unaffected children of the same age. This condition commonly affects 2- to 3-year-old children, but is rare in children older than age 7 years.

Clinical symptoms

Typically, a parent or older sibling lifts, swings, or pulls a child when the forearm is pronated and the elbow extended. The radial head is pulled in a distal direction and becomes loaded or wedged in the annular ligament, the structure that normally wraps around the more narrow neck of the radius.

With subluxation of the radial head, there is a lack of specific clinical and radiographic findings. Therefore, awareness of the condition and obtaining a history consistent for the injury are critical.

Immediately after the injury, the child will cry, but the initial pain quickly subsides. Thereafter, the child is reluctant to use the arm but otherwise does not appear to be in great distress. The extremity is held by the side with the elbow slightly flexed and the forearm pronated.

Tests

Exam
Tenderness over the radial head and resistance on attempted supination of the forearm are the only consistent clinical findings. Because young children are frightened of strangers, examination presents a challenge. Perform the exam sitting or kneeling below the child and with the child seated in the parent's lap. It is also helpful to examine the uninjured arm first and to begin by palpating the wrist and then moving toward the suspected site of injury.

Diagnostic
Radiographs are usually within normal limits and are only helpful to rule out other injuries. If the history and exam strongly support the diagnosis of radial head subluxation, it is reasonable to manipulate the elbow without obtaining radiographs. However, if the history is uncertain and if the exam is equivocal, then radiographs of the elbow should be obtained to rule out other injuries.

pediatric orthopaedics

Differential diagnosis

Early septic arthritis of the elbow (in the same age group)

Torus fracture of the distal radius (in the same age group)

Adverse outcomes of the disease

Recurrence is always a risk; however, this situation can be minimized by instructing the parents in the mechanism of injury and encouraging them to lift the child by the torso. Fortunately, even after multiple recurrences, the child eventually grows out of the problem without sequelae or need for reconstructive surgery.

Most subluxations of the radial head reduce spontaneously even when the injury is not recognized or treated. Without treatment, however, the child may experience continued discomfort. There have been only rare case reports of children who required open reduction when injury was not brought to medical attention for several weeks.

Treatment

Explain to the parents that the manipulation will momentarily cause pain. Place a thumb over the radial head, and with the other hand deftly supinate the patient's forearm. If that maneuver fails to produce the snap of reduction, then flex the elbow. Resistance may be perceived just before reaching full flexion. As one pushes through that resistance, the annular ligament will slip back into normal position and a snap will be perceived as the radial head reduces. If the reduction is successful, the child will begin to use the extremity normally in a few minutes.

Immobilization is probably not necessary, as parents report that slings are quickly discarded. The exception is the child who presents a day or two after injury. At this time, swelling of the annular ligament may obscure the snap that signals a successful reduction and also hinder immediate resumption of normal function. However, if the elbow has full flexion and supination, the radial head has been reduced. A sling can then be used with the expectation that full elbow function will resume in a few days. These children should be reexamined in a few days to ensure that reduction of the radial head has occurred.

For a child who has recurrent subluxations of the radial head, 2 to 3 weeks of immobilization in a long arm cast may be helpful; however, this treatment has not been proven and some recognized experts do not treat recurrent subluxation any differently than the primary event.

Adverse outcomes of treatment

Failure to recognize a persistent unreduced subluxation may, in rare instances, necessitate open reduction.

Referral decisions/Red flags

Failure of reduction and persistent irritability indicate the need for further evaluation.

Supracondylar Fracture of the Humerus

Synonyms

None

Definition

Fracture of the distal end of the humerus above the growth plate is the most common elbow fracture in children. This fracture typically occurs in children between the ages of 2 and 12 years. Most are extension fractures with the distal fragment displacing posteriorly.

Clinical symptoms

Patients have pain localized to the distal humerus. Swelling may be severe, but with nondisplaced fractures, swelling may actually be mild. Note that the child may not use the limb, and may actually hold it at the side with the elbow bent 30° or more.

Tests

Exam

Neurovascular injuries are more likely to occur in association with these fractures. Paralysis of the median, ulnar, or radial nerve may occur. The absence of pulses suggests compression or injury to the brachial artery. Compartment syndrome involving the forearm may develop in the first 24 to 48 hours. Disproportionate pain and the inability to extend the fingers are suggestive findings.

Diagnostic

Obtain AP and lateral radiographs of the elbow. In nondisplaced fractures, the radiograph may not reveal a fracture line. Point tenderness of the supracondylar area of the humerus and the presence of an abnormal fat pad sign on the radiograph (Figure 1) will confirm the diagnosis.

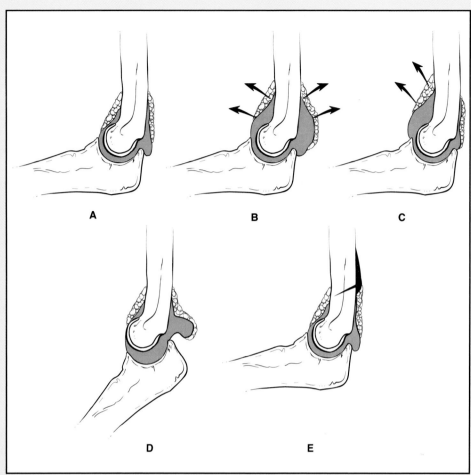

Figure 1
Normal relationships of the two fat pads (A); Displacement of both fat pads with an intra-articular effusion (B); In some cases, the effusion may displace only the anterior fat pad (C); In extension the posterior fat pad is normally displaced by the olecranon (D); and an extra-articular fracture may lift up the distal periosteum and displace the proximal portion of the posterior fat pad (E)

Differential diagnosis

Lateral condyle fracture of the humerus

Medial epicondyle fracture of the humerus

Subluxation of the radial head (in younger age group, normal posterior fat pad sign)

Transcondylar fracture of the distal humerus

Adverse outcomes of the disease

Compartment syndrome with Volkmann's ischemic contracture (necrosis of all muscles in the forearm), recurrent angulation, rotational or angular malunion, and deformity may develop.

pediatric orthopaedics

Treatment

Nondisplaced fractures are treated with a posterior splint and explicit instructions for the parents to call if the child cannot fully extend all fingers of the affected hand (check hourly for the first 24 hours), or if pain is unmanageable with over-the-counter pain medication. Splinting should continue for 3 weeks or longer if the fracture is still tender. Displaced fractures require closed reduction and pinning.

Adverse outcomes of treatment

These are the same as the adverse outcomes of the disease. In addition, postsurgical pin-tract infections may occur.

Referral decisions/Red flags

Any displacement or angulation, inability to extend all fingers of the affected hand, failure of over-the-counter medications to give pain relief, and absent or diminished radial pulse are all signs that further evaluation is necessary.

Acknowledgements

Figure 1 is adapted with permission from Murphy WA, Siegel MJ: Elbow fat pads with new signs and extended differential diagnosis. *Radiology 1977;*124:656–659.

pediatric orthopaedics

Syndactyly and Polydactyly

Synonyms

Multiple digits

Accessory digit

Webbed digit

Definition

Syndactyly is characterized by the lack of separation between fingers or toes (Figure 1). This can vary from a thin skin attachment to a bony attachment (synostosis) between parts of the phalanges.

Polydactyly is the presence of extra digits in the hand or foot. These usually affect the first (thumb or great toe) or fifth digit (Figure 2).

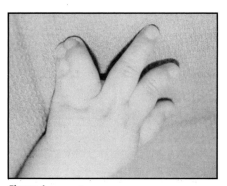

Figure 1
A fourth web space complex syndactyly

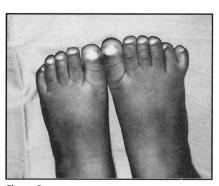

Figure 2
Postaxial polydactyly in the right foot

Clinical symptoms

None, other than the obvious deformity.

Tests

Exam
Watch for webbing of the fingers in syndactyly. Extra digits are typically incomplete and may be vestigial (no bony elements).

Diagnostic
Radiographs are needed only for surgical planning.

pediatric orthopaedics

Differential diagnosis

Other malformations occur in about 5% of patients.

Adverse outcomes of the disease

Because of the different lengths of the metacarpals and phalanges of the fingers, growth usually causes progressive deviation of the conjoined fingers. Therefore, syndactyly of the hand should be corrected. The growth differential of a syndactyly of the toes is typically insignificant. Therefore, correction of syndactyly of the toe is usually not required.

Polydactyly of the hand does not cause functional difficulties but is a significant cosmetic deformity. Accessory digits of the foot frequently cause difficulty in shoe wear.

Treatment

Vestigial digits can be ablated by suture technique. Otherwise, delay removal of extra digits until children are age 9 to 12 months. Syndactyly release is also performed at this age.

Adverse outcomes of treatment

Infection and skin sloughing may occur.

Referral decisions/Red flags

Any definite or questionable consideration of surgical reconstruction indicates the need for further evaluation.

Acknowledgements

Figure 1 is reproduced with permission from Kasser JR (ed): *Orthopaedic Knowledge Update 5.* Rosemont, IL, American Academy of Orthopaedic Surgeons, 1996, pp 295–309.

Tarsal Coalition

Synonyms

Calcaneal bar

Talocalcaneal bar

Peroneal spastic flatfoot

Rigid flatfoot

Definition

This condition is characterized by a connection of any two bones of the hindfoot that is initially fibrous or cartilaginous but that ossifies in adolescence and restricts motion of the hindfoot. Two locations are common. A bar connecting the calcaneus and navicular (calcaneonavicular coalition) typically becomes symptomatic in children between the ages of 9 and 13 years. A bar between the talus and calcaneus (talocalcaneal coalition) becomes symptomatic when children are older, with onset of symptoms typically occurring between the ages of 13 and 16 years. The condition is bilateral in approximately 50% of patients and may be found in other family members who are asymptomatic, but have no hindfoot motion.

Clinical symptoms

Symptoms are gradual in onset; pain is increased with activity and usually localized to the lateral aspect of the ankle. To restrict painful hindfoot motion, the peroneal muscles contract and produce a pes planus or flatfoot posture. Patients may limp or walk with the foot externally rotated.

Tests

Exam

Hindfoot (heel) motion (inversion and eversion) is markedly restricted. If the foot is quickly inverted, the peroneal muscles will spasm to restrict the painful movement. The foot is flat whether the patient is seated or standing; this is in contrast to a flexible flatfoot.

Diagnostic

AP, lateral, and oblique radiographs of the foot will clearly delineate a coalition between the calcaneus and navicular (calcaneonavicular bar) (Figure 1). Talocalcaneal bars are difficult to see on routine radiographs, even with special views, and a CT scan

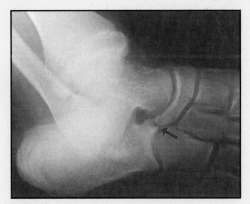

Figure 1
Oblique radiograph of a foot with a cartilaginous calcaneonavicular coalition (arrow)

may be necessary to confirm a diagnosis. Since two coalitions may be present in the same foot, this imaging study is routine before surgical treatment. If radiographs fail to demonstrate a coalition in a patient with restricted subtalar motion, an MRI scan may identify a fibrous coalition.

pediatric orthopaedics

Differential diagnosis

Accessory navicular (medial prominence)

Congenital vertical talus (in the neonate)

Flexible flatfoot (usually without pain)

Osteonecrosis of the tarsal navicular (in younger age group, tender medial navicular)

Adverse outcomes of the disease

Pain, stiffness, and arthritis of the hindfoot are all possible.

Treatment

Treatment options includes observation, short-term cast immobilization, resection of the coalition, and arthrodesis. Observation is appropriate for children who have minimal symptoms. Resecting the coalition will restore hindfoot motion and significantly reduce pain in most patients (about 85%) who meet the prerequisites for this procedure. Casting temporarily rests the foot. In some cases, the coalition then progresses to complete bridging of the joint. With loss of all motion, the pain also frequently abates.

Adverse outcomes of treatment

Complications include cast sores, postoperative infection, failure of fusion, recurrence of a resected coalition, arthritis, and skin slough.

Referral decisions/Red flags

Continued pain following conservative treatment indicates the need for further evaluation.

Acknowledgements

Figure 1 is reproduced with permission from Mosca VS: Flexible flatfoot and tarsal coalition, in Richards BS (ed): *Orthopaedic Knowledge Update: Pediatrics.* Rosemont, IL, American Academy of Orthopaedic Surgeons, 1996, pp 211–218.

Tibial Hemimelia

Synonyms

None

Definition

Tibial hemimelia is characterized by congenital absence of part or all of the tibia (Figure 1). This condition may occur in families and may be associated with hip dysplasia or upper extremity deformities. Other generalized associations include hernias, gonadal malformations, hypospadias, cleft palate, imperforate anus, and congenital heart disease.

Clinical symptoms

Patients have an obvious deformity.

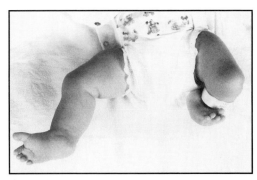

Figure 1A

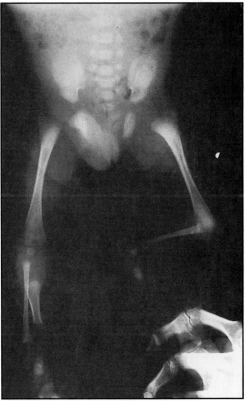

Figure 1B
Clinical appearance (A) and radiograph (B) of an infant with bilateral tibial hemimelia. The patient's right lower limb has incomplete absence of the tibia; on the left, the tibia is completely absent

pediatric orthopaedics

Tests

Exam
Examination reveals that the foot is deviated so that the sole faces the opposite leg. Evaluate whether the infant's knee joint can be fully extended and palpate for the presence of the proximal tibia. With the pelvis level and both legs in a straight line with the trunk, look for a leg length discrepancy.

Diagnostic
Obtain an AP radiograph of the lower extremities from hip to toes.

Differential diagnosis

Clubfoot

Metatarsus adductus (normal leg and ankle)

Adverse outcomes of the disease

Severe leg length discrepancy, inability to bear weight on the shortened, unstable limb, and severe gait disturbance are possible, as is the possibility of amputation.

Treatment

Usually, the foot and ankle are so deformed that amputation is necessary when children are between the ages of 1 and 2 years. If the proximal tibia is present and the quadriceps functional, an amputation through the ankle (Syme amputation) is possible. Otherwise, a knee disarticulation provides better long-term function.

Adverse outcomes of treatment

Complications include infection, painful scars, or neuromas.

Referral decisions/Red flags

Consultation is indicated immediately upon diagnosis.

Acknowledgements

Figure 1 is reproduced with permission from Kasser JR (ed): *Orthopaedic Knowledge Update 5*. Rosemont, IL, American Academy of Orthopaedic Surgeons, 1996, pp 437–451.

pediatric orthopaedics

Transient Synovitis of the Hip

Synonyms

Observation hip

Toxic synovitis (this implies an infectious arthritis and should not be used)

Definition

Transient synovitis of the hip is a sterile effusion of the hip that resolves without therapy or sequelae. The etiology is unknown, but mild trauma at an age when the socket or acetabulum is not fully developed may be the best theory. Another theory proposes a viral infection, but virologic cultures of the effusion have not identifed organisms.

Clinical symptoms

Although the reported age range is from 2 to 15 years, children age 2 to 5 years are most commonly affected, and boys are affected two to three times more often than girls. Typically, the child awakens with a limp.

Tests

Exam

Examination reveals mild restriction of motion, particularly abduction. If localized, the pain is typically in the groin or proximal thigh, but may be referred to the distal thigh. Some children deny pain, but the limp and restricted motion are obvious. Most children are afebrile.

Diagnostic

Since transient synovitis is a diagnosis of exclusion, the extent of the evaluation process depends on the degree and duration of symptoms. If a child is seen 1 to 3 days after onset, and is afebrile, does not appear ill, has only mildly restricted abduction (movement of at least 25° to 30°), and does not guard against movement in other planes, then radiographs and laboratory studies are not necessary. However, if the process has been going on for several days, but the child is afebrile and only has mildly restricted motion, then AP and frog lateral radiographs of the pelvis are appropriate. Obtain a CBC, erythrocyte sedimentation rate, blood cultures, and radiographs if the patient has very limited motion or a temperature greater than 99.5°F (37.5°C).

While radiographs are usually normal, they may show widening of the joint space. Ultrasound and bone scan are not usually necessary but often demonstrate changes. A joint effusion will be seen on ultrasound, and a bone scan may demonstrate mild increased uptake.

Aspirate the joint if there are findings compatible with septic arthritis.

pediatric orthopaedics

Differential diagnosis

Juvenile rheumatoid arthritis (multiple joints, persistent symptoms, systemic symptoms)

Legg-Calvé-Perthes disease (limp worse at the end of the day, present after 2 to 6 weeks of symptoms, abnormal radiographic finding)

Rheumatic fever arthralgias (progresses in a migratory fashion to other joints; systemic symptoms)

Septic arthritis of the hip (temperature > 99.5°F (37.5°C) and a sedimentation rate > 20 mm/hr)

Adverse outcomes of the disease

The symptoms and limp associated with this condition resolve within 3 to 14 days. Recurrence is possible, but uncommon. No sequelae have been associated with transient synovitis. Legg-Calvé-Perthes disease subsequently develops in 1% to 3% of patients, but whether the disorders have any association is unclear. A delay in diagnosing septic arthritis is also possible.

Treatment

Most children can be treated by bed rest at home, with the parents periodically checking the temperature. When the diagnosis is equivocal or if the patient is uncomfortable, hospitalization for observation and traction is indicated.

Adverse outcomes of treatment

None

Referral decisions/Red flags

A temperature higher than 99.5°F (37.5°C) or an erythrocyte sedimentation rate greater than 20 mm/hr suggests septic arthritis, in which case patients need further evaluation. Also, patients need additional evaluation if radiographs suggest Legg-Calvé-Perthes disease.

index

Medial compartment of the knee
  arthritis in, 324, *325*
Medial tennis elbow. *See also* Tennis
    elbow
  pain associated with, 126–127, 129
Medial tibial stress syndrome. *See*
    Leg, chronic pain in the
Median nerve
  compression (*See* Carpal tunnel syn-
    drome)
  entrapment at the wrist (*See* Carpal
    tunnel syndrome)
  percussion, *193*
Meniscus
  definition, 6*(def)*
  relationship to collateral ligaments,
    353
  tear of the
    adverse outcomes, 354
    clinical symptoms, 353
    definition, 353
    differential diagnosis, 354
    red flags, 355
    tests, 353–354
    treatment, 354
Meralgia paresthetica. *See* Femoral
    cutaneous nerve, entrapment of
    the lateral
Metacarpals, fracture of the
  adverse outcomes, 216
  clinical symptoms, 215
  definition, 215
  differential diagnosis, 216
  red flags, 217
  tests, 216
  treatment, 216–217, *218*
  types, *215*
Metacarpophalangeal (MP) joint
  dislocations of the thumb, 238
  flexor tendon injuries that affect,
    209
  injection, 179–180
Metacarpotrapezial degenerative
    arthritis, 181–183
Metaphysis, definition, 6*(def)*
Metatarsal
  fracture of the
    adverse outcomes, 430
    clinical symptoms, 429
    definition, 429
    differential diagnosis, 430
    red flags, 431
    tests, 429
    treatment, 430–431
  osteonecrosis of the head, 604–605
  stress fracture of the, 485–486
Metatarsalgia, 411–413
Metatarsal pad, application proce-
    dure, 400–401

Metatarsophalangeal (MP) joint
  arthritis of the, 444–446
  physical exam, *380*
Metatarsus adductus, 620–623
Metatarsus primus varus, 395–397
Metatarsus varus, 620–623
Midfoot
  fracture of the
    adverse outcomes, 434
    clinical symptoms, 432
    definition, 432
    differential diagnosis, 433
    red flags, 434
    tests, 432–433
    treatment, 434
  stress fracture of the, 485–486
Minor dystrophy, 54–56
Monteggia fracture-dislocation,
    600–601
Morton's neuroma
  adverse outcomes, 456
  application of a plantar pad, 457
  clinical symptoms, 455
  definition, 455
  differential diagnosis, 456
  injection procedure for, 458–459
  red flags, 456
  tests, 455–456
  treatment, 456
Motrin, dosage, *47*
Motrin IB, dosage, *47*
MP dislocation (finger), 235–238
MP joint. *See* Metacarpophalangeal
    (MP) joint; Metatarsophalangeal
    (MP) joint
MRI, principles of, 34–35
Mucoid cyst
  of the foot, 442–443
  of the hand and wrist, 254–256
Mucous cyst. *See* Finger, ganglion of
    the
Multiple digits, 661–662
Munchausen syndrome by proxy,
    567–570
Musculotendinous cuff rupture. *See*
    Rotator cuff, tear
Myelopathy, definition, 6*(def)*
Myeloproliferative disorders
  risk factor for osteonecrosis, 292
Myofascial low back pain, 528–530
Myofascial neck pain, 509–511

Nabumetone, dosage, *47*
Nalfon, dosage, *47*
Naprelan 375, dosage, *47*
Naprelan 500, dosage, *47*
Naprosyn, dosage, *47*

Naproxen, dosage, *47*
Naproxen sodium, dosage, *47*
Navicular fracture, 219–221
Neck. *See* Spine
Neck strain, 509–511
Necrosis, of muscle. *See*
    Compartment syndrome
Neer impingement sign (shoulder),
    75, 109
Neisseria gonorrhoeae, 648
Neonate, physical exam, 555
Neoplasm. *See* Bone tumor
Nerves, electrical conduction in, 137
Neurogenic claudication, 540–542
Neurologic disorders, pediatric ortho-
    paedics and, 551
Neuroma, interdigital, *381,* 455–456
Neuropathic foot, 416–419
Neuropathy, definition, 6*(def)*
Nonorganic physical findings
  clinical symptoms, 43
  definition, 43
  differential diagnosis, 45
  red flags, 45
  tests, 43–44
  treatment, 45
Nonsteroidal anti-inflammatory drugs
    (NSAIDs)
  choosing, 48
  compared to corticosteroids, 46
  dosages, *47*
  properties and mechanism of action,
    46
  side effects, 48
Nonunion, definition, 6*(def)*
NSAIDs. *See* Nonsteroidal anti-
    inflammatory drugs
Numbness, in the hand, 164
Nuprin, dosage, *47*
Nursemaid's elbow, 656–657

O

Observation hip, 667–668
Occupational arm pain, 21–23
Occupational cramp, 21–23
Occupational stress syndrome, 21–23
Olecranon
  anatomy of, *158*
  bursa aspiration, 158–159
  bursitis
    adverse outcomes, 157
    clinical symptoms, 156
    definition, 156
    differential diagnosis, 157
    pain associated with, 126–127
    red flags, 157
    tests, 156–157
    treatment, 157